Interfacing Bioelectronics and Biomedical Sensing

Hung Cao • Todd Coleman • Tzung K. Hsiai
Ali Khademhosseini
Editors

Interfacing Bioelectronics and Biomedical Sensing

Springer

Editors
Hung Cao
Electrical Engineering and
Computer Science
University of California, Irvine
Irvine, CA, USA

Tzung K. Hsiai
Department of Bioengineering
University of California Los Angeles
Los Angeles, CA, USA

Todd Coleman
Department of Bioengineering
University of California, San Diego
La Jolla, CA, USA

Ali Khademhosseini
California NanoSystems Institute
University of California Los Angeles
Los Angeles, CA, USA

ISBN 978-3-030-34469-6 ISBN 978-3-030-34467-2 (eBook)
https://doi.org/10.1007/978-3-030-34467-2

This Springer imprint is published by the registered company Springer Nature Switzerland AG
The registered company address is: Gewerbestrasse 11, 6330 Cham, Switzerland

Preface

Over the past decade, we have witnessed an unparalleled trajectory in the emerging flexible bioelectronics that transforms healthcare and biomedical research. While cutting-edge micro- and nano-fabrication techniques have provided the fundamental basis to miniaturize sensors and systems with superior performance, innovations in biomaterials, flexible electronics, and telecommunications have enabled both accurate and reliable biomedical devices at a reduced cost for high-throughput mobile health (m-Health) in the era of personalized- and tele-medicine. Furthermore, the integration with the rise of the Internet of Things (IoTs) and data science techniques, including machine learning and deep learning, has paved the way to translate bench discoveries to industry-grade biomedical systems for precision medicine.

In response to this era of electronic transformation, we have assembled a cadre of experts in the field of biomaterials, biosensors, and bioelectronics to provide fundamental knowledge, novel designs and preclinical implementation, potential challenges, and opportunities at the interface between bioelectronics and tissue. The principal audience will include university and industry researchers, regulatory agents and patent attorneys, clinicians, and engineers. Our vision is to inspire the young minds, to enrich the up-and-coming leaders, and to empower the next generation of clinicians, engineers, and scientists. Our mission is to promote cross-disciplinary research and to enable community engagement to solve the unmet clinical challenges. Finally, the authors would like to pay tribute to the giants in biosensors and bioelectronics for creating a better future for mankind.

Irvine, CA, USA — Hung Cao
La Jolla, CA, USA — Todd Coleman
Los Angeles, CA, USA — Tzung K. Hsiai
Los Angeles, CA, USA — Ali Khademhosseini

Contents

Challenges in the Design of Large-Scale, High-Density, Wireless Stimulation and Recording Interface 1
Po-Min Wang, Stanislav Culaclii, Kyung Jin Seo, Yushan Wang, Hui Fang, Yi-Kai Lo, and Wentai Liu

Advances in Bioresorbable Electronics and Uses in Biomedical Sensing 29
Michelle Kuzma, Ethan Gerhard, Dingying Shan, and Jian Yang

Inorganic Dissolvable Bioelectronics 73
Huanyu Cheng

Wirelessly Powered Medical Implants via Radio Frequency 101
Parinaz Abiri, Alireza Yousefi, Hung Cao, Jung-Chih Chiao, and Tzung K. Hsiai

Electrocardiogram: Acquisition and Analysis for Biological Investigations and Health Monitoring 117
Tai Le, Isaac Clark, Joseph Fortunato, Manuja Sharma, Xiaolei Xu, Tzung K. Hsiai, and Hung Cao

Flexible Intravascular EIS Sensors for Detecting Metabolically Active Plaque 143
Yuan Luo, Rene Packard, Parinaz Abiri, Y. C. Tai, and Tzung K. Hsiai

Epidermal EIT Electrode Arrays for Cardiopulmonary Application and Fatty Liver Infiltration 163
Yuan Luo, Parinaz Abiri, Chih-Chiang Chang, Y. C. Tai, and Tzung K. Hsiai

High-Frequency Ultrasonic Transducers to Uncover Cardiac Dynamics 185
Bong Jin Kang, Qifa Zhou, and K. Kirk Shung

Minimally Invasive Technologies for Biosensing 193
Shiming Zhang, KangJu Lee, Marcus Goudie, Han-Jun Kim, Wujin Sun, Junmin Lee, Yihang Chen, Haonan Ling, Zhikang Li, Cole Benyshek, Martin C. Hartel, Mehmet R. Dokmeci, and Ali Khademhosseini

Index 225

Challenges in the Design of Large-Scale, High-Density, Wireless Stimulation and Recording Interface

Po-Min Wang, Stanislav Culaclii, Kyung Jin Seo, Yushan Wang, Hui Fang, Yi-Kai Lo, and Wentai Liu

1 Introduction

Neural stimulation and recording have been widely applied to fundamental research to better understand the nervous systems as well as to the clinical therapy of a variety of diseases. Well-known examples include monitoring and manipulating neural activities in the brain to map the brain function [1–3], spinal cord implant to restore the motor function after spinal cord injury [4–6], gastrointestinal (GI) implant to monitor and treat GI motility disorders [7–9], and retinal prostheses to regain eyesight in the blind [10–12]. These applications require continuous technological advancement in designs of stimulation and recording electronics, electrodes, and the interconnections between the electronics and the electrodes.

Despite technological advances to date, there are still challenges to overcome in applications requiring large-scale, high-density, wireless stimulation and recording. Concretely, the study of the brain function through monitoring and manipulating neural activities in freely moving and behaving subjects requires decoding the complex brain dynamics by mapping brain activity with both large-scale and high spatial resolution. This decoding demands a large-scale, high-density electrode array with small electrode size, which imposes significant challenges in electrode array fabrication as well as its interconnect with electronics. In addition, the capability to record high channel number implies the need for wireless electronics with high

P.-M. Wang · S. Culaclii · Y. Wang · W. Liu (✉)
Department of Bioengineering, University of California, Los Angeles, Los Angeles, CA, USA
e-mail: wentai@ucla.edu

K. J. Seo · H. Fang
Department of Electrical and Computer Engineering, Northeastern University, Boston, MA, USA

Y.-K. Lo
Niche Biomedical Inc., Los Angeles, CA, USA

H. Cao et al. (eds.), *Interfacing Bioelectronics and Biomedical Sensing*,
https://doi.org/10.1007/978-3-030-34467-2_1

bandwidth. This need trades off with the low power consumption constraint set by the safety regulations of cortical implant. Furthermore, when a large-scale, high resolution neural interface is used to investigate the brain dynamics through simultaneous electrical stimulation and recording, the recorded signal suffers from severe stimulation artifact. The artifact, which is generated when stimulation is applied, can be several times larger than the recorded neural activities, which results in failure to detect neural response often by saturating the recording front-end. A recording scheme needs to integrate specific functions to cancel the stimulation artifact and thus capture the evoked neural response arising right after the stimulation. Furthermore, it is critical to achieve spatially precise stimulation. Performing selective and focalized stimulation to specific neurons would greatly ameliorate the stimulation efficacy by avoiding the undesired spread of electrical charge.

Overcoming these challenges is desperately needed to continue driving the technology toward large-scale, high-density wireless stimulation and recording interface. This chapter will review state of the art technologies that address the aforementioned design challenges. Section 2 presents the stimulation artifact cancellation recording schemas. Section 3 discusses the focalized stimulation. Section 4 describes fabrication of the high-density electrode array. Section 5 focuses on specifics of the high-data-rate wireless link. Section 6 presents a large-scale, high-density, wireless stimulation and recording system that integrates the above components.

2 Cancellation of Artifacts During Simultaneous Neural Stimulation and Recording

Next-generation neural interfaces, which investigate neural connectivity in diagnostics and research or enable implantable closed-loop therapy for neural diseases, will require the ability to simultaneously stimulate and record from the electrode arrays. But recording during stimulation incurs a challenge in the form of the stimulation artifact. The artifact signal is formed every time the stimulus is injected into the neural tissue; it also propagates to the recording electrode and contaminates the recording. The contamination can slightly distort the data if the artifact amplitude and duration are small compared to the neural signals recorded. The contamination can also completely confound the data of interest if the artifact is larger than the recording amplifier's input compliance limits or if the artifact's duration exceeds the timing and length of the neural response. The latter case is more likely in an ultrahigh-density neural interface where the stimulation electrode will be close to the recording electrodes due to shorter electrode-to-electrode pitch. The design of ultrahigh-density bidirectional neural interfaces thus must accommodate large stimulation artifacts while avoiding recording signal loss.

Many methods are developed in commercial FDA-approved and research-grade devices to reduce or eliminate the negative effects of stimulation artifacts in neural recording. Most designs solve the artifact problem either in the circuit domain or in the digital post-processing domain. The latter removes the artifact distortion in neural recordings where the artifact is small and does not saturate the amplifier. The focus of the former, instead, is to preserve amplifier linearity at the recording electrode during large artifact events.

2.1 Stimulation Artifact Cancellation by Circuit Design

Most common circuit design approaches employ amplifier blanking and signal filtering (Fig. 1a, b). The advantage of such designs is their simplicity, which allows a simple upgrade from a basic neural amplifier with insignificant additional area needed on the integrated circuit. A popular example of analog filtering for artifact cancellation is Medtronic's Activa RC + S device [13]. The device is designed to

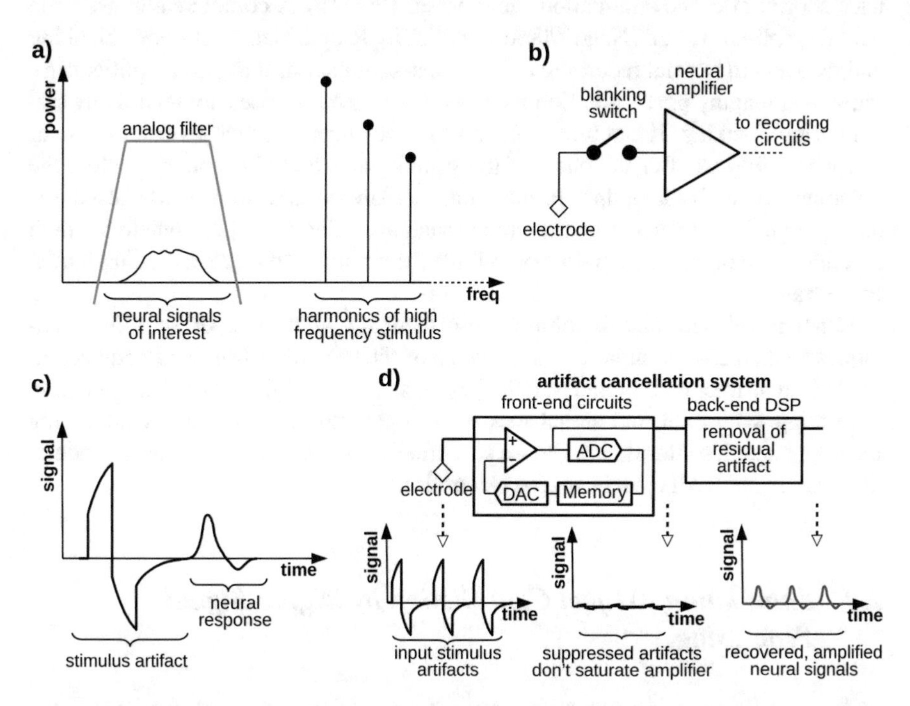

Fig. 1 Methods for stimulation artifact cancellation: (**a**) high-frequency stimulation tones are well separated in frequency domain from neural signals, allowing stimulation artifact removal by analog and digital filters. (**b**) Blanking switch disconnects signal input to the recording amplifier to avoid signal saturation during stimulus artifact. (**c**) Subtraction of averaged artifact template is effective when stimulus artifact is of similar amplitude to neural signals, and they do not overlap in time. (**d**) A complete artifact-free system design combines circuit techniques to reduce artifact and prevent amplifier saturation and uses digital signal processing to remove residual artifacts and reveal an intact neural signal, effectively removing very large artifacts that may overlap with short-latency neural responses

suppress stimulus artifacts specifically in high-frequency stimulation protocols, where low-frequency LFP signals are recorded. A high-order analog filter is used to separate the artifact from neural signals in the frequency domain (Fig. 1a). The filter by itself is not sufficient to suppress a very large artifact, so device users are recommended to constrain the electrode placements such that a bipolar stimulation is applied symmetrically at the opposite sides of the recording electrode. This allows the positive and negative stimulation artifacts to cancel in the middle at the recording site. The device is widely used for DBS in treatment of movement disorders with a low-channel-count depth electrode, but it is not suitable for high-density electrode arrays where a large number of recording sites are likely positioned non-symmetrically to the site of stimulation.

Alternatively, amplifier blanking circuits can be used to reduce the effects of artifacts on the recording. It is often implemented as a switch that momentarily disconnects the amplifier from the recording electrode during the period of the stimulation and optionally resets the amplifiers output to a zero value as it is reconnected back shortly after the stimulation event passes (Fig. 1b). A commercially available neural amplifier design (Nano/Micro 2 + Stim by Ripple Neuro) employs blanking and recovers to normal recording at 1 ms post-stimulation. It avoids amplifier saturation and lengthy post-saturation recovery, but inherently does not record any signal during blanking. It can thus miss short latency neural responses that can occur as early as 100 μs after the onset of the stimulation event [14] and end before the amplifier comes back online. In addition, blanking would omit neural responses during high frequency stimulation trains, such as during DBS [13], where the high repetition rate of the pulse train would force the amplifier to blank recording for the entire train.

Other novel pure-circuit solutions use feedback with chopper amplifier techniques to increase the input dynamic range of the recording front-end and accommodate an artifact with higher amplitude [15, 16]. However, they are insufficient in preventing saturation and signal loss with larger artifacts that can occur during recording near the stimulation site – a scenario likely to occur in high-density neural interface arrays with closely spaced electrodes.

2.2 *Stimulation Artifact Cancellation by Digital Signal Processing*

Signal post-processing is another approach to recover neural recording by removing the artifact using various algorithms. The algorithms recover artifact-free recordings by subtracting the representative artifact template, where the template is generated by modeling the artifact. One such simple and common approach creates the template by averaging a collection of periods of signal waveform during all stimulation periods [17, 18]. This approach excels especially when the stimulation artifact is of a similar size or smaller than the neural signal's features of interest and does not significantly overlap with those features (Fig. 1c). This is likely the case for any electrodes that are far from the stimulation site which cause the stimulation artifact to be highly attenuated in amplitude. This attenuation also reduces perceived variations from one artifact instance to another at the recording electrode, thus making the average template an accurate representation of all artifact instances.

When the artifact varies significantly between stimulation instances and the average template is subtracted, a residual waveform remains, which distorts the neural recording and possibly confounds the neural response to a stimulus. Then a more accurate template is needed to accurately represent each individual instance. More sophisticated artifact models and removal techniques accomplish this by polynomial curve fitting [19], Gaussian processes [20], wavelet transforms [21], and Hampel identifier filters [22]. Finally, in dense electrode array interfaces, an algorithm design can leverage coupling of an artifact to many electrodes simultaneously, with different attenuation at each. Here, principal component analysis (PCA) or independent component analysis (ICA) techniques can be added to an algorithm to accurately extract the artifact from the multichannel signal [23]. Still, a major constraint to the performance of any post-processing algorithm is the saturation of the amplifier, which must be avoided to recover a fully intact neural recording.

2.3 Stimulation Artifact Cancellation by a Complete System Design

To eliminate the large amplitude range of artifacts present in electrodes of the high-density neural interfaces and allow artifact-free recording during stimulation, both circuit design and digital processing techniques must work in synergy in a complete system solution. Circuit techniques must be used to keep the signal within amplifier's linear range, while back-end processing must separate intact recording from any artifact distortion. To realize the first task efficiently, the recording circuit must enable feedback, which will subtract a large artifact during recording. The second task can use any combination of signal processing techniques, working in real time with the recording circuits, to complete the artifact removal.

In one such system design [24], a switched capacitor front-end digitizes the input signal sample, and a back-end DAC subtracts that sample from the input by capacitive coupling, such that at the next clock, only the difference between the current and previous sample is amplified. This could eliminate saturation due to large artifacts, as only the delta between samples is amplified. The integration of the signal sums the deltas and theoretically allows the circuit to digitize full artifact signal without loss. In practice, however, the loss does occur when a large fast-edged artifact still saturates the signal path as the loop is challenged to prevent saturation. Then the technique instead accelerates recovery post-artifact. Here, the system design samples signals at 1 kS/s, slower than the duration of the artifact (<1 ms). The saturation and signal loss thus only happen for two samples, and the digital processing recreates them by a simple interpolation between neighboring data points.

Another complete system design [25] digitizes the first stimulus instance of the non-amplified artifact into memory using an ADC and then outputs the stored artifact points using a DAC into a subtraction node at the following artifact instances (Fig. 1d). The amplification happens after subtraction, and at all stimulation instances following the first one, the artifact is suppressed, and saturation is fully prevented. The design requires the post-subtraction artifact residue to be just small enough to meet the amplifier's input compliance limits. This relieves the design constraints on precision and thus resolution of the ADC and DAC in the artifact-suppressing loop. The back-end of the system digitally processes the resulting amplified signal and removes any residual artifact. The digital processing is accomplished by a combination of template averaging and adaptive filtering, which together removes the repetitive residual artifact and adjusts for its variation from instance to instance even when it fully overlaps with the neural signals. With this schema, the design can suppress and remove very large artifacts, which are 100 dB higher in amplitude relative to the neural signals recorded. Thus, unlike all other existing designs, this one is capable of stimulation and recording simultaneously at the same electrode where the amplitude of the observed artifact is the largest, making its removal most difficult. This flexibility allows the system to record intact neural signals from all electrodes of a high-density neural interface while stimulating at any arbitrarily chosen electrode. A performance demonstration of such a system is shown in a test-bench test (Fig. 2), where a periodic train of stimulation

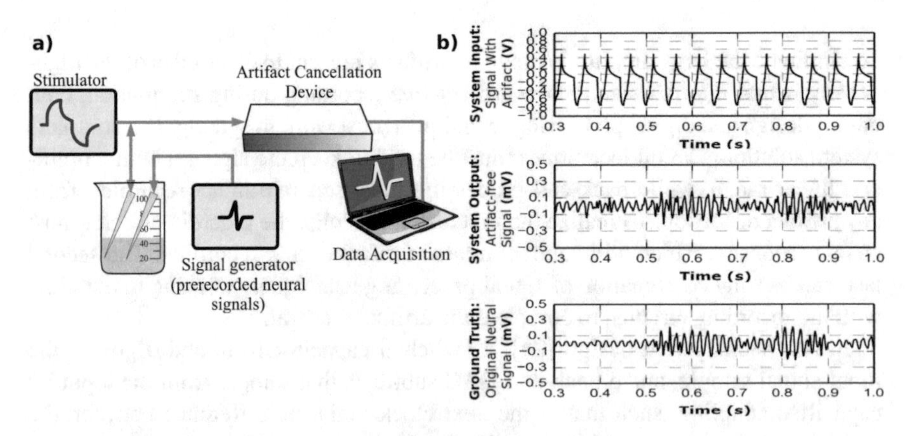

Fig. 2 Bench-top test results demonstrate the performance of an artifact cancellation system: (**a**) a periodic stimulus train with frequency of 20 Hz is injected into a saline solution through an electrode. A second electrode injects a prerecorded neural signal (human ECOG data containing high frequency oscillations during onset of a seizure) into the same solution. The artifact cancellation system is recording both the large stimulus artifact and the neural signal superimposed at the stimulating electrode. (**b**) The system's input contains stimulation artifacts (1.1 V amplitude), which is much larger than the ground truth neural signal (250 μV amplitude) superimposed on it. The system uses a combination of circuit techniques and digital processing to remove the artifact and recover the neural signal without recording loss. Some signal contamination by the system's noise is seen at the output but does not alter the main features of the recovered signal. (Human ECOG data is provided courtesy of Dr. Yue-Loong Hsin, Biomedical Electronics Translation Research Center at National Chiao Tung University, Taiwan. Copyright 2018 IEEE [25])

pulses is injected with an electrode resulting in 1.1 V artifacts. A prerecorded 250 μV neural signal is superimposed on the artifacts and recorded by the same stimulating electrode. The system records from this electrode, removes the artifacts online during recording, and recovers neural signals without signal loss. A standard neural amplifier would have saturated during this large stimulation pulse train, likely resulting in a complete loss of signal.

The two complete system design examples above are the only known works that have also successfully demonstrated full online artifact cancellation in vivo at the time this text is written. The future developments of high-density neural interfaces will inevitably need to continue exploring this design space to ensure fully bidirectional capability of the electrical neural interfaces.

3 Focalized Stimulation

An effective stimulation should be capable of selectively activating the desired neurons of interest while avoiding others, making selectivity and focality critical concerns in electrical stimulation systems. However, in the high-density stimulation interface with conventional stimulation paradigm, the resolution of stimulation is poor as the delivered charge from the stimulating electrodes would inevitably spread to nearby tissues. An in vitro study to perform and validate focalized stimulation using implantable microelectrode array and bioelectronics has been conducted in [10]. However, in order to further the stimulation focality/selectivity, it is essential to develop a stimulation approach that can focalize the electrical field in the desired region with ease.

For a high-density stimulation system, various algorithms are used to achieve focalized stimulation, including reciprocity principles [26], machine learning algorithms [27], and optimization of electric montage [28–30]. Different methods have their own strengths and weaknesses and may be suitable for different stimulation technologies. Thus, choosing an appropriate algorithm is one of the major challenges in achieving focalized stimulation. Here, we summarize some optimization algorithms to achieve the focalized stimulation in high-definition transcranial direct current stimulation (HD-tDCS) (Table 1). Least square [28] minimizes the secondary error term, which has strong focusing capability, but the produced e-field intensity is always small and insufficient to stimulate the brain effectively. Some methods, based on least square, are proposed to overcome its limitations. Weighted least square method [28] is designed to increase the e-field intensity by adding different weightings between target region and non-target region. Furthermore, researchers add a penalty term to the weighted least square method, which is the L1-norm on the stimulation current [30]. This improvement is made for targeting multiple brain regions. Maximum intensity [29], as the name suggests, maximizes the e-field intensity at the target region; however, the focusing capability is poor. Linearly constrained minimum variance (LCMV) [28] aims to achieve the exact desired intensity at the target region while minimizing the total energy, which is novel and is a great way to balance the trade-off between the intensity and the focusing capability.

Table 1 Some optimization methods for focalized stimulation.

Optimization algorithm	Cost function
Least square	$\min_S KS - e_d{}^2$ Subject to safety constraints
Weighted least square	$\min_S W^{1/2}(KS - e_d)^2$ Subject to safety constraints
Weighted least square with penalty term	$\min_S W^{1/2}(KS - e_d)^2 + \lambda\lvert S\rvert$ Subject to safety constraints
Max intensity	$\max_S e_0{}^T CS$ Subject to safety constraints
Linearly constrained minimum variance	$\min_S KS^2$ Subject to $CS = e_0$ and safety constraints

Notes: e_d is a vector that denotes the desired e-field distribution of the cortex, e_0 is desired e-field distribution of the target region on the cortex, S is the current of each electrode, K is the lead field matrix, which contains the relationship between the current of each electrode and the e-field intensity of each point on the cortex, C is the sub-matrix of K, which denotes contains the relationship between the current of each electrode and the e-field intensity of each point in the target region. W is the weighting matrix, which counter balances the asymmetry in the number of target and non-target elements

Nevertheless, it cannot guarantee that the highest intensity appears in the target region, and it may have no feasible solution due to the hard constraints. The obvious challenges are identifying the optimal object and selecting the optimal algorithm based on its advantages and drawbacks.

It should be noted that although those algorithms are designed and validated in noninvasive applications, we believe that the concept is also applicable to implantable devices. In [31], an optimized programming algorithm for the cylindrical and directional electrodes is proposed for DBS, which has the similar strategy of the max intensity method in HD-tDCS. Also, genetic algorithm is utilized to achieve close-loop optimization of DBS [32]. Machine learning algorithm [27] and optimization method [33] are also applied in spinal cord stimulation (SCS). Considering the development of those electric stimulation technologies, we envision that the next-generation electroconvulsive therapy (ECT) will be neuro-targeted, which requires multichannel stimulation device and algorithms to control the stimulation protocol.

4 High-Density Electrode Array

4.1 Scaling Trend of Neural Interfaces

Achieving neural interfaces with both large throughput and chronic biocompatibility is critically important. Reverse engineering the brain is extremely difficult, since there are over 10^9 neurons in mammalian cerebral cortices spaced 50 μm on average, and each one receives connections, i.e., synapses, from many other neurons [34]. It is also estimated that almost all neural circuits are composed of over thousands of neurons. Further complicating the challenge is the fact that the brain is plastic and changes over its lifetime in response to changes in activity. This synaptic plasticity is the neurological foundation for learning and memory. To reverse engineer the brain, it is therefore crucial that neural interface devices can map brain activity over the scale of thousands of neurons and over long periods of time. On the other hand, high functionality neuroprosthetics also require >1000 channels to possess advanced abilities such as fine motor control. Needless to say, in biomedical brain implants, the neural interface has to be chronically stable. For example, in neuroprosthetic limbs, the lifetime of the device in the brain needs to be as long as the brain's lifetime, i.e., decades [35].

The need for large-throughput neural recording has led to great strides in the development of high-density neural interfaces. Since the introduction of wires to measure the extracellular electrical activity of a few neurons in the 1950s, neural interface technologies have evolved into today's micro-fabricated electrode arrays with more than 1, 000 electrodes and have been the workhorse for modern neuroscience research and biomedical applications. The number of recorded neurons has increased exponentially since the first neural interface, approximately doubling every 7 years [36]. Recently developed silicon-based technology also allows hundreds of electrodes to be fabricated as arrays in a fully automated process.

4.2 Actively Multiplexed, Flexible Electrode Arrays

4.2.1 Rationale and Concept

While reduction of the electrode size and pitch increases the density and throughput of the arrays, it also raises significant problems with a high number of wiring, which results in interconnect difficulties and signal cross talk. To date, the largest-channel-count electrode arrays take advantage of large-scale integrated CMOS circuits to provide on-site time-division multiplexing, amplification, signal conditioning, and data conversion [2, 37–41]. On-site multiplexing is arguably the only viable way to achieve large-scale, high-density electrode arrays. With just a single level of multiplexing, the amount of interconnect wires needed would be as low as 20 for 100 electrodes, 64 for 1,024 electrodes, and 200 for 10,000 electrodes. In one of the simplest form of this approach, two transistors are used as a buffer and a multiplexer at each sensing site. The buffer will amplify the recorded signal from the adjacent tissue, and the multiplexer will act as a switch to decrease the number of wires that have to be connected to the system. Similar to the DRAM memory architecture, each buffered electrode will be in series with an on-site multiplexing transistor, which is controlled by a "word"-line connecting all transistors in a row. A "bit" line connecting all electrodes in the column will conduct neuronal signal or electrical stimulation current. This method has enabled two or three times increase of the number of the electrode per unit area compared to traditional passive neural probes.

Previously, the simultaneous recording electrode sites in passive arrays were mainly limited to 256 largely by the number of interconnect wires between neural interfaces and the recording system. Currently, active electrodes have demonstrated up to 1,356 channels in vivo, with remarkable promises for further scaling up [42].

4.2.2 Capacitively Coupled Arrays of Multiplexed Flexible Silicon Transistors for Chronic Electrophysiology

Long-term encapsulation from biofluids has been an important aspect in a chronic implant, especially when active electronics are used. As flexible devices are developed to target soft, curved, and moving surfaces of body organs, it becomes necessary to develop thin, flexible encapsulation layer that can provide sufficient protection yet not jeopardize the sensing interface. Ideally, this encapsulation material should cover all exposed surfaces of the electronics to prevent penetration of biofluids in all directions, including the following characteristics: (1) biocompatible molecular composition; (2) high electrical capacitance (for electrical interfaces); (3) low thermal conductivity and thermal mass (for thermal interfaces); (4) good optical transparency (for optical interfaces); (5) low areal density; (6) low flexural rigidity; (7) defect-free material perfection over large areas; (8) thermal and chemical compatibility with polymer substrates; and (9) lifetimes of multiple decades in electrolyte solutions at physiological pH and temperature under cyclic bending conditions. Despite more than a decade of research on this topic in academic and industrial groups around the world, there is currently no proven material system that offers these properties.

Conventional long-term encapsulation methods include thick, rigid layers of bulk metal or ceramics, which are not compatible with flexible systems. The use of thin flexible films is limited since they cannot be applied to active, semiconductor-based electronic platforms where operations involving continuous applied voltages and induced currents are needed. Using organic/inorganic-based multilayer is another option, but they cannot withstand harsh conditions in the body nor decades of timescale. Recently, a newly developed solution is to introduce ultrathin, pristine layers of silicon dioxide (SiO_2) thermally grown on silicon wafers, integrated with flexible electronic platforms [43]. This thermal SiO_2 layer provides reliable barrier characteristics of 70 years with thickness less than 1 μm, which also provides excellent mechanical flexibility.

Specifically, through an innovative device fabrication process, an ultrathin, thermally grown layer of SiO_2 covers the entire surface of the Si NM electronics. Thermal SiO_2 encapsulation on flexible electronics is extremely durable, due to an extremely slow hydrolysis process, $SiO_2 + 2H_2O \rightarrow Si(OH)_4$, while conventional inorganic and organic materials cannot provide long-term protection because of extrinsic defects such as pinholes and grain boundaries. The linear form of the lifetime of SiO_2 encapsulation and its zero intercept suggest that hydrolysis proceeds exclusively by surface reactions without any significant role of reactive diffusion into the bulk of the SiO_2 or of permeation through defect sites. At a pH of 7.4 and 37 °C, the dissolution rate for thermal SiO_2 is ~4×10^{-2} nm/day (Fig. 3a, b). Bending tests and soaking tests demonstrate mechanical flexibility and robustness of the

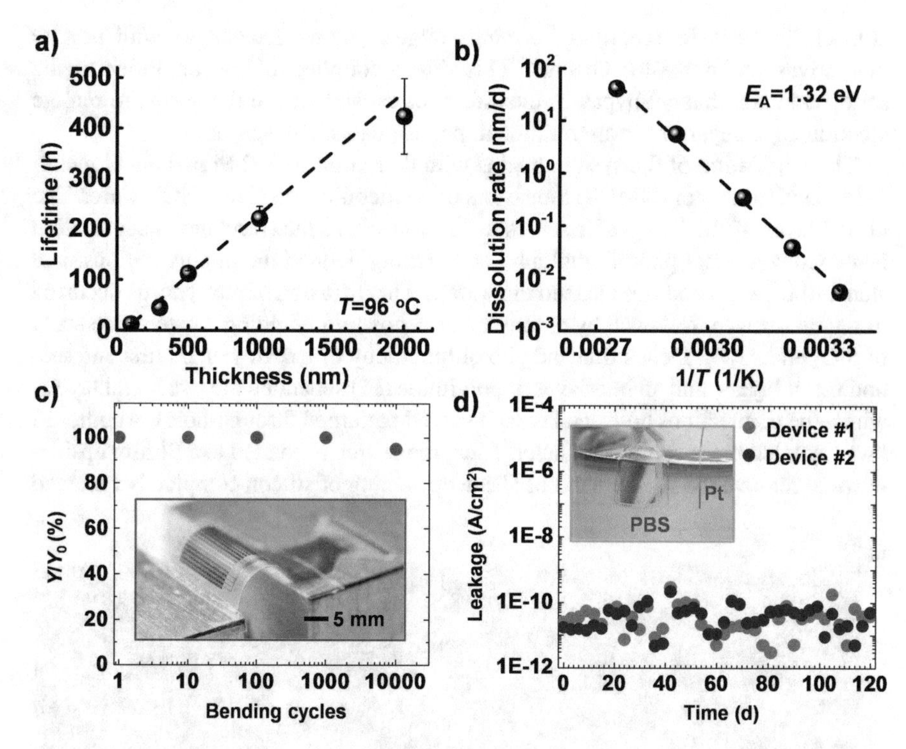

Fig. 3 Transferred, ultrathin thermal SiO_2 as long-term encapsulation layer for active flexible electronics. (**a**) Accelerated lifetime and (**b**) dissolution rate of thermal SiO_2 at elevated temperatures. (**c**, **d**) Demonstration of the bending robustness and soaking stability of the active flexible electronics covered with a 900-nm thermal SiO_2 layer, respectively. (Copyright 2016 P.N.A.S [43]. Copyright 2017 Nature Publish Group [37])

device (Fig. 3c, d). The performance of the device does not change up to 10,000 cycles of bending at a radius of 5 mm. We have achieved in vitro stability for over 120 days in PBS solution, three orders of magnitude longer than previous active flexible electronics. For a 900-nm-thick layer (which is sufficiently thin to meet the key requirements outlined above), this dissolution rate corresponds to a lifetime of nearly 70 years, exceeding the lifetime of most patients who might benefit from chronic flexible electronic implants.

The thermal SiO_2 encapsulation approach is versatile to many surfaces, including thin active electronics. A proof of concept capacitively coupled array consists of 396 multiplexed capacitive sensors (18 columns, 22 rows), each sensor having two transistors. The size of the sensor is 500 × 500 μm^2, with total sensing area of 9.5 × 11.5 mm^2 (Fig. 4a). An ultrathin layer of 900-nm-thick thermal SiO_2 covers the top surface. This layer serves as not only the dielectric for capacitive coupling of adjacent tissue to the semiconducting channels of the associated silicon nanomembrane transistors but also a barrier layer that prevents penetration of biofluids to the underlying metal electrode and associated active electronics. Each capacitive sensor consists of an amplifier and a multiplexer (Fig. 4b). The thermal SiO_2 layer contacts the tissue for recording, forming a large capacitor that couples with the gate that drives the transistor channel. This direct coupling of the amplifier to the semiconductor channel bypasses the effects of capacitance in the wiring to remote electronics, a departure from traditional, passive capacitive sensors.

The fabrication of the system begins with definition of 792 Si n-channel metal-oxide semiconductor (NMOS) transistors on a silicon-on-insulator (SOI) wafer. The buried layer of the SOI wafer serves as the capacitive interface and encapsulation layer. Conventional photolithography and etching defined the doping regions, and standard RCA procedures cleaned the wafers. The diffusion of phosphorus occurred in a tube furnace, followed by atomic layer deposition, yielding a gate oxide stack of SiO_2 and Al_2O_3. Deposition and photolithography of Cr/Au yielded first and second metal layers, and an interlayer of polyimide (PI) separated the two metal layers, where the connections between layers involved patterned through-holes. Another PI layer isolated the second metal layer. Then, the device is bonded to a PI film upside-down to remove the silicon handle. The deep etching of silicon completely removed

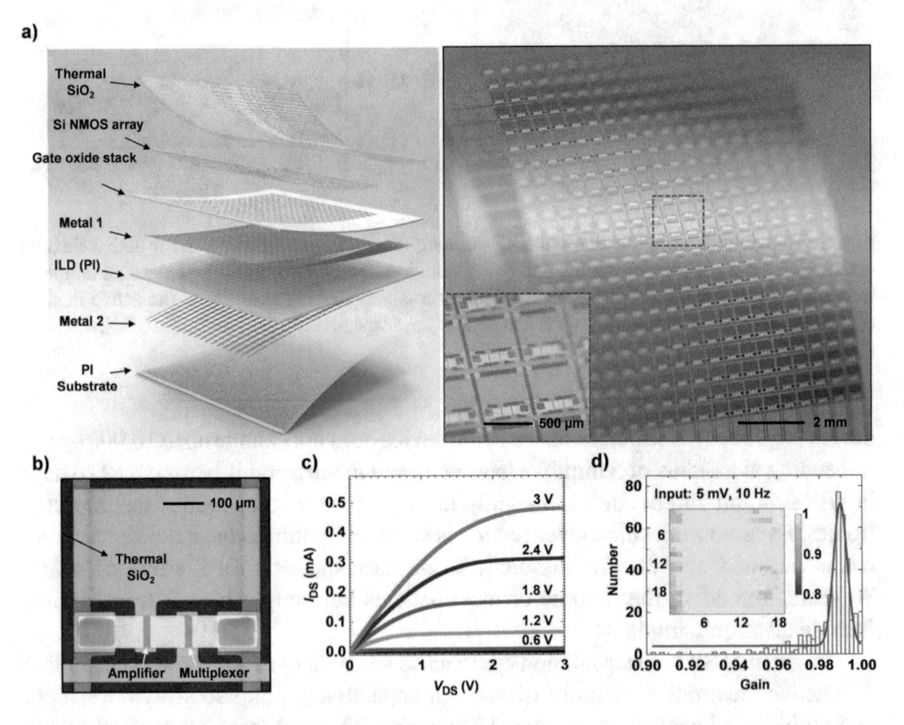

Fig. 4 Capacitively coupled silicon nanomembrane transistors (covered by a thermal SiO_2 layer) as amplified sensing nodes in an actively multiplexed flexible electronic system for high-resolution electrophysiological mapping. (**a**) A schematic of the device structure, highlighting the key functional layers (left) and a photograph of a completed capacitively coupled flexible sensing system with 396 nodes in a slightly bent state (right). Inset: a magnified view of a few nodes. (**b**) An optical microscope image of a unit cell, showing circuits with annotations of key components. (**c**) Output characteristics of capacitively coupled sensing using a test transistor. (**d**) Histogram of gain values from all 396 nodes with Gaussian fitting, indicating 100% yield and near-unity average gain. Inset: a spatial mapping of gain. (Copyright 2017 Nature Publish Group [37])

the bulk silicon substrate. Photolithography defined areas for forming opening for contact lead, and finally the device was peeled off from the handling substrate. The fabrication process presented here can be scaled up to any large silicon wafer, providing possibilities to cover most of the human body organs.

Figure 4c demonstrates the principle of the capacitive coupling to a Si NM transistor. Contacting SiO_2 layer and a droplet of phosphate buffered saline (PBS) solution causes coupling to the gate pad of the transistor, demonstrating the transfer characteristic of the effects of electrical potential generated from the contacting tissue. There is a good agreement of measurements from both direct biasing of the gate pad and the resulting electrical performance, validating the capacitively coupled sensing design. The capacitively coupled transistor shows an on/off ratio of ~10^7 and a peak effective electron mobility of ~800 cm^2 $(Vs)^{-1}$. The output characteristics show the high-performance transistors for the required high-fidelity amplification and fast multiplexing. A histogram plot of the gain values on all channels shows excellent uniformity in electrical responses across the device, demonstrating a 100% yield (Fig. 4d).

Implantation of the 64-channel device to the rat auditory cortex yielded successful long-term in vivo recording (Fig. 5a). Leakage current maintained the same level after 20 days, indicating the reliability of the thermal SiO_2 encapsulation layer (Fig. 5b). The spatial mapping from the rat primary cortex proves functionality of the system (Fig. 5c). The 64-channel system recorded for 20 days, demonstrating high-speed mapping of electrophysiology. Further engineering has demonstrated that a trilayer based on thermal SiO_2 with additional ultrathin Parylene C and HfO_2 improved the encapsulation performance by two orders of magnitude, with greatly enhanced barrier properties against mobile ions.

5 Gigabit Wireless Link

In this section, we first discuss several design considerations for wireless data link in high-density neural interfaces. Next, we summarize state-of-the-art high-data-rate wireless links that are promising candidates for cortical implants. Finally, we present a high-density gigabit wireless recording system.

5.1 Design Consideration

The design of wireless data link faces challenges in trade-offs between data rate, power consumption, and transmission distance. The situation is even more stringent in the design of large-scale high-density cortical implant, which needs a high-data-rate wireless link with limited power budget set by safety regulations. Moreover, the transmission distance must be long enough to ensure the subjects can freely move. In this subsection, we will discuss the design consideration in terms of data rate, power consumption, and transmission distance.

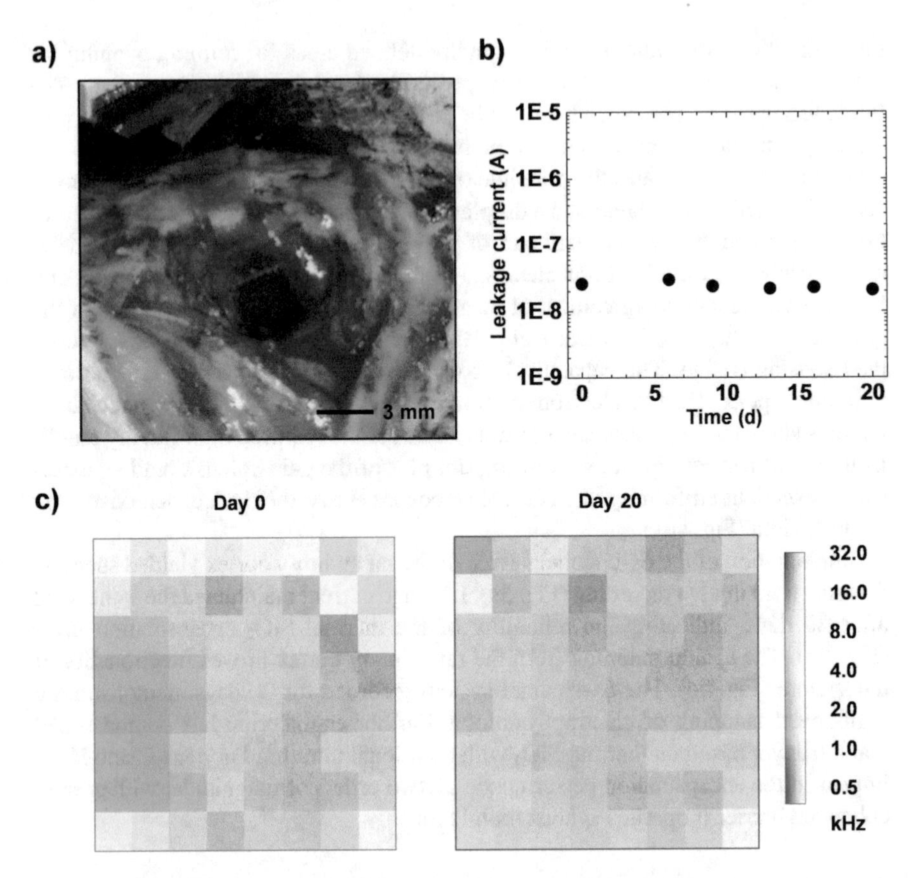

Fig. 5 In vivo recording of rat auditory cortex using a flexible, capacitively coupled, actively multiplexed sensing matrix with 64 nodes. (**a**) A photograph of the device on the rat auditory cortex. (**b**) Leakage current measured from the array as a function of time, up to 20 days. (**c**) Spatial mapping of best frequency revealed the dorsorostral-ventrocaudal gradient of representation in rat primary auditory cortex on day 0 (left) and day 20 (right), indicating stable sensing capabilities throughout the study. (Copyright 2017 Nature Publish Group [37])

5.1.1 Bandwidth/Data Rate Requirement

Several researches have demonstrated large-scale cortical recording in behaving animals with recording sites of ~500 to 1,024 for the study of brain activity and realization of brain-machine interface (BMI) [44–47]. It is further postulated that a BMI targeting the restoration of limb movement requires recording of 5,000 to 10,000 neurons, while recording of 100,000 neurons is essential for the control of whole body movement [45]. Recording with such a high channel count necessitates a wireless link with a data rate in Gb/s range. Using a sampling rate of 30 kHz to capture the neural spike within 10 kHz bandwidth [48] and a resolution of 10 bit to digitize the spike signal, the amount of data would be 0.3 Gb/s for 1,000 channels,

3 Gb/s for 10,000 channels, and 30 Gb/s for 100,000 channels, respectively. This significant amount of data can be reduced by ~10× if state-of-the-art compressed sensing techniques are applied [49, 50]. However, the use of compressed sensing might make the power constraint of the wireless link even more stringent, when the power consumption of the compressed sensing hardware is larger than the power saving contributed by the data rate compression. We will discuss this issue in Sect. 5.1.2.

5.1.2 Power Constraint

The power constraint of the cortical implants is set by the safety guideline that the temperature elevation in the brain cannot be greater than 1 °C [51, 52]. The temperature elevation is mainly attributed to the power dissipated by the electronics as well as the electromagnetic (EM) fields induced by the wireless telemetries in the implant [53]. The power of both heat sources that induce 1 °C increase has been quantified using 2-D and 3-D head models [52, 54]. It has also been shown that the temperature rise resulting from EM field induced by RF frequency smaller than ~6 GHz is negligible compared to the temperature rise resulting from power dissipation of electronics [54]. Here, we only consider the power dissipation effect for power constraint estimation for simplicity.

Figure 6 shows the energy budget of one recording channel used for the power constraint estimation for the high-data-rate wireless link [55]. A digital processing unit (DSP) with a 20× data compression capability is utilized. According to [52], assuming the chip size is 8 mm × 8 mm, the power consumption of the chip should be kept below 37 mW to ensure the temperature rise in the brain is less than 1 °C. Based on Fig. 6, each recording channel exclusive of the transmitter consumes 12.6 μW. The required energy cost per bit for the wireless data link in order not to exceed the upper power limit thus can be calculated as 1.63 nJ/bit for 1,000 channel recording and 10.6 pJ/bit for 2,900 channel recording, respectively. This stringent energy efficiency specification can be relieved by using advanced low-power recording circuitry. Yet an energy-efficient wireless link with a performance of <2.5 pJ/bit is still essential to realize a 100,000 channel wireless recording system for a BMI targeting whole body movement.

It should be noted that the compressed sensing technique does not always help relieve power budget. For example, [49] demonstrated a compressed sensing circuitry that is capable of 8× compression for spike signal with a power consumption of ~5 μW per recording channel. Based on Fig. 6, the required data rate of the wireless link is decreased from 300 kb/s to 37.5 kb/s with an 8× data compression. If a wireless link with an energy efficiency of <10 pJ/bit is adopted [56–60], the power savings contributed by the data compression is <2.625 μW, which is smaller than the power consumption of data compression circuitry. It is thus not worthy to apply compressed sensing technique in this case.

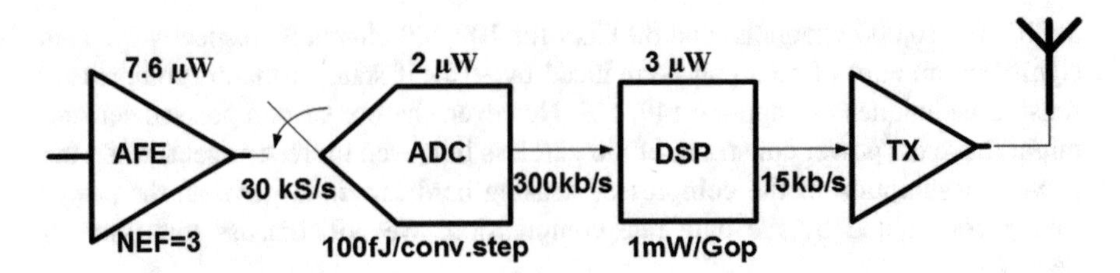

Fig. 6 Energy cost and specification of functional bocks in a multichannel recording system [55]

Although the temperature elevation resulting from EM field is negligible when the RF frequency is smaller than 6 GHz, it should be noted that the radiated power absorbed by the tissue increases with the higher frequency. It is thus necessary to consider the temperature elevation due to EM field when millimeter (mm) wave telemetry is used. Researchers have investigated the biological effects of low-intensity mm waves on human skin cells [61, 62]. Nevertheless, the upper power limit of the mm wave telemetry in the cortical implant has not been characterized yet. Taking both the radiated power limit and the low-power requirements of the electronics into consideration would make the design of mm wave wireless link even more challenging.

5.1.3 Transmission Distance

The transmission distance of a wireless link depends on the signal path loss and reflection loss, delivered power from the transmitter, the data rate and signal bandwidth, and the sensitivity of the receiver. Unfortunately, those factors trade off with the transmission distance in the realization of long-distance, high-data-rate wireless link in the cortical implant. For example, a strong transmitted power favors the long-distance transmission, yet it conflicts with the low energy cost per bit requirement of the transmitter. At the receiver side, the increased noise floor due to the required large data bandwidth leads to the degradation of the sensitivity, thus shortening the transmission distance. On the other hand, the signal loss is determined by the characteristic of the tissue, position of the antenna, as well as carrier frequency. The signal loss is larger if a higher carrier frequency or large bandwidth is used in order to support higher data rate.

Received signal intensity can be calculated according to different antenna locations inside the brain. Figure 7 shows several options for the placement of transmitter antenna for a cortical implant: (a) underneath the skull, (b) underneath the skin, and (c) uncovered by the skin. According to Beer-Lambert law [63], the intensity of an EM wave inside a material decays exponentially from the surface of the source. Assuming the transmitted power from the antenna is P_t, the power at a distance z from the antenna P_z within the same material is

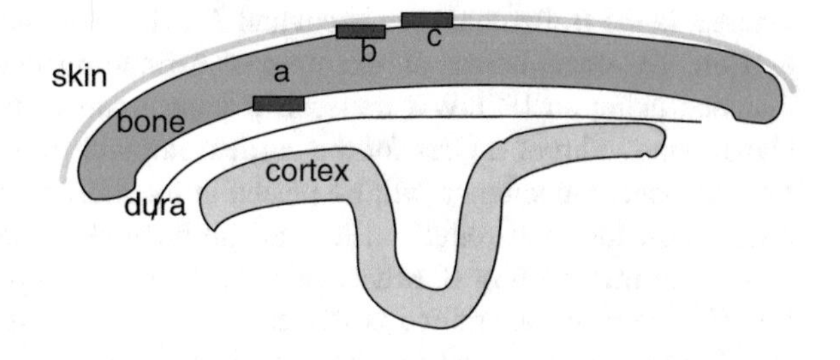

Fig. 7 Potential location for the transmitter antenna inside the brain: (**a**) underneath the bone, (**b**) underneath the skin, and (**c**) uncovered by the skin. (Copyright 2015 IEEE [56])

$$P_z = P_t \times e^{-z/\delta}, \tag{1}$$

where δ is the penetration depth of the material with different values for different tissues at various frequencies. The reflection loss can be estimated by the refractive index of the material. Here we assume a unity-gain antenna and the incident EM signal is perpendicular to the skin-air boundary with an incident angle of 0° for simplicity. According to Fresnel equations [63], the reflection coefficient is

$$\frac{P_{\text{reflective}}}{P_{\text{incident}}} = \left(\frac{1-\beta}{1+\beta}\right)^2 = R, \tag{2}$$

where β is the ratio of the refractive index of the material at the incident and emergent side. Based on (1) and (2), the received EM power at the received antenna in the air is

$$\begin{aligned} P_{\text{received,dB}} &= 20 \times \log P_t + 20 \times \log\left(e^{-\frac{z_1}{\delta_{\text{bone}}}}\right) + 20 \times \log\left(1 - R_{\text{bone-skin}}\right) + \\ & 20 \times \log\left(e^{-\frac{z_2}{\delta_{\text{skin}}}}\right) + 20 \times \log\left(1 - R_{\text{skin-air}}\right) + 20 \times \log\left(\frac{4\pi d}{\lambda}\right) + G_T + G_R, \end{aligned} \tag{3}$$

where z_1 and z_2 are the thickness of the cortex bone and skin, respectively; δ_{bone} and δ_{skin} are the penetration depth of bone and skin at the transmission frequency; $R_{\text{bone-skin}}$ and $R_{\text{skin-air}}$ are the reflection coefficients at the bone-skin interface and skin-air interface, respectively; d is the distance between the receiver and signal source exiting the skin in free space; λ is the free space wavelength of the EM signal; and G_T and G_R are the antenna gain at the transmitter and receiver, respectively. For simplicity, the tissue layers between the skin and bone are neglected in our estimation. The required receiver sensitivity is set by the signal to noise ratio (SNR) at a bit error rate (BER) under a specific modulation scheme, receiver noise figure (NF), and the integrated thermal noise in signal bandwidth (BW)

$$\text{Sensitivity} = 10 \times \log\left(k \times T \times BW\right) + NF + SNR \tag{4}$$

where k is the Boltzmann's constant and T is the absolute temperature.

Here we demonstrate an example of estimating the achievable transmission distance using an IR-UWB transceiver with a carrier frequency of 4 GHz and a signal bandwidth of 2 GHz for the cortical implant [64]. For low power consumption purpose, the antenna will be placed at location b or c in Fig. 7 to bypass the attenuation loss and reflection loss at the bone-skin interface. The transmitter is capable of transmitting 10 Mb/s at a low power consumption of 100 μW, i.e., 10 pJ/bit. The average transmitted power can be calculated as −32 dBm from the measured output power spectrum. The receiver sensitivity is reported as −79 dBm to achieve a BER of 10^{-3}. Skin loss and reflection loss can be estimated as −11.5 dB based on the scalp skin of 8 mm thickness. The transmission distance (d) can be calculated based on the power delivered from the scalp (P_T), the received signal intensity (P_R), and the free space path loss based on (3 and 4).

$$d = \frac{\lambda}{4\pi} \times 10^{(P_T - P_R)/20}. \tag{5}$$

The transmission distance d is estimated as ~0.3 m and ~0.74 m for antenna position b and c in Fig. 7, respectively. This <1 m transmission distance might only be suitable for small animal recording inside a cage. The distance would be even shorter if a higher carrier frequency or a larger bandwidth is used in order to support higher data rate. A novel technique that can increase the transmission distance of the wireless link is thus in an urgent need.

5.2 State-of-the-Art Gigabit Wireless Telemetry

State of the art wireless links achieve a data rate in 0.135–6 Gb/s range with an energy efficiency of 2–10 pJ/b (Table 2). Based on the discussion in Sect. 5.1.1, this data rate capability corresponds to 450–20,000 channels of neural spike recording. Most of the works shown in Table 2 utilize impulse radio ultra-wideband (IR-UWB) technology that provides a high data rate with low power consumption. Among various modulation schemes, the most popular one is on-off keying (OOK), which favors the design of low power transmitter as no signal is transmitted for a digital logic 0.

Figure 8 shows an example of gigabit wireless telemetry that achieves a data rate of 6 Gb/s and an energy efficiency of 2 pJ/b [56]. The transmitter (Tx) is composed of an on-off keying (OOK) modulator and a 60 GHz crossed-couple voltage controlled oscillator (VCO) (Fig. 8a). A transformer is utilized to drive the current generating transistors (M1 and M2) in the VCO, while the serialized data controls the switching transistors (M3 and M4) to turn on/off the current to achieve the OOK modulation. A corresponding receiver (Rx) contains a 60 GHz amplifier and a self-mixer that amplifies and down converts the received signal, respectively (Fig. 8b). A transformer couples the signal from the antenna to the 60 GHz amplifier

Table 2 State-of-the-art wireless data links

	[56]	[57]	[58]	[59]	[60]
Frequency (GHz)	60	2.6–5.6	3.1–7	3–5	4
Modulation scheme	OOK	BPSK	OOK	PPM	OOK
Data rate (Mb/s)	6,000	800	500	135	1,000
Power consumption (mW)	12	5.36	3.5	1.4	5
Energy/bit (pJ/b)	2	6.7	7	10.4	5
Antenna position	Air	Underneath the skin	Underneath the bone	Air	Air
Transmission distance (m)	0.012	N/A	0.006	0.12	0.05

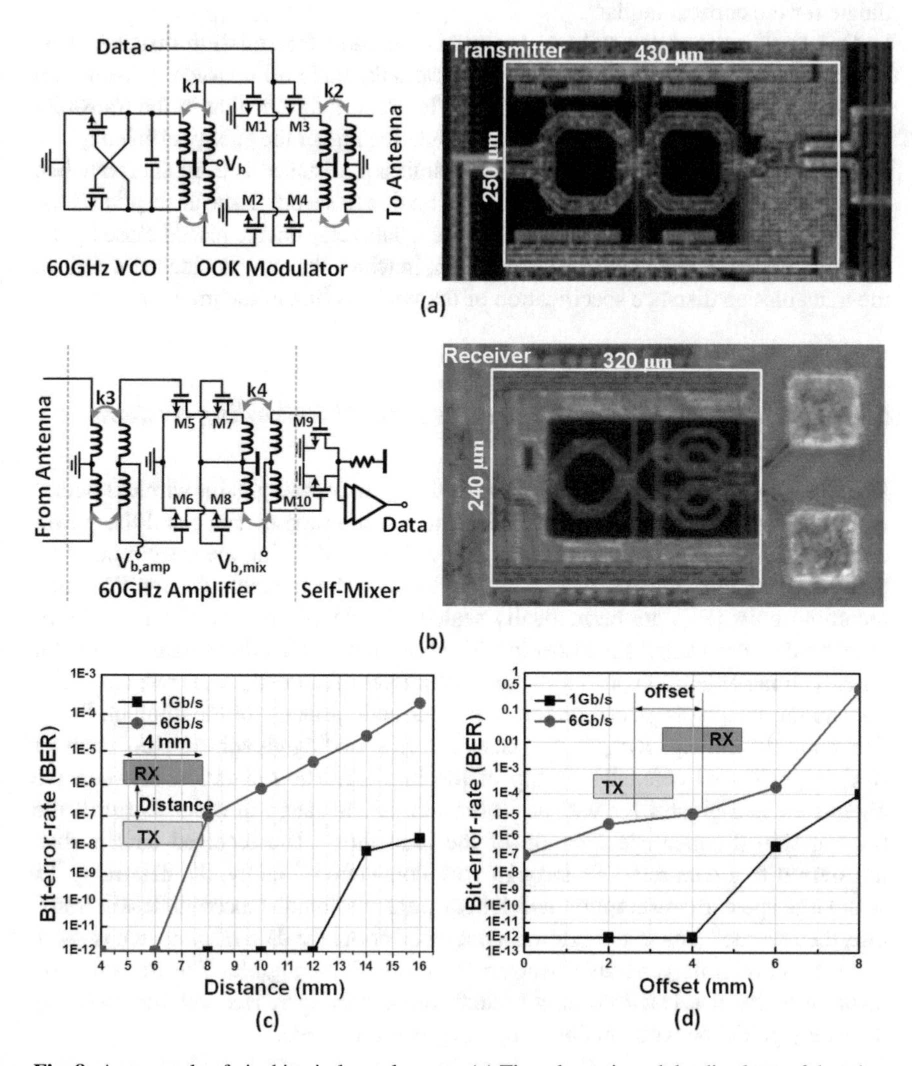

Fig. 8 An example of gigabit wireless telemetry. (**a**) The schematic and the die photo of the wireless transmitter. (**b**) The schematic and the die photo of the wireless receiver. (**c**) Measured BER versus different Tx-Rx distance. (**d**) Measured BER versus different Tx-Rx offset. (Copy-right 2015 IEEE [56])

(M5-M8). A second transformer then takes the amplifier output to drive the self-mixer (M9 and M10) to down convert the 60 GHz signal to a baseband data. The measured results indicate that a bit error rate (BER) of 10^{-12} is achieved at both 1 Gb/s and 6 Gb/s data rates under a transmission distance of <6 mm (Fig. 8c). The transmission distance can be increased to 12 mm at 6 Gb/s data rate with a BER of 10^{-5}. In order to mimic the situation that the transceiver pair is not placed coaxially, the BER versus offset between Tx and Rx is characterized under a transmission distance of 8 mm (Fig. 8d). A BER of $\sim 10^{-4}$ is demonstrated with an 8 mm offset and a 6 mm offset for a data rate of 1 Gb/s and 6 Gb/s, respectively. This gigabit wireless link with a high data rate and a high energy efficiency is a promising candidate for the cortical implant.

One challenge needed to be overcome is the short transmission distance. It is interesting to point out that Table 2 demonstrates the trade-off between transmission distance, energy efficiency, and data rate. The trend indicates that in the transmission medium of air, the higher the data rate and the higher the energy efficiency, the shorter the transmission distance. The transmission distance is even shorter when the signal needs to pass through skin and bone due to severe tissue absorption. One possible alternative to this challenge is to use a data relay device placed close by the implant [56], as shown in the next subsection, in which the relay device helps relieve the transmission distance specification of the wireless link in the implant.

5.3 High-Density Gigabit Wireless Neural Recording System

Figure 9 shows the conceptual diagram of the high-density gigabit wireless neural recording system. The trade-off between the transmission distance and the power consumption of the wireless system is relieved by partitioning the system into two parts (the implant and the data relay device). In the implant, the multichannel recording units (RU) are hermetically sealed and placed on the cortical region of interests through a small skull opening. The central unit (CU) that contains a gigabit wireless transmitter (Tx) processes and transmits the recorded neural response. The transmitter antenna is placed on top of the hermetic package of the implant. In the data relay device, e.g., a cap in this case, a corresponding gigabit wireless receiver (Rx) is positioned coaxially on top of the implant. The separation between the gigabit Tx and gigabit Rx will be in the mm to cm range in order to fulfill the transmission distance specification of the transmitter. The received data is then transferred to a data-relay Tx through wire connection. Finally, the data-relay Tx with larger power consumption and longer transmission distance wirelessly transmits the recorded data to the external data receiver. As the data relay device is away from the cortex, its generated heat on the cortex is negligible. The transmission distance of the wireless recording system thus can be increased without violating the safety guideline while maintaining the gigabit data rate.

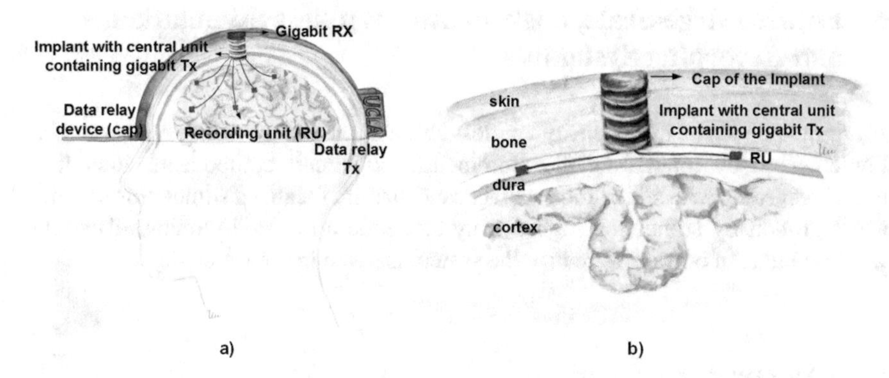

Fig. 9 Conceptual diagram of the high-density gigabit wireless neural recording system. (Copy-right 2015 IEEE [56])

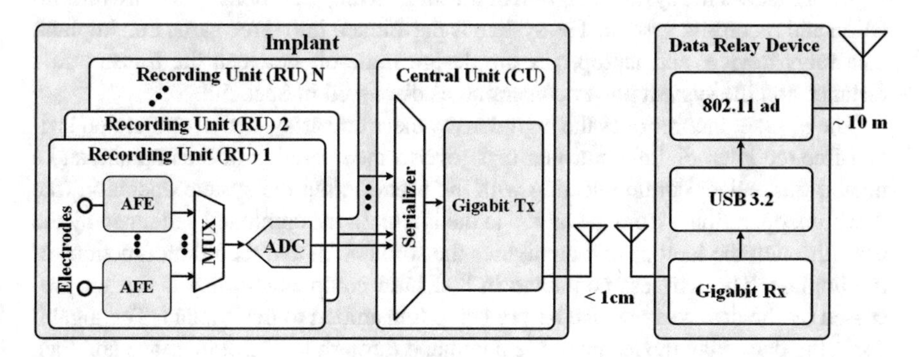

Fig. 10 Block diagram of the gigabit wireless recording system consisting of an implant and a data relay device

The block diagram of the gigabit wireless recording system that comprises an implant and a data relay device is shown in Fig. 10. The RU in the implant captures, amplifies, and digitizes the neural signal. The digitized data is then transferred to the CU to perform data serialization. Subsequently, the gigabit Tx modulates the serialized data and sends the data to the gigabit Rx in the data relay device through the antennas. The gigabit Rx downconverts and demodulates the received signal to a single-bit data stream. The design of the gigabit wireless link has been discussed in detail in Sect. 5.2. The data stream is then transferred to 802.11ad wireless transceiver through USB 3.2 interface that can support 20 Gb/s in Super Speed+ mode. Finally, the wireless 802.11ad transceiver with a data rate of 7 Gb/s and a transmission distance of 10 m transmits the recorded data to an external receiver. The presented architecture achieves a safe and long-distance wireless gigabit neural recording system that has the power hungry long-distance transmitter away from the cortex. The transmission distance specification of the wireless link in the implant thus can be relieved.

6 Future Large-Scale, High-Density Wireless Stimulation and Recording System

A future large-scale, high-density wireless stimulation and recording system can be implemented by integrating the aforementioned circuit components, including neural recording with stimulation artifact cancellation, focalized stimulation scheme and high-density stimulator, high-density electrode array, and ultrahigh-data-rate wireless link. An outlook to realize the system is given in the following.

6.1 System Architecture

Figure 11 shows the system diagram of the large-scale, high-density wireless stimulation and recording system. The system is partitioned into three parts, i.e., implant, data relay device, and laptop, per the design trade-off between the transmission distance and the system power constraint, as discussed in Sect. 5.3.

The implant incorporates the high-density electrode array, neural stimulator, artifact-free recorder, digital controller that governs the operation of the implant, and a gigabit transceiver communicating with the user. During the system operation (the downlink operation – from the laptop to the implant), the command generated by the user (through the laptop) first configures the stimulation and recording functions of the implant. The wireless command in 802.11ad compliant format is further processed by the data relay device before being transmitted to the implant. The gigabit Tx in the data relay device takes the command through USB 3.2 interface and then wirelessly transmits it to the gigabit Rx in the central unit (CU) in the implant. The command programs the digital controller in the CU to generate the desired focalized stimulation parameters, and the stimulation parameters along with the command that sets the recording function are sent to control the stimulation and recording units (SRUs). The high-density stimulator in the SRU then generates the desired stimulation pattern based on the received command. In order to reduce the number of wiring between the high-density electrode array and the high-density stimulator, de-multiplexers (DEMUXs) are utilized to perform time-division de-multiplexing, which is similar to the idea of multiplexing that is discussed in Sect. 4.2. The design of high-density stimulator and DEMUX has been discussed in detail in [10, 65]. The data relay device can be attached to a cap, as shown in Fig. 9, while the gigabit data receiver is placed coaxially on top of the implant with a separation of <1 cm.

On the other hand, during the uplink operation (from the implant to the laptop), the high-density electrode array acquires the evoked neural response right after the stimulation. The multiplexers (MUXs) are used to reduce the number of wiring between the electrodes to the recording circuitry. The captured neural response and stimulation artifact are then recorded by the stimulation artifact-free recorders. A detailed description of the high-density implementation of artifact-free recorder can be found in [25]. The neural signals are then digitized, serialized, and sent to the laptop via two-step wireless transmission, as the process presented in Sect. 5.3.

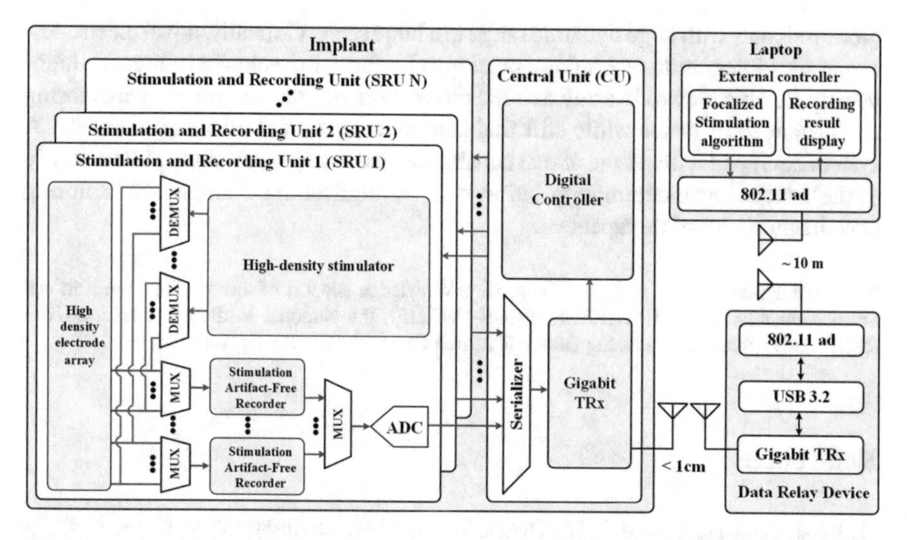

Fig. 11 Block diagram of the future large-scale, high-density wireless stimulation and recording system

6.2 Outlook

The wireless, high-density neural interface has been growing rapidly and has achieved numerous progress under the trend of investigating complex brain dynamic and causality. Specifically, the results from capacitively coupled high-density arrays, as presented in Sect. 4, demonstrate a promising route toward safe, robust, and high-performance flexible electrode array for high-density brain mapping in both clinical and research settings. We see no fundamental hurdles, for example, in achieving systems with thousands or even hundreds of thousands of channels utilizing the same materials. Using the ultrathin layer of thermally grown SiO_2 as long-lived encapsulation, the active flexible electrode array system can be used in long-term in vivo recordings. In the future, the mechanical rigidity of these devices can be further reduced. Rather than just flexibility, stretchable neural interfaces might allow better coverage of soft and irregular curvilinear surfaces of the brain, as well as minimize the implant-tissue mechanical mismatch and the resulted foreign body responses.

In terms of high-data-rate wireless link, the state of the arts have demonstrated a data rate of ~Gb/s range and an energy efficiency of <10 pJ/b [56, 57, 60] with a transmission distance in the cm range. It is feasible to perform ~20,000 channels of neural spike recording if using 6 Gb/s data link [56]. Future efforts might be focused on the integration of the high-data-rate wireless transceiver and the sensitive neural recording and stimulation circuitries. Moreover, though it has been demonstrated that the stimulation artifact can be fully cancelled in vivo by combining both the circuit design and digital processing techniques, it is of great benefit to develop a low-noise, low-power stimulation artifact-free high-density recording front-end to pick up signals with large dynamic range in a large scale. Critically, novel electrodes-electronics integration technology must also be further developed to link the high-density flexible electrode array and the electronics without significantly increasing the system form factor while still maintaining robust mechanical connectivity. A system equipped with all the above capabilities will bring a paradigm shift not only to the neuroscience community but also to the clinical applications for exploring new diagnostics and therapeutics.

Acknowledgments Seo, K. J. and Fang, H. acknowledge support of this work by the National Science Foundation (NSF CAREER, ECCS-1847215), the National Institutes of Health (NIH R21EY030710) and the Samsung Global Research Outreach (GRO) program.

References

1. Khodagholy, D., Gelinas, J. N., Thesen, T., Doyle, W., Devinsky, O., Malliaras, G. G., & Buzsáki, G. (2015). NeuroGrid: Recording action potentials from the surface of the brain. *Nature Neuroscience, 18*(2), 310.
2. Viventi, J., Kim, D.-H., Vigeland, L., Frechette, E. S., Blanco, J. A., Kim, Y.-S., Avrin, A. E., Tiruvadi, V. R., Hwang, S.-W., Vanleer, A. C., Wulsin, D. F., Davis, K., Gelber, C. E., Palmer, L., Van der Spiegel, J., Wu, J., Xiao, J., Huang, Y., Contreras, D., Rogers, J. A., & Litt, B. (2011). Flexible, foldable, actively multiplexed, high-density electrode array for mapping brain activity *in vivo*. *Nature Neuroscience, 14*(12), 1599.
3. Yin, M., Borton, D. A., Komar, J., Agha, N., Lu, Y., Li, H., Laurens, J., Lang, Y., Li, Q., Bull, C., Larson, L., Rosler, D., Bezard, E., Courtine, G., & Nurmikko, A. V. (2014). Wireless neurosensor for full-spectrum electrophysiology recordings during free behavior. *Neuron, 84*(6), 1170–1182.
4. Lo, Y.-K., Kuan, Y.-C., Culaclii, S., Kim, B., Wang, P.-M., Chang, C.-W., Massachi, J. A., Zhu, M., Chen, K., Gad, P., Edgerton, V. R., & Liu, W. (2017). A fully integrated wireless SoC for motor function recovery after spinal cord injury. *IEEE Transactions on Biomedical Circuits and Systems, 11*(3), 497–509.
5. Shahdoost, S., Frost, S., Guggenmos, D., Borrell, J., Dunham, C., Barbay, S., Nudo, R., & Mohseni, P. (2016). A miniaturized brain-machine-spinal cord interface (BMSI) for closed-loop intraspinal microstimulation. In: *IEEE biomedical circuits and systems conference (BioCAS),* Shanghai, China (pp. 364–367).
6. Liu, W., Wang, P.-M., & Lo, Y.-K. (2017). Towards closed-loop neuromodulation: A wireless miniaturized neural implant SoC. In: *SPIE defense+ security, International Society for Optics and Photonics*, Anaheim, California, USA (pp. 1019414–1019418).
7. Lo, Y.-K., Wang, P.-M., Dubrovsky, G., Wu, M.-D., Chan, M., Dunn, J. C., & Liu, W. (2018). A wireless implant for gastrointestinal motility disorders. *Micromachines, 9*(1), 17.
8. Arriagada, A., Jurkov, A., Neshev, E., Muench, G., Andrews, C., & Mintchev, M. (2011). Design, implementation and testing of an implantable impedance-based feedback-controlled neural gastric stimulator. *Physiological Measurement, 32*(8), 1103.
9. Deb, S., Tang, S.-J., Abell, T. L., McLawhorn, T., Huang, W.-D., Lahr, C., To, S. F., Easter, J., & Chiao, J.-C. (2012). Development of innovative techniques for the endoscopic implantation and securing of a novel, wireless, miniature gastrostimulator (with videos). *Gastrointestinal Endoscopy, 76*(1), 179–184.
10. Lo, Y.-K., Chen, K., Gad, P., & Liu, W. (2013). A fully-integrated high-compliance voltage SoC for epi-retinal and neural prostheses. *IEEE Transactions on Biomedical Circuits and Systems, 7*(6), 761–772.

11. Noorsal, E., Sooksood, K., Xu, H., Hornig, R., Becker, J., & Ortmanns, M. (2012). A neural stimulator frontend with high-voltage compliance and programmable pulse shape for epiretinal implants. *IEEE Journal of Solid-State Circuits, 47*(1), 244–256.
12. Monge, M., Raj, M., Honarvar-Nazari, M., Chang, H.-C., Zhao, Y., Weiland, J., Humayun, M., Tai, Y-C., & Emami-Neyestanak, A. (2013). A fully intraocular 0.0169 mm 2/pixel 512-channel self-calibrating epiretinal prosthesis in 65nm CMOS. In: *IEEE international solid-state circuits conference digest of technical papers (ISSCC),* San Francisco, California, USA (pp. 296–297).
13. Stanslaski, S., Afshar, P., Cong, P., Giftakis, J., Stypulkowski, P., Carlson, D., Linde, D., Ullestad, D., Avestruz, A.-T., & Denison, T. (2012). Design and validation of a fully implantable, chronic, closed-loop neuromodulation device with concurrent sensing and stimulation. *IEEE Transactions on Neural Systems and Rehabilitation Engineering, 20*(4), 410–421.
14. Sekirnjak, C., Hottowy, P., Sher, A., Dabrowski, W., Litke, A. M., & Chichilnisky, E. (2008). High-resolution electrical stimulation of primate retina for epiretinal implant design. *Journal of Neuroscience, 28*(17), 4446–4456.
15. Yang, Z., Xu, J., Nguyen, A. T., Wu, T., Zhao, W., & Tam, W.-K. (2016). Neuronix enables continuous, simultaneous neural recording and electrical microstimulation. In: *IEEE 38th annual international conference of the Engineering in Medicine and Biology Society (EMBC),* Orlando, Florida, USA (pp. 4451–4454).
16. Chandrakumar, H., & Marković, D. (2017). A high dynamic-range neural recording chopper amplifier for simultaneous neural recording and stimulation. *IEEE Journal of Solid-State Circuits, 52*(3), 645–656.
17. Hashimoto, T., Elder, C. M., & Vitek, J. L. (2002). A template subtraction method for stimulus artifact removal in high-frequency deep brain stimulation. *Journal of Neuroscience Methods, 113*(2), 181–186.
18. Wichmann, T. (2000). A digital averaging method for removal of stimulus artifacts in neurophysiologic experiments. *Journal of Neuroscience Methods, 98*(1), 57–62.
19. Wagenaar, D. A., & Potter, S. M. (2002). Real-time multi-channel stimulus artifact suppression by local curve fitting. *Journal of Neuroscience Methods, 120*(2), 113–120.
20. Mena, G. E., Grosberg, L. E., Madugula, S., Hottowy, P., Litke, A., Cunningham, J., Chichilnisky, E., & Paninski, L. (2017). Electrical stimulus artifact cancellation and neural spike detection on large multi-electrode arrays. *PLoS Computational Biology, 13*(11), e1005842.
21. Yochum, M., & Binczak, S. (2015). A wavelet based method for electrical stimulation artifacts removal in electromyogram. *Biomedical Signal Processing and Control, 22*, 1–10.
22. Allen, D. P., Stegemöller, E. L., Zadikoff, C., Rosenow, J. M., & MacKinnon, C. D. (2010). Suppression of deep brain stimulation artifacts from the electroencephalogram by frequency-domain Hampel filtering. *Clinical Neurophysiology, 121*(8), 1227–1232.
23. O'Shea, D. J., & Shenoy, K. V. (2018). ERAASR: An algorithm for removing electrical stimulation artifacts from multielectrode array recordings. *Journal of Neural Engineering, 15*(2), 026020.
24. Zhou, A., Santacruz, S. R., Johnson, B. C., Alexandrov, G., Moin, A., Burghardt, F. L., Rabaey, J. M., Carmena, J. M., & Muller, R. (2017). WAND: A 128-channel, closed-loop, wireless artifact-free neuromodulation device. *arXiv preprint arXiv*:170800556.
25. Culaclii, S., Kim, B., Lo, Y.-K., Li, L., & Liu, W. (2018). Online artifact cancelation in same-electrode neural stimulation and recording using a combined hardware and software architecture. *IEEE Transactions on Biomedical Circuits and Systems, 12*(3), 601–613.
26. Fernandez-Corazza, M., Turovets, S., Luu, P., Anderson, E., & Tucker, D. (2016). Transcranial electrical neuromodulation based on the reciprocity principle. *Frontiers in Psychiatry, 7*, 87.
27. Sui, Y., & Burdick, J. (2014). Clinical online recommendation with subgroup rank feedback. In: *Proceedings of the 8th ACM conference on recommender systems,* Foster City, California, USA (pp. 289–292).
28. Dmochowski, J. P., Datta, A., Bikson, M., Su, Y., & Parra, L. C. (2011). Optimized multi-electrode stimulation increases focality and intensity at target. *Journal of Neural Engineering, 8*(4), 046011.

29. Sadleir, R., Vannorsdall, T. D., Schretlen, D. J., & Gordon, B. (2012). Target optimization in transcranial direct current stimulation. *Frontiers in Psychiatry, 3*, 90.
30. Huang, Y., Thomas, C., Datta, A., & Parra, L. C. (2018). Optimized tDCS for targeting multiple brain regions: An integrated implementation. *Conference Proceedings: Annual International Conference of the IEEE Engineering in Medicine and Biology Society, 2018*, 3545.
31. Anderson, D. N., Osting, B., Vorwerk, J., Dorval, A. D., & Butson, C. R. (2018). Optimized programming algorithm for cylindrical and directional deep brain stimulation electrodes. *Journal of Neural Engineering, 15*(2), 026005.
32. Feng, X.-J., Greenwald, B., Rabitz, H., Shea-Brown, E., & Kosut, R. (2007). Toward closed-loop optimization of deep brain stimulation for Parkinson's disease: Concepts and lessons from a computational model. *Journal of Neural Engineering, 4*(2), L14.
33. Alo, R., Alo, K., Ilochonwu, O., Kreinovich, V., & Nguyen, H. P. (1998). Towards optimal pain relief: Acupuncture and spinal cord stimulation. In: *Proceedings of the 2nd International Workshop on Intelligent Virtual Environments*, Xalapa, Veracruz, Mexico (pp. 16–24).
34. Herculano-Houzel, S. (2009). The human brain in numbers: A linearly scaled-up primate brain. *Frontiers in Human Neuroscience, 3*, 31.
35. Carmena, J. M. (2013). Advances in neuroprosthetic learning and control. *PLoS Biology, 11*(5), e1001561.
36. Stevenson, I. H., & Kording, K. P. (2011). How advances in neural recording affect data analysis. *Nature Neuroscience, 14*(2), 139–142.
37. Fang, H., Yu, K. J., Gloschat, C., Yang, Z., Song, E., Chiang, C.-H., Zhao, J., Won, S. M., Xu, S., Trumpis, M., Zhong, Y., Han, S. W., Xue, Y., Xu, D., Choi, S. W., Cauwenberghs, G., Kay, M., Huang, Y., Viventi, J., Efimov, I. R. & Rogers, J. A. (2017). Capacitively coupled arrays of multiplexed flexible silicon transistors for long-term cardiac electrophysiology. *Nature Biomedical Engineering, 1*(3), 0038.
38. Viventi, J., Kim, D.-H., Moss, J. D., Kim, Y.-S., Blanco, J. A., Annetta, N., Hicks, A., Xiao, J., Huang, Y., Callans, D. J., Rogers, J. A., & Litt, B. (2010). A conformal, bio-interfaced class of silicon electronics for mapping cardiac electrophysiology. *Science Translational Medicine, 2*(24), 24ra22–24ra22.
39. Du, J., Blanche, T. J., Harrison, R. R., Lester, H. A., & Masmanidis, S. C. (2011). Multiplexed, high density electrophysiology with nanofabricated neural probes. *PLoS One, 6*(10), e26204.
40. Yu, K. J., Kuzum, D., Hwang, S.-W., Kim, B. H., Juul, H., Kim, N. H., Won, S. M., Chiang, K., Trumpis, M., Richardson, A. G., Cheng, H., Fang H., Thompson, M., Bink, H., Talos, D., Seo, K. J., Lee, H. N., Kang, S.-K., Kim, J.-H., Lee, J. Y., Huang, Y., Jensen, F. E., Dichter, M. A., Lucas, T. H., Viventi, J., Litt, B., & Rogers, J. A. (2016). Bioresorbable silicon electronics for transient spatiotemporal mapping of electrical activity from the cerebral cortex. *Nature Materials, 15*(7), 782.
41. Jun, J. J., Steinmetz, N. A., Siegle, J. H., Denman, D. J., Bauza, M., Barbarits, B., Lee, A. K., Anastassiou, C. A., Andrei, A., Aydın, Ç., Barbic, M., Blanche, T. J., Bonin, V., Couto, J., Dutta, B., Gratiy, S. L., Gutnisky, D. A., Häusser, M., Karsh, B., Ledochowitsch, P., Lopez, C. M., Mitelut, C., Musa, S., Okun, M., Pachitariu, M., Putzeys, J., Rich, P.D., Rossant, C., Sun, W.-L., Svoboda, K., Carandini, M., Harris, K. D., Koch, C., O'Keefe, J., & Harris, T. D. (2017). Fully integrated silicon probes for high-density recording of neural activity. *Nature, 551*(7679), 232.
42. Raducanu, B. C., Yazicioglu, R. F., Lopez, C. M., Ballini, M., Putzeys, J., Wang, S., Andrei, A., Rochus, V., Welkenhuysen, M., Helleputte, N. V., Musa, S., Puers, R., Kloosterman, F., Van Hoof, C., Fiáth, R., Ulbert, I., & Mitra, S. (2017). Time multiplexed active neural probe with 1356 parallel recording sites. *Sensors, 17*(10), 2388.
43. Fang, H., Zhao, J., Yu, K. J., Song, E., Farimani, A. B., Chiang, C.-H., Jin, X., Xue, Y., Xu, D., Du, W., Seo K. J., Zhong, Y., Yang, Z., Won, S. M., Fang, G., Choi, S. W., Chaudhuri, S., Huang, Y., Alam, M. A., Viventi, J., Aluru, N. R., & Rogers, J. A. (2016). Ultrathin, transferred layers of thermally grown silicon dioxide as biofluid barriers for biointegrated flexible electronic systems. *Proceedings of the National Academy of Sciences, 113*(42), 11682–11687.

44. Berényi, A., Somogyvári, Z., Nagy, A. J., Roux, L., Long, J. D., Fujisawa, S., Stark, E., Leonardo, A., Harris, T. D., & Buzsáki, G. (2013). Large-scale, high-density (up to 512 channels) recording of local circuits in behaving animals. *Journal of Neurophysiology, 111*(5), 1132–1149.
45. Schwarz, D. A., Lebedev, M. A., Hanson, T. L., Dimitrov, D. F., Lehew, G., Meloy, J., Rajangam, S., Subramanian, V., Ifft, P. J., Li, Z., Ramakrishnan, A., Tate, A., Zhuang K. Z. & Nicolelis M. A. L. (2014). Chronic, wireless recordings of large-scale brain activity in freely moving rhesus monkeys. *Nature Methods, 11*(6), 670.
46. Shobe, J. L., Claar, L. D., Parhami, S., Bakhurin, K. I., & Masmanidis, S. C. (2015). Brain activity mapping at multiple scales with silicon microprobes containing 1,024 electrodes. *Journal of Neurophysiology, 114*(3), 2043–2052.
47. Rajangam, S., Tseng, P.-H., Yin, A., Lehew, G., Schwarz, D., Lebedev, M. A., & Nicolelis, M. A. (2016). Wireless cortical brain-machine interface for whole-body navigation in primates. *Scientific Reports, 6*, 22170.
48. Harrison, R. R. (2007). A versatile integrated circuit for the acquisition of biopotentials. In: *IEEE custom integrated circuits conference,* San Jose, California, USA (pp. 115–122).
49. Liu, X., Zhang, M., Xiong, T., Richardson, A. G., Lucas, T. H., Chin, P. S., Etienne-Cummings, R., Tran, T. D., & Van der Spiegel, J. (2016). A fully integrated wireless compressed sensing neural signal acquisition system for chronic recording and brain machine interface. *IEEE Transactions on Biomedical Circuits and Systems, 10*(4), 874–883.
50. Zhang, J., Suo, Y., Mitra, S., Chin, S. P., Hsiao, S., Yazicioglu, R. F., Tran, T. D., & Etienne-Cummings, R. (2014). An efficient and compact compressed sensing microsystem for implantable neural recordings. *IEEE Transactions on Biomedical Circuits and Systems, 8*(4), 485–496.
51. Kim, S., Tathireddy, P., Normann, R. A., & Solzbacher, F. (2007). Thermal impact of an active 3-D microelectrode array implanted in the brain. *IEEE Transactions on Neural Systems and Rehabilitation Engineering, 15*(4), 493–501.
52. Silay, K. M., Dehollain, C., & Declercq, M. (2008). Numerical analysis of temperature elevation in the head due to power dissipation in a cortical implant. In: *IEEE 30th annual international conference of the Engineering in Medicine and Biology Society, (EMBS),* Vancouver, British Columbia, Canada (pp. 951–956).
53. Lazzi, G. (2005). Thermal effects of bioimplants. *IEEE Engineering in Medicine and Biology Magazine, 24*(5), 75–81.
54. Ibrahim, T. S., Abraham, D., & Rennaker, R. L. (2007). Electromagnetic power absorption and temperature changes due to brain machine interface operation. *Annals of Biomedical Engineering, 35*(5), 825–834.
55. Chen, F., Chandrakasan, A. P., & Stojanovic, V. M. (2012). Design and analysis of a hardware-efficient compressed sensing architecture for data compression in wireless sensors. *IEEE Journal of Solid-State Circuits, 47*(3), 744–756.
56. Kuan, Y.-C., Lo, Y.-K., Kim, Y., Chang, M.-C. F., & Liu, W. (2015). Wireless gigabit data telemetry for large-scale neural recording. *IEEE Journal of Biomedical and Health Informatics, 19*(3), 949–957.
57. Rezaei, M., Bahrami, H., Mirbozorgi, A., Rusch, L. A., & Gosselin, B. (2016). A short-impulse UWB BPSK transmitter for large-scale neural recording implants. In: *IEEE 38th annual international conference of the Engineering in Medicine and Biology Society (EMBC),* Orlando, Florida, USA (pp. 6315–6318).
58. Mirbozorgi, S. A., Bahrami, H., Sawan, M., Rusch, L. A., & Gosselin, B. (2016). A single-chip full-duplex high speed transceiver for multi-site stimulating and recording neural implants. *IEEE Transactions on Biomedical Circuits and Systems, 10*(3), 643–653.
59. Elzeftawi, M., & Theogarajan, L. (2013). A 10pJ/bit 135Mbps IR-UWB transmitter using pulse position modulation and with on-chip LDO regulator in 0.13 μm CMOS for biomedical implants. In: *2013 IEEE topical conference on biomedical wireless technologies, networks, and sensing systems (BioWireleSS),* Austin, Texas, USA (pp. 37–39).

60. Crepaldi, M., Angotzi, G. N., Maviglia, A., Diotalevi, F., & Berdondini, L. (2018). A 5 pJ/pulse at 1-Gpps pulsed transmitter based on asynchronous logic master–slave PLL synthesis. *IEEE Transactions on Circuits and Systems I: Regular Papers, 65*(3), 1096–1109.
61. Chahat, N., Zhadobov, M., Le Coq, L., Alekseev, S. I., & Sauleau, R. (2012). Characterization of the interactions between a 60-GHz antenna and the human body in an off-body scenario. *IEEE Transactions on Antennas and Propagation, 60*(12), 5958–5965.
62. Zhadobov, M., Nicolaz, C. N., Sauleau, R., Desmots, F., Thouroude, D., Michel, D., & Le Dréan, Y. (2009). Evaluation of the potential biological effects of the 60-GHz millimeter waves upon human cells. *IEEE Transactions on Antennas and Propagation, 57*(10), 2949–2956.
63. Feynman, R. P., Leighton, R. B., & Sands, M. (2005). *The Feynman lectures on physics including Feynman's tips on physics: The definitive and extended edition*. Reading: Addison Wesley.
64. Zou, Z. (2011). *Impulse radio UWB for the internet-of-things: A study on UHF/UWB hybrid solution*. Doctoral dissertation, KTH Royal Institute of Technology.
65. Chen, K., Yang, Z., Hoang, L., Weiland, J., Humayun, M., & Liu, W. (2010). An integrated 256-channel epiretinal prosthesis. *IEEE Journal of Solid-State Circuits, 45*(9), 1946–1956.

Advances in Bioresorbable Electronics and Uses in Biomedical Sensing

Michelle Kuzma, Ethan Gerhard, Dingying Shan, and Jian Yang

1 Introduction to Bioresorbable Electronics

1.1 Motivation and Classification

More than 46 million tons of electronic waste (e-waste) are produced every year worldwide [67]. E-waste produced by the USA alone exceeds five million tons with similar amounts produced by other developing and established markets [67, 103]. Given the fast turnover of technology and increasing use of electronics in daily life, the rate of e-waste growth was startlingly three times higher than that of any other waste stream in 2016 with only 15–20% being recycled [4, 15, 67, 119]. Healthcare-related electronics furthermore contribute to the growing e-waste epidemic. Unprecedented advanced systems, such as infusion pumps, enhanced bioimaging, wireless communication, dialysis machines, electronic healthcare records, and long-term, implantable electronic devices, such as the cardioverter defibrillator, ventricular assist devices, and brain-instrument interfaces, became emerging essentials to patient care over the course of the twentieth century [15, 37, 106]. Despite the vast strides made in clinical care due to these rising technologies, soft, *biodegradable* electronic implants remained a clear, unmet need, with only flexible non-electronic biodegradable implants or non-resorbable electronic implants that not only added to e-waste accumulation, but contributed to adverse patient outcomes including tissue obstruction due to rigid designs, limited sensitivity from poor device tissue contact, and foreign body responses in addition to iatrogenic

M. Kuzma
Department of Materials Science and Engineering, Materials Research Institute,
The Pennsylvania State University, University Park, PA, USA

E. Gerhard · D. Shan · J. Yang (✉)
Department of Biomedical Engineering, Materials Research Institute, The Huck Institutes
of The Life Sciences, The Pennsylvania State University, University Park, PA, USA
e-mail: jxy30@psu.edu

H. Cao et al. (eds.), *Interfacing Bioelectronics and Biomedical Sensing*,
https://doi.org/10.1007/978-3-030-34467-2_2

complications from surgical removal, like infection, tissue damage, psychological distress, and superfluous costs widely available [15, 59].

It was not until Hwang and colleagues reported a state-of-the-art proof-of-concept electronic device composed of thin layers of silk with magnesium (Mg), magnesium oxide (MgO), silicon dioxide (SiO_2), and monocrystalline silicon nanomembranes (Si-NMs) that the first entirely transient electronic device was realized (Fig. 1) [36]. The device demonstrated successful engineering of both soluble *active components* (i.e., transistors) and [49] *passive components* (i.e., resistors, inductors, capacitors). The fabricated device was capable of energy storage, wireless heating, and imaging, demonstrating that this technology could have versatile and unprecedented utility in many biomedical applications [36, 49]. The bioresorbable materials were used in concert to comprise active and passive components with corresponding interconnects placed onto a silk substrate that visibly degraded in as little as 5 minutes into nontoxic and non-immunogenic by-products [35, 36]. Furthermore, this study established the critical finding that the device and degradation by-products were safe in vivo, sparking inspiration to revolutionize implantable electronics for acute, clinical scenarios while reducing healthcare-related e-waste [36].

Following this breakthrough, the class of *transient electronics* came to fruition. Although transient electronic systems often incorporate materials used in conventional, nondegradable electronics, the components in transient devices are nanometers to micrometers in dimension, which enables complete degradation within

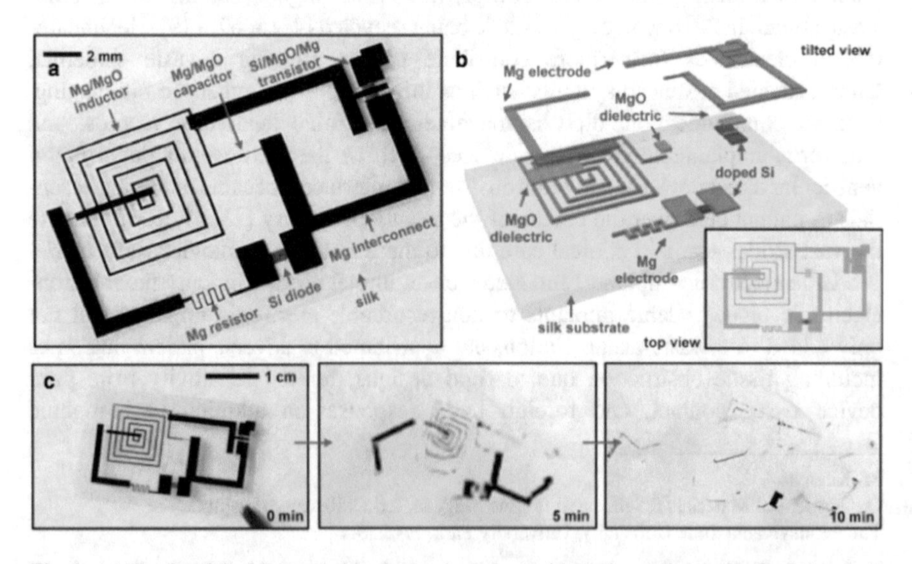

Fig. 1 (**a**) The first fully bioresorbable electronic system with engineered active and passive components using thin films of magnesium (Mg), magnesium oxide (MgO), and silicon (Si) on a silk substrate (**b**) Exploded schematic of the resorbable device and an inset containing a top view. (c) Demonstrated transience of the device upon immersion in DI water within 10 minutes. (From Hwang et al. [36]. Reprinted with permission from AAAS)

minutes to days under benign conditions [59]. The bulk materials used in traditional silicon-based electronics, on the contrary, hydrolyze at negligible rates on the order of nanometer's per day [59]. Therefore, distinct from traditional built-to-last electronics, transient electronics encompass electronic devices capable of disappearing by degradation or dissolution in a controllable manner over relevant timeframes in benign environmental conditions [6, 78, 116].

The terms transient electronics and biodegradable electronics are used to describe degradable or resorbable devices with applications in and beyond the biomedical field (i.e., environmental sensing, solar cells, self-degrading data security systems, and photodetectors) [59, 78]. Meanwhile, *bioresorbable electronics* frequently refers to the subdivision of transient electronics that safely disappear in the body [59, 62, 119]. Likewise, in this chapter, the same classification of bioresorbable electronics will be adopted to describe the biomedical cohort of transient and biodegradable electronic systems.

1.2 Background

Prior to the advent of the first completely biodegradable electronic device, increased attention was devoted to small-scale material deposition techniques and flexible organic electronics [23, 48, 102] and the development of organic light-emitting devices (OLEDs), thin-film transistors (TFTs), and thin film organic photovoltaic cells, among other flexible microdevices at the end of the Twentieth century [23, 48, 102]. Furthermore, successes in transfer printing in 2006 enabled electronic devices with components that are sensitive to traditional, harsh integrative circuit (IC) fabrication conditions (i.e., vacuum or high temperature) possible [86, 119]. These milestones in flexible electronic production were imperative to the creation of electronic implants that conform to curvilinear shapes of body tissues without resulting in mechanical injury to the surrounding tissue or the device [62]. Meanwhile, water-soluble electronics were being developed during this time. Reported partially soluble devices included organic thin-film transistors (OTFTs), biodegradable organic field-effect transistors (OFETs), stretchable complementary metal-oxide-semiconductor (CMOS) circuits on plastic substrates, and elastic monocrystalline silicon (mono-Si) electronic mesh designs [5, 48, 61].

Furthermore, nonelectronic, resorbable technology (i.e., surgical sutures, cardiovascular stents, tissue scaffolds, drug delivery vehicles, and orthopedic equipment) were emerging at the turn of the twentieth century [15, 119]. However, biodegradable implants possessing electronic capabilities were limited by the absence of active resorbable components capable of sensing, processing, communication, and actuation [15, 37]. The first implantable electronic device with partial bioresorption was reported circa 2010, a single crystal (monocrystalline, single crystalline) silicon-based n-channel metal oxide semiconductor (nMOS) transistor, which incorporated insoluble electrical components on a fully resorbable silk substrate [62]. Although this device functioned successfully in an aqueous environment, was non-immunogenic, and only possessed finite amounts of insoluble materials, it still necessitated a secondary surgery to remove the insoluble components [62]. It was the collective findings of Si-NMs and metal thin films soluble in water in appreciable timeframes and flexible Si electronics that led to the first fully bioresorbable Si electronic device using CMOS technology in 2012.

Successive innovations in production of transient, soft electronics established utility biomedical engineering applications including photothermal therapy, on-demand drug release, multifunctional vascular stents, intracranial pressure monitoring, temperature monitoring, electrical stimulation, and biosensing [14, 39, 66, 97, 119]. Despite the many advances since the first bioresorbable device, research is ongoing to achieve multifunctional bioresorbable electronics, refine on-demand transience, and develop economical and scalable manufacturing processes for commercialization as well as to study long-term physiological and environmental effects of by-products [14, 15, 34, 72, 119].

The main types of materials used in bioresorbable electronics include thin layers of biodegradable polymers, dielectrics, semiconductors, and conductors [119]. Depending on the material properties, each material can serve different roles or multiple roles from a physical foundation to a protective layer for electronic components or contribute to charge transfer. Many of the properties, such as dissolution rate, electrical performance, and biocompatibility of materials used in transient electronics, have been studied under various conditions [59, 119]. The basic components and properties of respective materials that have developed into current, mainstream transient electronics will be outlined in detail (Sect. 2).

To date, mechanisms of transience in biodegradable electronics include dissolution, degradation, hydrolysis, corrosion, and enzymatic breakdown in response to stimuli including heat, water moisture, aqueous solutions, acid, enzymes, and light [14, 62, 105, 119]. The main types of degradation mechanisms for complete dissolution of implantable bioresorbable electronics are hydrolysis and dissolution in aqueous media [27]. Rate of transiency can be regulated through materials selection and processing, device design, and environmental conditions [119].

2 Overview and Advancements of Constituent Resorbable Materials

Evaluation of the fundamental bioresorbable materials and ensuing innovations moving transient electronics toward practical advanced degradable medical devices are explored in this section in terms of device components (i.e., conductor, semiconductor, dielectric, substrate, and encapsulating layers). Core materials comprising each component are inorganic and/or organic in nature exhibiting conducting, semiconducting, dielectric, and/or insulating properties (Fig. 2) [14, 27, 35, 78, 119]. Material choice for each component is governed by the desired properties (i.e., mechanical, thermal, magnetic, optical, electric, or chemical) needed to generate

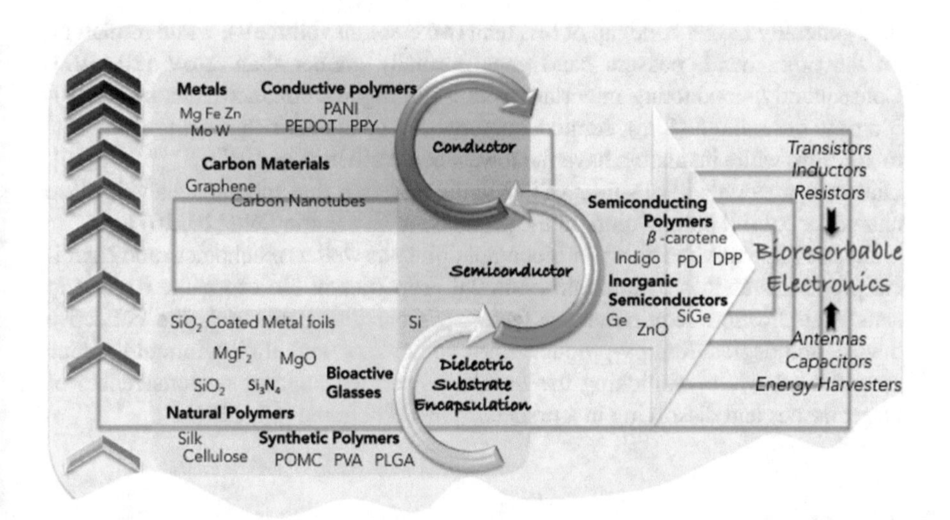

Fig. 2 Key materials used to create functioning components that constitute bioresorbable electronic devices. Materials are divided into three main categories: insulators (i.e., dielectric, substrate, and encapsulating materials), semiconductors, and conductors, with increasing levels of conductivity directed from the bottom to the top of the figure. Silicon (Si) is represented between the semiconductor and insulator windows due to its successful employment as a semiconductor and an encapsulating layer, respectively. (Adapted with permission from Hwang et al. [42]. Copyright 2015 American Chemical Society)

necessary cooperative performance for translatable applications [10]. Further modification through functionalization, engineering, and processing inspired by the established principles reviewed here guides progress in this field for medical advances (Sect. 3) [14].

A property vital to any functioning electronic device is *electrical conductivity*, a second rank tensor that relates the current density to the electric field intensity according to Ohm's law (Eq. 1) [10, 107]:

$$J = \sigma \xi, \tag{1}$$

where J is the current density, σ is the electrical conductivy, and ξ is the electric field intensity [10]. Electrical conductivity is the degree to which electrically charged particles move (i.e. electron, hole, ion, or charged vacancy) within a material in response to an applied electric field [10, 107]. Movement of charged particles, in turn, provokes and dictates the magnitude of an electric current through the material [10, 107]. How easily electrons are available for conduction with respect to energy can be visualized through energy band diagrams [10, 107]. Simply put, conducting materials typically have overlapping valence and conduction bands so that electrons are readily available for conduction, whereas in solid semiconducting and insulating materials, there is a distinct energy barrier electrons must overcome to become involved in conduction (i.e., the energy band gap) [10, 107]. Semiconducting materials generally have a bandgap of less than two electron volts (eV), while insulators, on the other hand, possess band gaps normally greater than 2 eV [10, 107]. Consequently, conducting materials exhibit the highest conductivities of 10^4–10^7 Siemens per minute (S/m), semiconductors have conductivities ranging from 10^{-6} to 10^4 S/m, while insulators have the lowest conductivities of 10^{-10}–10^{-20} S/m [10]. Dielectric materials are a subset of insulating materials that spontaneously displace charge, or exhibit polarization, in the presence of an electric field [10, 107].

It is as imperative to consider biocompatibility as well as mechanical and chemical properties as it is to evaluate electrical properties in bioresorbable devices to satisfy application requirements (Table 1). Constituent material, the collective device, and degradation by-products must be nontoxic and elicit minimal immune response all while mimicking the flexibility, softness, and/or responsiveness of target tissues and dissolving in a predetermined timeframe [62, 119].

2.1 *Conductor*

Both conductive inorganic and organic materials have been used in transient electronics as functional materials for interconnects and electrodes [27, 119]. The inorganic and organic materials used as conducting components in bioresorbable electronics accompanied by pertinent developments and properties are described here.

2.1.1 Inorganic

Like conventional silicon electronics, metals and their alloys serve as the inorganic conducting materials in bioresorbable electronics [27]. Unique to metallic materials are the delocalized electrons that attribute to high thermal and electrical conductivities, low resistivities, and opacity to visible light [10, 107]. Electron mobilities exhibited by elemental metals are generally near 10 $cm^2/V \cdot s$ [107]. Furthermore, metals have high packing densities of atoms relative to other material types resulting in multiple slip planes attributing to exhibited notable ductility and strength [107]. Dissolvable metals used in bioresorbable electronics have low electrical resistance, predictable properties, and are biocompatible. The commonly employed metals in transient electronics include Mg, iron (Fe), zinc (Zn), molybdenum (Mo), and tungsten (W); each of which have inherent physiologic roles with the exception of W [14]. The metal alloy, AZ31B combines aluminum (Al) and Zn with trace amounts of other elements, and has been used in construction of a bioresrobable battery [27, 34, 116, 119].

Upon hydrolysis of the aforementioned metals, products formed include respective metal ions, oxides, or hydroxides [119]. Corrosion of the metals is driven by changing to a state that exhibits a lower Gibbs free energy Gibbs (G). The change in Gibbs free energy is directly related to the standard electrode potential through the relationship (Eq. 2) [119]:

Table 1 Key device components, common constituent materials, and corresponding reported dissolution behaviors

Device component	Purpose	Material	Dissolution mechanism	Dissolution conditions	Dissolution rate (nm/hr)[a]
Conductor	To transfer charge, commonly used in conductive components, like electrodes, interconnects, and antenna wires	Mg	$Mg + 2H_2O \rightleftharpoons Mg(OH)_2 + H_2$	SBF, 37 °C, pH 7.4	0.05–500
		Fe	$Fe + 2H_2O \rightleftharpoons Fe(OH)_2 + H_2$	SBF, 37 °C, pH 7.4	~2
		Zn	$Zn + 2H_2O \rightleftharpoons Zn(OH)_2 + H_2$	In vivo PBS	5
		Mo	$2Mo + 2H_2O + 3O_2 \rightleftharpoons 2H_2MoO_4$	Hank's solution, 37 °C, pH 7.4	0.3–1.1
		W	$2W + 2H_2O + 3O_2 \rightleftharpoons 2H_2WO_4$	In vivo	11.4
Semiconductor	Functions to manage charge flow and directionality; commonly found in transistors and diodes	Mono-Si (p-and n- type (10^{17} cm^{-3}))	$Si + 4H_2O \rightleftharpoons Si(OH)_4 + 2H_2$	0.1 M buffer solution, 37 °C, pH 7.4	0.132
		Mono-Si (p-type 10^{20} cm^{-3})	$Si + 4H_2O \rightleftharpoons Si(OH)_4 + 2H_2$	0.1 M buffer solution, 37 °C, pH 7.4	0.010
		Mono-Si (n-type 10^{20} cm^{-3})	$Si + 4H_2O \rightleftharpoons Si(OH)_4 + 2H_2$	0.1 M buffer solution, 37 °C, pH 7.4	0.021
		Mono-Si	$Si + 4H_2O \rightleftharpoons Si(OH)_4 + 2H_2$	Bovine serum	4.2
		a-Si	$Si + 4H_2O \rightleftharpoons Si(OH)_4 + 2H_2$	Buffer solution, 37 °C, pH 7.4	0.17
		p-Si	$Si + 4H_2O \rightleftharpoons Si(OH)_4 + 2H_2$	Buffer solution, 37 °C, pH 7.4	0.14
		Ge	$Ge + O_2 + H_2O \rightleftharpoons H_2GeO_3$	Buffer solution, 37 °C, pH 7.4	0.13

(continued)

Table 1 (continued)

Device component	Purpose	Material	Dissolution mechanism	Dissolution conditions	Dissolution rate (nm/hr)[a]
		SiGe	–	Buffer solution, 37 °C, pH 7.4	4×10^{-3}
		ZnO	$ZnO + H_2O \rightleftharpoons Zn(OH)_2$	DI H_2O, RT	13.3
Dielectric	Can aid in transferring charge to other device components, store charge and redistribute charge, and insulate electrical materials	SiO_2 (PECVD)	$SiO_2 + 2H_2O \leftrightharpoons Si(OH)_4$	Buffer solution, 37 °C, pH 7.4	–[b]
		SiO_2 (e-beam)	$SiO_2 + 2H_2O \leftrightharpoons Si(OH)_4$	Buffer solution, 37 °C, pH 7.4	–[b]
		SiO_2 (thermal)	$SiO_2 + 2H_2O \leftrightharpoons Si(OH)_4$	Buffer solution, 37 °C, pH 7.4	–[b]
		Si_3N_4 (LPCVD)	$Si_3N_4 + 12H_2O \rightleftharpoons 9Si(OH)_4 + 4NH_3$	Buffer solution, 37 °C, pH 7.4	–[b]
		Si_3N_4 (PECVD-LF)	$Si_3N_4 + 12H_2O \rightleftharpoons 9Si(OH)_4 + 4NH_3$	Buffer solution, 37 °C, pH 7.4	–[b]
		Si_3N_4 (PECVD—HF)	$Si_3N_4 + 12H_2O \rightleftharpoons 9Si(OH)_4 + 4NH_3$	Buffer solution, 37 °C, pH 7.4	–[b]
		MgO	$MgO + H_2O \rightleftharpoons Mg(OH)_2$	–	–
Substrate	Provides a foundation for functional components	Mg foil	$Mg + 2H_2O \rightleftharpoons Mg(OH)_2 + H_2$	aCSF, 37 °C	~167
		Fe foil	$Fe + 2H_2O \rightleftharpoons Fe(OH)_2 + H_2$	PBS, 37 °C, pH 7.4	3
		Zn foil	$Zn + 2H_2O \rightleftharpoons Zn(OH)_2 + H_2$	PBS, 37 °C, pH 7.4	14.5
		Mo foil	$2Mo + 2H_2O + 3O_2 \rightleftharpoons 2H_2MoO_4$	PBS, 37 °C, pH 7.4	0.8

Device component	Purpose	Material	Dissolution mechanism	Dissolution conditions	Dissolution rate (nm/hr)[a]
		W foil	$2W + 2H_2O + 3O_2 \rightleftharpoons 2H_2WO_4$	PBS, 37 °C, pH 7.4	6.25
		Silk	Enzymatic degradation	–	–[c]
		Cellulose derivatives	Dissolves in water	–	–
		POC	Hydrolysis of ester bonds	–	–
		POMC	Hydrolysis of ester bonds	–	–
		PLGA	Hydrolysis of ester bonds, enzymatic degradation	–	–
		PLA	Hydrolysis of ester bonds, enzymatic degradation	–	–
		PVA	Dissolves in water	–	–
		PGS	Hydrolysis of ester bonds, enzymatic degradation	–	–
Encapsulation	Serves as packaging, or encapsulating layers, to enhance mechanical robustness, protect device components, and prolong operational lifetime of the device	–[d]	–	–	–

Adapted under CC BY 4.0 License [79] and adapted with permission from Yin et al. [116]. Copyright 2014 John Wiley and Sons [10, 39, 49, 55–57, 9178, 93107, 119, 124, 125]

[a]Please note that degradation rates are reported from different study designs and cannot be compared directly; however, provided rates may instead be used as a general guide for dissolution trends. The affects of intrinsic and extrinsic factors on dissolution rates are reviewed in the text (Sect. 2)

[b]Only relative rates, not absolute rates, are reported for respective conditions [55]

[c]Complete dissolution of polymeric substrate films occurs on the order of minutes to years depending on processing conditions and environmental factors (Sect. 2.4.2)

[d]The common materials used for encapsulation are also used in other device components provided in the table (i.e., mono-Si, SiO_2, Si_3N_4, polymeric substrate materials) (Sect. 2.5)

$$\Delta G^{\circ} = -nFE^{\circ}, \tag{2}$$

where ΔG° is the change in the standard Gibbs free energy, n is the number of electrons transferred in the reaction, F is the Faraday's constant (96,486 J·(V·mol)$^{-1}$), and E° is the standard electrode potential in volts (V) [10, 119]. The electrode potential of metals are compared to the standard hydrogen electrode—an arbitrarily selected control cell for oxidation-reduction potentials with an understood E° of 0 V at 25 °C. Metals with higher reactivity than that of hydrogen in aqueous media are appealing candidates for metals in bioresorbable electronics for reliable dissolution in aqueous media [119]. Under nonstandard conditions, the electrode potential is calculated using the Nernst equation (Eq. 3) [119]:

$$E = E^{\circ} - \frac{RT}{nF}\ln(Q), \tag{3}$$

where E is the nonstandard electrochemical potential, R is the universal gas constant (8.314 J (mol·K)$^{-1}$), T is temperature in Kelvin, and Q is the reaction quotient determined from concentrations of the product and reagent ions [10, 119]. Using the electrode potential, Pourbaix diagrams can be constructed to determine the equilibrium state of each metal (i.e., metal, metal oxide) as a function of pH [119].

It should be noted that metals used in transient electronics are in thin film form consequently leading to possible variations in dissolution properties from the bulk material of the same metal [116]. Unlike bulk materials, defects in the material, such as pin holes, as well as grain size, will dramatically affect the rate of dissolution of thin films [15]. Furthermore, the dissolution of each metal in water or biofluids occurs at variable rates dependent on several intrinsic and extrinsic factors. Intrinsic factors correspond to the material itself. Examples include reactivity, surface chemistry, crystalline structure, microstructures, porosities, defects, and metal thickness. For example, the metal oxides that form upon corrosion are typically less soluble in water and degrade at slower rates than the parent metal in aqueous solution [78]. Extrinsic factors are dependent on the surrounding environment including solution pH, surrounding temperature, ion and protein concentration, O_2 levels, and processing conditions [27, 78, 116, 119]. Chloride ions in biological fluids expedite dissoltuion rates of some metals, such as Mg, by reacting with the parent metals or hydroxides to produce a metal chloride and hydrogen gas relative to environments with little to no chloride content [77]. Mg, Zn, and respective oxides dissociate more rapidly and in a less controlled manner than materials composed of Mo, Fe, W, and their oxides in aqueous solutions, which can help guide materials selection when considering the required functional lifetime of the medical device being fabricated [14, 27, 116, 119].

Magnesium Mg is the fourth most common mineral in the human body with highest abundance in skeletal tissue and intracellular fluid [24, 110]. This trace element activates over 600 enzymes with indispensable roles in calcium and phosphate homeostasis, activation of cell metabolism, regulation of intracellular antioxidants, and stabilization of nucleic acids [24, 110].

To satisfy functional lifetime requirements of biomedical devices, it is crucial to assess the dissolution rates of bioresorbable materials. The standard electrode potential of Mg is −2.363 V, which is lowered, as expected, by the formation of surface hydroxide layers in aqueous media indicating the reduction of dissolution of the hydroxide species in water [10, 114, 119]. Due to high conductivity, existence in conventional electronics, established biocompatibility, and high dissolution rates, Mg was chosen as the conductive material in the first completely soluble silicon-based bioelectronic device [36, 117] [27]. In this device, Mg was incorporated into the inductor, capacitor, transistor, interconnects, and resistor pieces [36]. Upon dissolution, the hydroxide of Mg, $Mg(OH)_2$, is formed (Eq. 4) [116, 119]:

$$\mathrm{Mg} + 2\mathrm{H_2O} \rightleftharpoons \mathrm{Mg}(OH)_2 + \mathrm{H_2}. \quad (4)$$

Thereafter, in body fluids, $Mg(OH)_2$ will further react with chlorine ions to form the highly soluble salt, $MgCl_2$ [35]. Despite the formation of oxide and hydroxide features at the surface, the film thickness of Mg thin films has been shown to decrease linearly with time at rates dependent on the initial thickness of the film [77, 114, 116] with a reported weight loss rate of 19–44 mg/cm^2/day in Hank's solution (pH 6, 37 °C) as well as a surface area normalized dissolution rate of 0.05–0.5 μm per hour in simulated body fluid (SBF)under physiologic conditions (i.e., pH 7.4 and 37 °C) [98, 116]. The reported electrical dissolution rate (EDR)was higher than the mass loss dissolution rate for Mg at 4.8 ± 2.2 μm/hour in Hank's solution [117]. Aluminum (Al) can be added to Mg to reduce the dissolution rate [116]. Similarly, the Mg alloy, AZ31B, has a reduced EDR in Hank's solution of 2.6 ± 2.1 μm/hour [116]. Other magnesium films have incorporated silicon carbide (SiC), silicon dioxide (SiO_2), and polycaprolactone (PCL) to slow the dissolution of Mg as well [35]. Aside from alloying, Mg dissolution rates can be altered with distinct processing methods and/or coating layers [114].

Iron Over 60% of Fe is bound to hemoglobin present in red blood cells for delivery of oxygen, to body tissues while roughly 25% is stored until needed [43], and the remaining amount is predominantly bound to myoglobin in muscle cells or serves as a cofactor for enzymes needed for cellular function.

Fe has a positive $E°$ of +0.800 V signifying it is less readily soluble in water than Mg. When Fe thin films are immersed in water, the hydroxide by-product, $Fe(OH)_2$, results (Eq. 5) [116, 119]:

$$Fe + 2\mathrm{H_2O} \rightleftharpoons Fe(OH)_2 + \mathrm{H_2}. \quad (5)$$

In simulated body fluid at physiological conditions, Fe has a reported corrosion rate of 0.02 μm/hour [116]. The reported EDR of Fe in Hank's solution at pH 7.4 and 37 °C is 7×10^{-3} μm/hour, much lower than that of Mg in similar conditions [116].

Zinc Zn is a trace element with structural roles in protein folding and serves as a regulatory mediator in synaptic signaling, apoptosis, gene regulation, and catalysis [45]. Zn is a cofactor to over 70 enzymes, has antioxidant properties, and stabilizes endothelial membranes making it useful in the prevention of atherogenesis in cardiovascular stents and tissue scaffolds [8, 45].

The degradation products of Zn in aqueous solution include zinc hydroxide ($Zn(OH)_2$) and hydrogen gas (Eq. 6) [116, 119]. Further characterization has revealed that wires also break down into ZnO and Zn carbonate ($ZnCO_3$) in vivo [8].

$$Zn + 2H_2O \rightleftharpoons Zn(OH)_2 + H_2. \tag{6}$$

Zn degrades more slowly than Mg in body fluid and has an $E°$ of −0.763 V [10, 35]. Zn wire abdominal aorta stents reveal that Zn wires undergo uniform corrosion in rats for 3 months followed by localized, rapid corrosion behavior after 4.5 months with a corrosion rate ranging from 12 to 50 μm per year over the first sixth months [8, 35]. Another study reported an in vivo stent mass loss dissolution rate of 5×10^{-3} μm per hour [116]. The reported EDR was found to be 0.3 ± 0.2 μm per hour in Hanks' solution [116]. In regard to processability, Zn has a higher stability in organic solvents than Mg, which is more desirable for low-cost solution-based printing [35].

Molybdenum Mo is known to serve as a cofactor for various oxidases and catabolize sulfur-containing amino acids and heterocyclic compounds [35, 44, 109]. The dissolution of Mo results in the formation of molybdic(VI) acid (H_2MoO_4) (Eq. 7) [116, 119]:

$$2Mo + 2H_2O + 3O_2 \rightleftharpoons 2H_2MoO_4. \tag{7}$$

The $E°$ of Mo is −0.2 V. Unlike other bioresorbable metals, the EDR of Mo did not rise in acidic environments compared to that in water supported by an observed positive rise in $E°$ with increasing hydrochloric acid (HCl) concentrations [27, 99, 111]. Conversely, the solubility of Mo will increase with oxygen and chloride content [27, 35, 116]. The EDR of Mo under physiologic conditions is reported to be approximately $7 \pm 4 \times 10^{-4}$ μm/hour in Hanks solution (pH 7.4, 37 °C) [116].

Tungsten In contrast to the other metals used in bioresorbable electronics, W is not inherently found in the body; however, W thin films successfully deteriorate in aqueous solutions without eliciting toxic effects in vivo at concentrations less than 50 μg/mL [15, 116]. The hydrolysis of W produces tungstic(VI) acid (H_2WO_4) (Eq. 8) [116, 119]:

$$2W + 2H_2O + 3O_2 \rightleftharpoons 2H_2WO_4. \tag{8}$$

W has a E° of +0.1 V and a dissolution rate that increases with pH. Furthermore, W deposited via chemical vapor deposition (CVD) dissolves at a rate that is an order of magnitude slower than W that is sputtered [27, 111]. CVD-deposited W had a reported EDR of approximately 2×10^{-3} μm/hour at physiological conditions, while sputter-deposited W had a determined EDR and mass loss corrosion rate of 0.02 μm/hour and 0.02–0.06 μm/hour, respectively [116].

2.1.2 Organic

Conjugated carbon systems support movement of delocalized electrons through a material and are therefore critical to conductive polymers [107]. Unlike inorganic conducting materials, besides electronic conduction, organic conductive materials are also capable of ionic conductivity, a common conduction pathway in the body involving the motion of ions through a material upon an applied electric field [47, 107]. Furthermore, electronic materials constituted of organic and carbon-based components (i.e., graphene and carbon nanotubes (CNTs)) have easily modified surface chemistries, can be designed to be flexible, and can undergo more economical manufacturing methods (i.e., solution processing) than conventional silicon electronics [119]. To date, main limitations include incompatibility with the conventional electronic fabrication methods with established manufacturing plants as well as inferior performance compared to metallic electrodes [119]. Scalable manufacturing methods and increasing conductivity of polymers to be comparable to that of metals for use in bioresorbable electronics are an area of ongoing research [22, 119].

One of the first organic electronic devices used melanin as the conductive component to make a resistive switching element, and more recently a derivative, eumelanin, has been used in biodegradable supercapacitors for energy storage [47, 68]. Another biodegradable organic material capable of conduction is chitosan, the deacetylated form of chitin, a material found in the exoskeleton of arthropods and in the cell wall of mushrooms [20, 47]. Chitosan thin films have been successfully fabricated on paper substrates [47]. Biocompatible, conductive polymers include polyaniline (PANI), polypyrrole (PPY), and poly(thiophenes) like poly(3,4-ethylenedioxythiophene) (PEDOT) and PEDOT-doped polyanion poly(styrene sulfonate) (PEDOT:PSS) [47, 119]. In PEDOT:PSS, PEDOT exhibits anionic conduction, while PSS facilitates movement of cations, like calcium (Ca^{2+}), sodium (Na^{+}), potassium (K^{+}), and acetylcholine (ACh) [47]. PEDOT nanotubes have been used for neural recording and PEDOT:PSS for in vivo biosensing and for high-performance electrodes in electrocorticography (ECoG) [35, 47].

Carbon based materials (i.e., graphene and carbon nanotubes (CNTs)) have found useful as conductive materials with graphene having reported electron mobilities as high as 200,000 $cm^2/V{\cdot}s$ [70, 107]. Oxidation of carbon nanotubes (CNTs), tubular structures of graphene, results in carboxylation of the surface facilitating degradation via oxidative enzymes [35]. Silkworms can be fed CNTs and graphene to create a conductive silk composite with improved mechanical strength due to reinforcement from the carbon-based features [35, 120]. Graphene has been used as the conductive component in gelatin-based transient energy harvesters [70, 107].

2.2 *Semiconductor*

Semiconductors have intermediate conductivity between conductors and insulators [10, 49, 107]. To augment movement of charge carriers through *intrinsic* semiconductors, impurities coined *dopants* are added through a process called *doping* to alter the material into what are considered *extrinsic* semiconductors [10, 49]. Extrinsic semiconductors with electron-rich impurities, or donors, are referred to as n-type semiconductors, whereas p-type semiconductors are created by the addition of electron accepting dopants known as acceptors. Common examples of donor and acceptor impurities are phosphorous and boron, respectively [49]. Semiconducting materials are employed in transistors and diodes, namely, to govern charge flow and directionality [49].

2.2.1 Inorganic

Biocompatible semiconductors inlcude Si-NMs (i.e., polycrystalline (poly-Si), amorphous (a-Si), nanoporous (np-Si), and monocrystalline (mono-Si)), silicon-germanium (SiGe), germanium (Ge), and ZnO [18, 35, 39, 57]. Amorphous silicon (a-Si) and polycrystalline silicon (p-Si) can be found in solar cells and in the active matrix of flat panel displays, while SiGe is implemented in communication devices as well as high-frequency heterojunction bipolar transistors [57]. ZnO is a direct wide bandgap semiconductor (3.37 eV) that serves as a semiconductor in thin-film transistors and photodetectors. Furthermore, having optical transparency, high electron mobility, and piezoelectric properties, ZnO is a desirable material for mechanical energy harvesters and actuators [39, 50, 55]. Electronic mobilities for conventional inorganic semiconductors can vary between 1 $cm^2/V{\cdot}s$ to the order of 10^4 $cm^2/V{\cdot}s$ [107]. The dissolution reactions of each of these semiconducting materials lead to hydroxide or oxide products (Eqs. 9, 10, and 11) [119]. Currently, the most commonly used semiconducting material in bioresorbable electronics is Si as it is fundamental to conventional electronics with robust performance and an established biocompatibility profile [40, 59].

Orthosilicic (silicic) acid ($Si(OH)_4$), the by-product of Si hydrolysis, is a naturally occurring compound in the body with concentrations ranging from 14×10^{-6} to 39×10^{-6} M in serum [35, 59]. Silicon dissolves in a spatially uniform manner exhibiting linear kinetics indicating surface erosion rather than reactive diffusion with water [40, 59].

$$Si + 4\mathrm{H_2O} \rightleftharpoons Si\left(OH\right)_4 + 2\mathrm{H_2}, \tag{9}$$

$$Ge + O_2 + H_2O \rightleftharpoons H_2GeO_3, \quad (10)$$

$$ZnO + H_2O \rightleftharpoons Zn(OH)_2. \quad (11)$$

Environmental and intrinsic parameters can be used to tune degradation of semiconductors in order to meet lifetime requirements of desired bioresorbable devices. The dissolution rate of each of material is accelerated with temperature, which may be prevalent in the case of fever, local injury, or immune reactions [55]. Ge in buffer solution at physiological conditions dissolves at 3.1 nm/day—similar to mono-SiNM—and will increase drastically with increasing pH. SiGe appears resistant to changes until a pH of 8, and deteriorates about 30 times slower than mono-SiNM (buffer solution, pH 7.4, 36 °C) [55, 79]. Although relatively resistant to changes in pH, SiGe dissolutions increase by approximately 185 times when placed into bovine serum compared to PBS at 37 °C, whereas Si degradation rates only increase 30–40 times under the same conditions [55]. Dissolution rates of mono-SiNMs can vary from 0.5 to 624 nm/day depending on surface chemistry, doping type and levels, temperature, and ion concentration in the surrounding solution, allowing for nanomembranes of silicon to disappear in a matter of days to weeks [35, 40, 59, 78]. The dissolution rate of Si-NMs increases with temperature and ion concentrations in aqueous solution (i.e., Ca^{2+} and PO_4^-), and exponentially, with pH as silicic acid, formation is catalyzed by hydroxyl ions [35, 119]. Conversely, dissolution rates of Si are decreased by doping levels that exceed 10^{20} cm^{-3}, the presence of silicic acid in solution, and proteins in solution [59, 74]. When in similar conditions (buffer solution at 37 °C and pH 7.4), the morphology of Si alone can affect dissolution rates where a-Si degrades faster rate than p-Si, due to enhanced diffusion, followed by mono-SiNM [59, 79]. For ZnO, a 20 nm-thick film fully dissolves at room temperature in deionized (DI) water in 15 hours [119].

2.2.2 Organic

Biodegradable organic semiconducting compounds have conjugated carbon systems like in conducting polymers. Examples include β-carotene, indigo, perylenediimide (PDI), chlorophyll, and derivatives of anthraquinone, indanthrene brilliant orange RF, and indanthrene yellow G [35]. Appealing attributes of organic as opposed to inorganic semiconductors include low-cost, high-throughput low-temperature fabrication techniques (i.e., large-area, solution-based processing), high mechanical flexibility, and diverse surface functionalization [59, 76].

Unfortunately, however, many natural, organic semiconductors used in transient electronics exhibit intrinsically low carrier mobilities compared to traditional semiconducting materials used in currently fabricated transient electronic devices, which is a limitation for facilitating interfacial charge transport [22, 46, 59]. Mono-Si typically exhibits an electron mobility of 1400 cm^2·(V·s)$^{-1}$, whereas β-carotene and indigo have much lower mobilities (4×10^{-4} cm^2/(V·s) and 1×10^{-2} cm^2/ (V·s), respectively) [35, 48, 59]. Diketopyrrolopyrrole (DPP)can be conjugated with p-phylenediamine (PPD) to synthesize the bioresorbable, semiconducting polymer, PDPP-PD. Degradation begins initially degrading by acid catalysis of the imine bonds followed by hydrolysis of the lactam group present in the DPP monomers [59, 76]. PDPP-PD exhibited a carrier mobility as high as 0.21 ± 0.03 cm^2/(V·s) when used with an aluminum oxide (Al_2O_3) dielectric, gold (Au) electrodes and gate, and cellulose substrate [59].

Other organic compounds investigated as biodegradable semiconductors include perylenediimide (PDI), a chromophore for red cosmetics, used in an edible OFET, and 5,5′-bis(7-dodecyl-9*H*-fluoren-2-yl)-2,20-bithiophene (DDFTTF) explored to assemble OTFTs made with a poly(vinyl alcohol) (PVA) dielectric, Au and silver (Ag) electrodes, and poly(lactic-co-glycolic acid) (PLGA) substrate to accomplish carrier mobilities of ~0.25 cm^2/(V·s)$^{-1}$ [5, 46]. Electron mobilities for existing polymer semiconductors can extend up to 10 cm^2/(V·s) [107]. Translating higher mobilities exhibited by conventional organic materials to safe and practical organic semiconductors for bioresorbable devices is an area of research that necessitates further exploration.

2.3 Insulator: Dielectric

In capacitors, dielectric materials store and transport electrical charge and/or can be used to isolate an electrical charge from the surrounding components in electronic devices [10, 107]. Higher dielectric constants in parallel plate capacitors signify an increase capacitance for a given area [10]. In bioresorbable electronics, gate and interlayer dielectrics as well as passivation coatings are comprised of dielectric materials [55, 119]. Moreover, thin film dielectric materials mitigate short circuiting and are resistant to water penetration making them especially instrumental in hydrolytic protection (Sect. 2.5) [55].

2.3.1 Inorganic

Inorganic biocompatible and biodegradable dielectric materials include SiO_2, MgO, silicon nitride (Si_3N_4) magnesium fluoride (MgF_2), and spin on glass (SOG) [48, 56, 59, 78, 121]. SiO_2 and Si_3N_4 are widely adopted dielectric materials in current microelectronics. Amorphous SiO_2 has a dielectric constant of 3.9 at 1 MHz with high electrical resistivities making it a suitable material for insulation and passivation [35, 107]. Si_3N_4 exhibits a high dielectric constant of around 7 and has been used as the gate dielectric in TFTs and for charge storage in nonvolatile memory devices in microelectronics [35, 94]. MgO is commonly utilized as a gate dielectric in transistors and as the dielectric layer in capacitors [35]. Thin-film MgO exhibits optical transparency, high electrical resistivity, chemical and thermal stability, and high dielectric permittivity values ranging from 8 to 10 depending on temperature and frequency, with a dielectric constant of 9.65 at 1 MHz [107, 119]. MgF_2 has a band-gap of 11.3 eV and a dielectric constant ranging from 4.87–5.45 at 1 MHz. MgF_2 is the dielectric material in the bioresorbable resistive random-access memory (RRAM) device for memory storage developed by Zhang et al. [121, 123].

The dissolution reaction for SiO_2 yields silicic acid ($Si(OH)_4$) as a by-product (Eq. 12) [55]:

$$SiO_2 + 2H_2O \leftrightharpoons Si(OH)_4. \tag{12}$$

After a two-step dissolution reaction of Si_3N_4 (Eqs. 13 and 14), the net products are $Si(OH)_4$ and ammonia (NH_3) (Eq. 15) [55]:

$$Si_3N_4 + 6H_2O \rightleftharpoons 3SiO_2 + 4NH_3, \tag{13}$$

$$SiO_2 + 2H_2O \rightleftharpoons 3Si(OH)_4, \tag{14}$$

$$Si_3N_4 + 12H_2O \rightleftharpoons 9Si(OH)_4 + 4NH_3. \tag{15}$$

The dissolution kinetics of dielectric materials are affected by the morphology, density, and chemical stoichiometry [59]. The type of deposition process can control the density of a dielectric film. Kang et al. studied the dissolution behaviors of SiO_2 grown by wet and dry thermal growth with oxygen gas and water vapor, respectively, plasma-enhanced chemical vapor deposition (PECVD), and electron beam (e-beam) deposition in buffer solutions between pH 7.4 and 12 at room temperature as well as at 37 °C. In the same study, dissolution behavior of Si_3N_4 deposited by low-pressure chemical vapor deposition (LPCVD), low-frequency PECVD (PECVD-LF), and high-frequency PECVD (PECVD-HF) was investigated at the same pH range and temperatures as SiO_2 [55]. Dissolution rates were determined by ellipsometry and atomic force microscopy to evaluate surface topography changes with dissolution over time [55]. It was found the dissolution rates of SiO_2 under the same conditions were fastest when grown via e-beam deposition and slowest through thermal growth consistent with e-beam deposition exhibiting fragmentation and having lowest densities and highest diffusivities relative to PEVCD-deposited materials and thermally grown materials that yield the most dense and uniform growth of the three methods [55]. The LPCVD-deposited Si_3N_4 possessed the highest density and lowest dissolution rate, followed by the PECVD-LF-deposited Si_3N_4, and PECVD-HF-deposited Si_3N_4 which had the lowest density and the highest dissolution rate [55]. The dissolution was uniform for each type of growth process, and the dissolution rates increased in bovine serum compared to buffer solution along with temperature, ion concentration, and pH as hydroxyl ions (OH^-) are understood to initiate the dissolution reaction [55].

In the dissolution of MgO, the dissolution yields a $Mg(OH)_2$ by-product (Eq. 16) [35]:

$$MgO + H_2O \rightleftharpoons Mg(OH)_2. \tag{16}$$

The dissolution rate of MgO trends inversely with pH as the reaction is driven by protons (H^+) [21, 35]. MgO was determined to have a morphological dissolution rate of 50 nm per hour in deionized water at room temperature [119].

2.3.2 Organic

Most bioresorbable devices investigated used inorganic dielectric materials; however, various organic molecules have been explored as dielectric materials in transient transistors, energy storage devices, and pressure sensors. Organic materials investigated span from nucleotides, polysaccharides, caffeine, and egg albumen to silk fibroin, keratin, and gelatin along with synthetic and flexible polymers, such as poly(vinyl alcohol) (PVA), poly(glycerol sebacate) (PGS), and poly(lactic-co-glycolic acid) (PLGA) [14, 33, 35, 59, 65, 79, 105, 113, 119]. At 1 kHz, dielectric constants of these organic semiconductors range from ~3.85 to ~7.6 [46].

2.4 Insulator: Substrate

The substrate serves as the foundational supporting material for the electrical components during processing, handling, and/or operation. In bioresorbable electronics, biodegradable polymers typically serve as the substrate material. In response to low thermal stability, water permeability, and sensitivity to standard IC photolithography and etching fabrication methods of most biodegradable polymers, metallic foils and bioactive glasses have been realized to overcome this limitation [14, 22, 56]. Substrate layers typically have a thickness on the order of microns and can be engineered to fully degrade at rates from almost instantly to over several years [22, 63, 119]. Many substrate materials also function as encapsulation layers and dielectrics in transient devices [119].

2.4.1 Inorganic

Metallic substrates are mechanically robust with high dimensional stability compared to biodegradable polymeric substrates that are sensitive to standard electronic fabrication techniques and swell in solution [56, 78]. Metal foils with a thickness of ~10 μm exhibit low bending stiffness while retaining mechanical toughness and reasonable dissolution times for bioresorbable electronics [56]. Metal thin foils of biocompatible metals (Fe, Mo, Zn, W, and Mg) have served as substrates in n-channel metal-oxide-semiconductor field-effect transistors (n-MOSFETs), Si PIN diodes, capacitors, and inductors.

Degradation rates of the thin metal foils vary depending on thickness of the foil, grain morphology, and surface structure as well as solution temperature and composition [14, 56, 119]. Metal thin foils without a SiO_2 coating were submerged into solution under various degradation conditions. In PBS solution (pH 7.4 and 37 °C), the Mo and Fe foils exhibited similar rates of dissolution (0.02 and 0.08 μm/day, respectively) with W having a slightly faster rate (0.15 μm/day) followed by Zn foils (3.5 μm/day) [56]. When used in a bioresorbable electronic brain sensor, the Mg foil was found to degrade at a high rate of 4 μm/day in artificial cerebrospinal fluid (ACSF)at 37 °C [58]. Contrary to metal thin films, Mo and Fe metal foils dissolve faster in Hank's solution compared to PBS. This conflicting behavior was attributed to morphology differences of foils compared to thin films and possible differences in solutions between each study.

Adnan et al. developed a biodegradable borate bioactive glass that was completely soluble in simulated body fluids to use as a substrate for a spiral, thin film resistor, inductor, capacitor (RLC) resonator device [2]. Bioactive glasses have been used in tissue engineering and are considered safe in vivo. The intention to use a glass material as a substrate was to retain structural integrity without swelling or disintegrating to prolong lifetime of a transient electronic device while withstanding IC processes to enable direct fabrication [2]. The bare borate glass substrates were dissolved in simulated body fluid at 37 °C. Time to complete dissolution was about 40 hours having a dissolution rate of 5.5 mg/hour with a surface area of 4.35 cm^2 [2]. Although a borosilicate glass was tested for this study, the composition of the glass can be varied to acquire desired properties for device function [2]. Glasses primarily composed of silicate dissolve at a slower rate than borosilicate glasses, ranging from 10% weight loss per day to 50% weight loss per day, respectively [2]. Bioactive glasses can also contain calcium and sodium. In these cases, sodium dissolves into solution, while calcium binds to extracellular phosphates causing eventual formation of porous hydroxyapatite, a mineral found in native bone tissue [2], which may not be desirable for fully resorbable devices.

2.4.2 Organic

Organic substrates possess a wide range of properties, degradation products, and dissolution behaviors. Despite limitations in conventional electronic fabrication, polymer substrates are redeemed by tunable properties, ease of chemical functionalization, Food and Drug Administration (FDA) approval for biomedical uses, and benign and economical processing compared to that of metals [56]. The primary degradation mechanisms of current biodegradable polymers in the body occur through hydrolysis of carbonyl functional groups or enzymatic degradation [35]. Additional degradation mechanisms empoloyed by transient substrates include oxidation and UV radiation exposure [22].

Polymers with slow swelling rates are desirable to eschew premature device disintegration during dissolution [78]. Functional lifetime and corrosion behavior can be adjusted by altering composition, crystallinity, surface chemistry, and thickness of the polymer. Hydrophilic polymers undergo accelerated dissolution in aqueous solutions compared to hydrophobic counterparts. Meanwhile, slower dissolution rates occur with increasing crystallinity [22, 35]. Degradation mechanisms of polymers can also affect dissolution rates. Namely, polymers that undergo dissolution, rather than biodegradation, frequently reveal faster transience. Surface erosion and enzymatic degradation typically transpire at slower rates compared to bulk degradation [22, 119].

Key organic substrates used in transient devices can be subdivided into *natural* organic materials and *synthetic* organic and carbon-based materials. Key natural materials reviewed herein are silk, cellulose, and cellulose ethers, and rice paper, and synthetic materials reviewed below include poly(lactic acid) (PLA), poly(glycolic acid) (PGA), poly(lactic-co-glycolic acid) (PLGA), polycaprolactone (PCL), poly(octamethylene citrate) (POC), poly(octamethylene maleate citrate) (POMC), poly(glycerol sebacate) (PGS), and poly(3-hydroxybutyrate-co-3-hydroxyvalerate) (PHB/V).

Natural Silk was used as the substrate in early partial and fully bioresorbable electronic devices [36, 62]. Silk originates from silk worms, being the strongest and toughest natural fiber known to date, with tunable optical and mechanical properties [105]. Silk isolated from the cocoons of silkworm larvae is composed of the proteins fibroin and sericin [14, 63, 119]. While sericin induces an inflammatory response in humans, silk fibroin is immunologically benign and is therefore purified from raw silk for use in bioimplant applications [105]. Silk fibroin has increased conformability to tissues with decreased thickness. Film thickness and crystallinity, tuned by controlling β-sheet formation, can control degradation times on the order of minutes to years along with mechanical properties [22, 119]. Main processing conditions of silk are mild occurring in water-based, neutral pH solutions at room temperature [105]. Silk has been processed into many forms including hydrogels, ultrathin films, thick films, conformal coatings, three-dimensional porous and solid matrices, and fibers for use in tissue engineering, wound dressings, surgical sutures, implantable devices, and drug delivery [14, 105, 119].

Another resilient polymer from nature is cellulose and cellulose-derived materials. Water soluble derivatives including methyl cellulose, sodium carboxymethyl cellulose (Na-CMC), and cellulose nanofibril paper, are natural, low-cost, and nontoxic insulating materials [27, 78]. Cellulose is the most abundant organic polymer on Earth composed of β-sheets of D-glucose monomers primarily found in the cell walls of plants serving as a structural material and is commonly sourced from tree bark [22, 101, 119]. Due to the high thermal and chemical stability of cellulose, it is possible to fabricate electronic devices onto cellulose substrates directly for potential roll-to-roll fabrication in addition to transfer printing methods used for other organic substrates with success in printed and paper-based electronics [22, 119]. Nanocellulose paper is a transparent material that can be used as a substrate and has been implemented into OFETs [27]. A water-soluble cellulose ether, methylcellulose, has been used for thermo-responsive transient electronics due to its low critical solution temperature (45 °C), while another cellulose ether, carboxymethyl cellulose, has served as a substrate that can dissolve in water within 10 minutes for bioresorbable printed circuit boards, heterojunction bipolar transistors (HBTs), and screen and laser printed bioresorbable conductors [27, 119]. For enhanced flexibility and faster time to complete degradation compared to thicker films, cellulose films on the order of 800 nm have been created via functionalizing cellulose with trimethylsilyl groups with successive hydrolytic deprotection [22].

Rice paper is an alternative low-cost option for substrate materials [41, 78, 119]. However, rice paper swells rapidly upon contact with water, causing functional components of the device to rupture quickly thereby limiting its use. A CMOS inverter device using a rice paper film substrate with 200 μm thickness disintegrated within 48 hours in DI water at 37 °C [41]. Other natural substrate materials less commonly experimented with include hydrogels like alginate and gelatin, cheese and seaweed, sugar-based epoxy, and charcoal (i.e., an allotrope of carbon) used alone or synthesized with other polymers [35, 59, 119].

Synthetic Biodegradable aliphatic polyesters commonly used in US Food and Drug Administration (FDA)-approved medical devices include the poly(α-hydroxy) acids: PLA, PGA, PLGA, and PCL [14]. These and the elastomers, POC, POMC, PGS, and PHB/V, degrade primarily through hydrolysis of ester bonds in aqueous solution [6, 7, 14, 69, 93]. Degradation rates may vary with time in solution, molecular weight, and polymer composition [14]. The substrates composed of PLA, PLGA, PGA, PCL, and PHB/V, and others are commonly fabricated via spin casting, hot melting, and electrospinning [119].

PLA and the copolymer, PLGA, formed by the ring opening polymerization reaction of lactide and/or glycolide is amongst the most widely used commercially available polymers for biomedical use in pharmaceuticals, tissue engineering, and medical devices[14, 83]. The polymer is composed of lactic acid (LA) and/or glycolic acid (GA) units. The presence of the methyl moiety in lactic acid gives PLA higher crystallinity and hydrophobicity relative to PLGA resulting in longer degradation times at the tradeoff of flexibility compared to similar molecular weight PGA [78, 119]. Likewise, PLGAhas tunable degradation rates from weeks to months being prolonged by increasing LA content and molecular weight [14, 22, 27, 30]. Moreover, the enantiomers of lactic acid can be used to modify degradation rates of PLA or PLGA with the l-isomer yielding a higher crystalline polymer (PLLA) and degradation rates compared to similar molecular weight polymers composed of the racemic mixture of LA (PDLA) [14]. PLA and PLGA polymers are highly processable, being fabricated into forms like films, scaffolds, hydrogels, and micro- and nanospheres [14, 83].

Poly(caprolactone) is an aliphatic polyester comprised of hexanoate monomers synthesized by polycondensation of 6-hydroxyhexanoic acid or, more commonly, via the ring opening polymerization of ε-caprolactone [69]. PCL typically has a slower degradation rate and higher hydrophobicity relative to PLGA and PGA due to a high degree of crystallinity (up to 69%) [35, 69]. The mechanical properties of PCL including tensile strength, Young's modulus, and elongation at break range from 4 to 785 MPa, 0.21 to 0.44 GPa, and 20 to 1000%, respectively, due to variations in molecular weight, crystallinity, and dissolution environment, which also affect rate of degradation [69]. Additionally, copolymerization with other polymers like PLGA, PLA, and PGS leads to enhanced properties, such as shape memory, flexibility, and elasticity [35, 69, 88]. Besides hydrolytic cleavage, thermal stimulation can achieve transience in devices containing PCL. At temperatures higher than the melting temperature (60 °C), the polymer quickly wrinkles and shrinks in a matter seconds [29]. An applied voltage can induce heating in a transient device as by Gao and colleagues [29].

Polyhydroxyalkanoates, like PHB/V, are another type of polyester family formed by bacteria or through copolymerization; for example, the copolymerization of 3-hydroxybutanoic acid and 3-hydroxypentanoic acid produces PHB/V [126]. Polyhydroxyalkanoates are biocompatible and shown to have nontoxic by-products after degradation via hydrolysis [95, 119]. Although 50 μm films of PHB/V were reported to fully degrade over 50 weeks in sea water at 15 °C, a device with an encapsulating layer made of PHB/V lasts tens of minutes once sweat covers the PHB/V and enters its surface [6].

In many circumstances, flexibility and elasticity of implantable materials are essential to fit to the curvilinear and dynamic anatomical structures in the body [22, 27]. In these cases, a lightly cross-linked network polymer with viscoelastic properties, or an *elastomer*, serves as an ideal insulator for substrate and encapsulating layers [27]. As mentioned above, POC, POMC, and PGS are three well-established elastomers employed in transient electronics [27, 35, 59, 78]. POC, a polymer composed of two nontoxic monomers (citric acid and 1,8-octanediol), is synthesized via a scalable, facile, one-pot polycondensation reaction without the use of catalysts [115]. POC has versatile and widely recognized use in tissue engineering and drug delivery [42, 115]. Similar to other synthetic polymers, the mechanical and degradation properties of POC can be finely tuned depending on monomer ratios, cross-linking conditions, molecular weight, compositing, and functionalization of the polymer [115]. The rate of dissolution is further altered by temperature, solute concentration, and pH of the dissolution media [27]. The Young's modulus for POC can range from 2.84 to 6.44 MPa, and the polymer has an elongation at break varying from 253% to 375% [35]. In PBS, at pH 10 at room temperature, POC degrades over several weeks [27]. Hwang et al. have used POC as a substrate and encapsulating material for CMOS MOSFET technology to create a electrocardiography monitor (EKG) and pH sensor [42]. The POC film used in this device could be stretched to roughly 30% [42].

Not only does POC have appealing degradation and mechanical properties, it can be easily formulated with other monomers creating a biodegradable, citrate-based material platform for versatile biomedical applications [100]. An example is the addition of the maleic anhydride to citric acid and 1,8-octane diol resulting in a photocross linkable elastomer, POMC, which has been used in an flexible, bioresorbable strain and pressure sensor for postsurgical tendon repair monitoring; meanwhile, the addition of amino acids, like cysteine and serine, manifests photoluminescent properties creating biodegradable photoluminescent polymers (BPLPs) for applications in bioimaging, drug delivery, and tissue engineering [85, 100].

Furthermore, BPLP has been successfully doped with aniline tetramer to synthesize electroactive, biodegradable elastomers that have controllable elasticity, conductivity, and degradation with dual-imaging capabilities [96]. To increase the mechanical strength of POC and BPLP, *hexamethylene diisocyanate* (HDI) can be reacted with each to result in cross-linked urethane-doped polyester (CUPE) and urethane-doped biodegradable photoluminescent polymers (UBPLPs), respectively [19, 85, 100].

PGS is another elastomer developed in 2002 by Wang et al. and is synthesized via the polycondensation reaction of glycerol and sebacic acid [6, 113]. PGS exhibits a Young's modulus ranging from 0.05 to 2 MPa and an elongation at break of greater than 260%. PGS has been used as a substrate in a piezoelectric pressure sensor device demonstrating no significant change in viscoelastic behavior after 7 weeks in PBS at 37 °C [6]. Like polyanhydrides, PGS undergoes surface erosion rather than bulk degradation as do the poly(α-hydroxy) acids, explaining a long functional lifetime [6, 14, 93, 119]. Only 17% of PGS was reported to degrade when incubated in agitated PBS at 37 °C for 60 days in vitro, while in vivo surface degradation revealed to be between 0.2 and 1.5 mm per month [6, 35].

Polyanhydrides are versatile, biocompatible copolymers of methyl vinyl and maleic anhydride groups that can be aliphatic, aromatic, or a mixture of both [31, 95]. Moreover, acrylated anhydrides capable of UV and visible light induced cross-linking have been designed to permit alternative processing regimes [95]. Polyanhydrides have been extensively studied in tissue engineering, medical devices, and controlled-release drug delivery [60, 95]. Polyanhydrides have reported tensile strengths of 25–27 MPa with a Young's modulus around 1.3 MPa and elongation at break between 14% and 85% [95]. To improve mechanical properties, polyanhydride can be copolymerized with polyimide as tried in orthopedic applications [95]. Polyanhydrides typically exhibit controlled surface erosion, rather than bulk degradation [31, 60, 119]. The degradation products, diacid and linear methacrylic acid molecules, of polyanhydrides, dissolve in water [95, 119]. The rates of degradation can be tuned from days to weeks in biofluids by increasing the degree of crystallinity and anhydride composition [28, 31, 95, 119]. Combination with poly(ethylene glycol) (PEG) has been reported to increase degradation rates [28, 53].

Substrate layers composed of poly(vinyl alcohol) PVA and poly(vinylpyrrolidone) (PVP) are nontoxic and water-soluble, demonstrating reversible dissolution occurring within minutes at room temperature at modifiable rates according to solution temperature and degree of polymerization [27, 54, 119]. These polymers are soluble in polar solvents allowing their use in printing inks, food products, pharmaceuticals, and cosmetics [27, 119]. PVA has been composited to adjust degradation rates and mechanical properties [1, 10]. PVA composites with gelatin and sucrose prolonged and expedited dissolution in water, respectively, compared to PVA alone. The decrease in the dissolution rate and swelling of the PVA/gelatin composite were attributed to collective hydrogen and ester bonding believed to form a stable, triple-helix structure, whereas the increase in the dissolution rate of the PVA/sucrose composite was rationalized by the high solubility of sucrose in water [1, 119].

2.5 Insulator: Encapsulation Layer

Encapsulation, or encapsulating, layers prolong lifetime in addition to increasing mechanical robustness of transient electronic devices. When using an encapsulating layer on a bioresorbable device, two-phase dissolution kinetics are classically observed with the initial dissolution of the encapsulation layer being slow followed by an expedited dissolution of the metal components resulting in rapid functional decline [36, 38, 56, 119]. This outer layer is critical to translatable bioresorbable electronics since electrical failure ensues once electronic materials come into contact with water, shortening the final lifetime before complete dissolution of the transient electronic components [119]. Typically swelling or nonuniform dissolution due to pinhole defects, pitting corrosion, or inconsistent composition is more pronounced in single-layer materials compared to bulk counterparts of substrate and encapsulating layers contributing to premature water diffusion of thin pieces in transient electronics. Recent developments to increase efficacy of encapsulating layers include compositing (Sect. 2.4.2), designing air pockets between semipermeable layers for indirect passivation between layers, and using various deposition techniques to reduce water permeable defects [1, 31, 55, 59, 78, 119]. Polymeric materials and film oxides like silk, polyanhydrides, PVA, PCL, PLGA, MgO, SiO_2, and Si_3N_4 are commonly used alone or in combination to form encapsulating layers with more recent investigation of p-type Si-NMs for this use [59, 74, 90, 119]. Novel encapsulation designs are of high interest as this layer provides a great opportunity to accomplish triggered degradation and precisely control functional lifetimes of transient devices [16, 36, 119].

2.5.1 Inorganic

The superior thermal and mechanical stability of inorganic materials relative to organic pieces inspires investigation of inorganic materials for protective coatings in bioresorbable electronics [35, 119]. In the first fully reported transient N-MOSFET device, a bilayer of MgO and highly crystalline silk protected the electrical components of the device for as much as 90 hours [36]. An 800 nm encapsulating layer of MgO alone on an inverter permitted protection against degradation for approximately 7 hours; meanwhile, in another study, a MgO layer (1–3 μm) provided a functional lifetime between 4 and 7 hours [39, 56]. A metal insulator metal (MIM) capacitor and Mg spiral inductor using metallic glass substrates were encapsulated with a 3 μm of layer of MgO and demonstrated stable operational times of approximately 5 and 6 hours in DI water at room temperature, respectively [56].

Recently, mono-Si-NM encapsulating layers (1.5 μm) have been studied for encapsulation for Mg thin films degrading at a rate of ~4.8 nm/day to keep the Mg film intact for over 60 days in PBS at 37 °C. In contrast, the same Mg film degrades within minutes using PLGA (5 μm) or e-beam-deposited SiO_2 (200 nm) encapsulating layers in PBS at room temperature. Boron-doped Si-NM encapsulating layers degraded at rates of 1–60 nm per day depending on degradation conditions. Lifetime of Si-NM-protected Mo/Si electrodes is expected to be roughly 50 days at 37 °C [75].

Silicate-based SOGs are appealing for insulation, and encapsulation as SOGs have tunable dissolution rates in aqueous solution, demonstrated in vitro cytocompatibility, established use in conventional electronics, and provide deposition that is confroms tightly with its substrates. SOGs were cured at different temperatures to control dissolution rates ranging from 6 nm/day and 50 nm/day when cured at 800 °C and 300 °C, respectively (PBS, nitrogen purge, pH 7.4, 37 °C). Less Si-O-Si bridges that provide stability against hydrolysis form at the glass surface at low curing temperatures compared to high temperatures [56].

It was shown that dissolution times can be prolonged synergistically by depositing alternating layers of SiO_2 and Si_3N_4 by PECVD to minimize exposed defects [55]. Furthermore, uniform deposition methods (i.e., atomic layer deposition (ALD)) result in less defects than PECVD. Therefore, ALD can be used to coat PECVD-deposited materials to lengthen dissolution times compared to PECVD-deposited materials without ALD-deposited coatings [55].

2.5.2 Organic

Although not as mechanically or chemically durable compared to inorganic materials, organic encapsulation layers are useful in transient devices to provide flexibility and can increase device robustness by reducing stress accumulation at interconnects [58, 73]. Many of the same organic materials used as substrates and dielectrics are also used as the packaging materials. It should be considered that many organic polymers are permeable to water and can swell causing rapid deterioration of the functional device components [119]. Adjusting crystallinity, encapsulating layer thickness, degree of polymerization, and monomer ratios of constituent polymers, as well as compositing, can alter water resistance, swelling, and degradation time of polymeric encapsulating layers [17, 27, 78].

PLA was used as an encapsulating layer for a piezoelectric force sensor in a resorbable intra-organ pressure monitoring device that performed unchanged for up to 16 days in vivo [17]. In a study conducted by Tsang et al., PCL (5 μm) coated biodegradable iron/magnesium batteries resulted in battery performance of 30 μW of power for 100 hours. This was reported to be sufficient to operate low-power transient implants, such as pacemakers, low-power neurostimulators, or flow rate sensors, for a matter of days [108]. Polyanhydrides have also been used as an encapsulating layer in bioresorbable electronics due to their low permeability to water and surface erosion behavior to minimize swelling and resulting potential damage to electronic components [28, 58, 73]. An 120-μm-thick polyanhydride encapsulation layer permitted stable operation of an intracranial pressure (ICP) sensor in vivo for a maximum of 72 hours. Another hydrophobic polyanhydride was synthesized to encapsulate a bioresorbable microsupercapacitor (MSC), a small-scale energy storage device composed of metal electrodes and electrolyte hydrogel [58, 59]. Similarly, Lee and colleagues used a thick hydrophobic polyanhydride encapsulation layer to protect MSC components while reducing water evaporation from the electrolyte solution and keeping it from contact with the surrounding media to prolong device lifetime compared to the use of a thinner PLGA encapsulating layer (Sect 3.1.3) [73]. Thermal sealing was used for PVA packaging of a transient lithium battery as well as for a multilayer silk encapsulated transient split-ring resonator antenna where noncrystalline silk layers coated the device and were then thermally sealed with an outer layer of crystalline silk [9, 26, 119]. The air pocket formed between the semipermeable crystalline and noncrystalline layers fills with water vapor instead of fluid to prolong diffusion of water through the encapsulating layer. Further increasing amount of silk layers and pockets lengthened time to device failure [9].

3 Applications in Biomedical Engineering

Through advancements in flexible electronics, deposition and fabrication techniques, degradable CMOS, wireless communication, and memory storage technologies, bioresorbable electronics have foreseeable potential for translation to biomedically relevant devices possessing complete transience, conformal contact to tissues, and enhanced precision relative to conventional diagnostic, treatment, and monitoring tools [14, 65, 79, 97, 121]. Application of transient electronic implant in clinical settings evades the necessity for a secondary removal procedure and potential complications while minimizing the risk for foreign body response, infection, device migration, and physical damage to surrounding tissue with essentially no e-waste production once the device is no longer needed [27]. Herein, key representative devices composed of materials previously discussed (Sect. 2) critical to advancing resorbable power sources, biosensing devices, and therapeutic systems are reported (Fig. 3). Table 2 provides a comprehensive reference guide to bioresorbable systems including and beyond the devices that are discussed in this section.

3.1 Energy Supply

A considerable restraint to realizing practical use of many transient devices is the need of external power sources for successful operation. Self-sustainable, wireless energy supplies are therefore crucial for the progression of transient electronics in order to function independently and extend operational lifetimes [81]. Representative examples of current energy sources include batteries, mechanical energy harvesters (MEHs), and microsupercapacitors (MSCs) [34]. Bioresorbable radiofrequency antennas and power scavengers have also been developed for wireless power and communication [119].

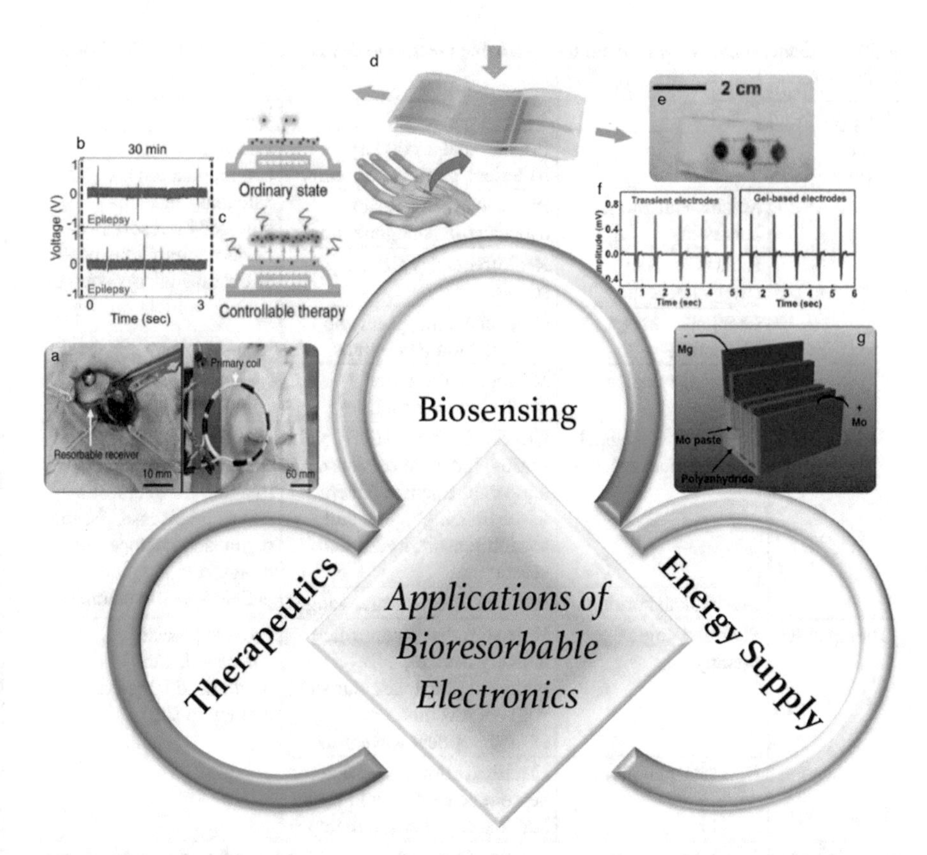

Fig. 3 Main classes and respective representative devices of clinically translatable bioresorbable electronics (**a**) Postoperative wireless stimulation by a bioresorbable device equipped with a radiofrequency power harvester connected to an electrical interface provided to a transected rat sciatic nerve for enhanced peripheral nerve regeneration. (Reprinted by permission from RightsLink Permissions Springer Nature Customer Service Centre GmbH: Springer Nature Nature Medicine [66] Copyright 2018). (**b–d**) A resorbable dual monitoring and treatment device powered by a triboelectric nanogenerator (TENG) for epilepsy management. (Adapted with permission from Zhang et al. [120]. Copyright 2018 John Wiley and Sons). (**b**) Detection by a drug-loaded device of epileptic signals (top) compared to an empty control (bottom) (**c**) A schematic of the inactive device (top) and activated device releasing drug due to resistor-mediated thermal stimuli triggered by epileptic signals (**d**) a dual-functioning, flexible pressure sensor and strain gauge foreseen to be used to guide patient-specific treatment for postoperative tendon repair. (Reprinted by permission from RightsLink Permissions Springer Nature Customer Service Centre GmbH: Springer Nature Nature Electronics [7]. Copyright 2018). (**e, f**) an electrophysiological (EP) sensor adhered to a POC substrate for EKG monitoring. (Adapted with permission from Hwang et al. [42]. Copyright 2015 American Chemical Society.) (**e**) The EP sensor placed on a human chest (**f**) comparable EKG readings from the bioresorbable device (left) compared to commercially available gel-based electrode readings (right) (**g**) a primary battery made of stacked cells of Mg and Mo tested in a PDMS dish containing PBS electrolyte solution. Reported output voltages of 1.5–1.6 V and discharge current density of 0.1 mA·cm^{-2} was sustained over 6 hours. (Reproduced with permission from Yin et al. [117]. Copyright 2014 John Wiley and Sons)

Table 2 General overview of main bioresorbable electronic devices and intended implications in healthcare

Purpose	Designs	Key devices	Implications
Energy supply	Battery	Primary and secondary batteries[a], photovoltaic cells[b]	Enable autonomous communication (i.e., RF antennas) and self-powered bioresorbable devices without the need of external energy sources
	Mechanical energy harvesting	Piezoelectric harvester, triboelectric nanogenerator[c]	
	Radiofrequency energy harvesting	Radiofrequency (RF) scavenger[d]	
	Graphene energy harvesting	Graphene flake, graphene oxide gelatin biocomposites[e]	
	Supercapacitance	Microsupercapacitor[f], organic supercapacitor[g]	
Biosensing	Electrophysiological and chemical sensing	EKG[h], ECoG and EEG[i], pH sensor[j], biochemical sensor[k], temperature sensor[l], deep tissue imaging[m], flow and motion sensor[n], hydration sensor[o]	Deliver direct and sensitive detection and imaging for diagnostics and acute disease, tissue, organ, and postoperative monitoring to guide patient-specific therapies
	Mechanical sensing	Pressure sensor, strain gauge[p]	
Therapeutics	Pharmacologic therapy	Heat-stimulated-controlled delivery device[q], multifunctional vascular stent with RRAM memory storage[r], optical waveguide[s]	Provide precise, controlled, and on-demand targeted therapy in situ
	Non-pharmacologic therapy	Electrical stimulus device for nerve regeneration and tissue stimulation[t], thermal therapy[u]	

[a][13, 25, 26, 34, 51, 52, 64, 108, 117]; [b][81]; [c][17, 18, 80, 120, 122]; [d][38, 72]; [e][70]; [f][73]; [g][68, 112]; [h][6, 42]; [i][58, 118]; [j][42, 58]; [k][3, 42, 84, 89]; [l][3, 58, 71]; [m][3]; [n][58, 97]; [o][89]; [p][6, 7, 17, 36, 58, 82]; [q][12, 71, 97, 104, 120]; [r][97]; [s][3]; [t][66, 122]; [u][36, 104]

3.1.1 Batteries

Transient primary and secondary batteries (i.e., non-rechargeable and rechargeable voltaic batteries, respectively) have been designed with and without demonstrated biocompatibility to date [38, 72]. Kelvin Fu and colleagues outlined five key criteria for successful design and performance of transient batteries including complete transience in the body; controllable desorption rates; mechanical properties and performance that meet device demands over specified timeframes; appropriate device dimensions; and compatibility with bioresorbable devices [13, 25, 34, 51, 52, 108, 117].

Yin and colleagues fulfilled these fundamental criteria with a fully resorbable, biocompatible primary battery ($3\ cm^3 \times 1.3\ cm^3 \times 1.6\ cm^3$, ~3.5 g) composed of Mg foil anodes, Mo cathodes, and porous polyanhydride packaging that was permeable to hydrogen evolved from the battery [27]. Single cells of Mg/Fe, Mg/W, and Mg/Mo batteries had output voltages of about 0.75 V, 0.65 V, and 0.45 V, respectively, in PBS electrolyte solution over 24 hours. To achieve comparable performance to conventional primary batteries, the bioresorbable batteries were assembled with four stacks of Mg-Mo cells to increase output voltage to 1.5–1.6 V and achieve a discharge current density of 0.1 $mA \cdot cm^{-2}$ sustained over 6 hours in PBS electrolyte solution (Fig. 3h) [117]. Potential pitting corrosion of metal foils or collective leakage of water through the porous polyanhydride overtime was used to explain the diminished stable lifetime of the stacked cells compared to the single cell batteries [35, 117].

Kim et al. developed a proof-of-concept edible sodium battery intended to directly supply power to self-sufficient, noninvasive devices for direct sensing of heart rate, core body temperature, chemicals, and metabolic activity. It was also envisaged to provide power for stimulation for gastrointestinal (GI) motility disorders and electrically mediated chemical release throughout the GI tract [117]. The battery was fabricated with flexible, rectangular electrodes (~2 mm × 0.4 mm) composed of silver nanowires dispersed through flexible poly(glycerol-co-sebacate)-cinnamate polymer, an activated carbon anode, and MgO_2 cathode. The battery was bent in half and placed into a gelatin capsule where current flow commenced once electrolyte solution reached the inside of the capsule [64]. Onset of activity can be adjusted by modifying encapsulating materials. In a 0.5 M isothermal solution of Na_2SO_4 at 37 °C, the battery supplied current to recording stainless steel electrodes to mimic an attached device up to 100 mA and then decreased gradually over the course of 5 hours. The energy density of the device could be improved by altering the cathode/anode mass ratios, increasing the amount of sodium content on the anode, and increasing the electrode mass density [64]. The battery is anticipated to be safe in vivo according to previously established biosafety and biocompatibility studies of each device component [64].

3.1.2 Mechanical Energy Harvesters

A bioresorbable mechanical energy harvester (MEH) using ZnO as the piezoelectric material was developed in 2013 by Dagdeviren et al. Piezoelectric energy harvesters produce energy by converting mechanical energy into electrical energy for low-power devices, such as microelectronics. The movement of the body can be used to power the energy supply [64]. Each mechanical energy harvester had a capacitor-type structure with a ZnO layer between bottom and top Mg electrodes (50 μm × 2 mm). There were six groups of ten capacitor ensembles connected in series by Mg interconnects creating a ~1 cm × 1 cm capacitor stage in the center of a silk fibroin substrate that was a few cm in length [18]. Conversion efficiency of the device was measured to be 0.28% under dry conditions but varies with mechanical load, physical dimensions, and elasticity of the substrate and piezoelectric components [18]. Lifetime of the energy harvester was not reported; however, a thin-film transistor with similar components without encapsulation reported in the same study exhibited steady function for over 3 hours and fully disappeared within 15 hours in DI water at room temperature [18].

The use of bioresorbable triboelectric generators for implantable devices has been reported by multiple authors. A voltage is produced in these generators by fluctuating the distance between two dissimilar materials capable of electron exchange. The arrangement includes an inner coating of metal electrodes followed by an insulating spacer in between the two metal coatings. Upon mechanical cycling, an electric potential is produced, which is capable of inducing an AC current between the metal electrodes [18]. In the study conducted by Zheng et al., a completely biodegradable and biocompatible triboelectric generator was developed using layers of commercially available polymers, like PLGA, PHB/V, PCL, and PVA, with a polymeric spacer in the center and Mg electrode layers. The nanogenerator (2 cm × 3 cm) was reported to have open-circuit voltage outputs ranging from 10 to 40 V with simulated biomechanical motion (1 Hz) depending on the polymer layers used. Further modification of output voltage could be executed by roughening the surfaces of the polymers through etching via immersion into sodium hydroxide (NaOH, 2 M) over a specified time frame (0–10 minutes). Output voltage increase with increasing etching time. Also, polymer selection dictates voltage output of the device, as each polymer possesses a different capacity of transferring or retaining electrons upon mechanical stimuli (i.e., triboelectric potential). In vivo studies using PLGA (75:25) and PVA encapsulating layers showed stable output for 14 days and less than 1 day, respectively. To demonstrate medical utility, the nanogenerator was linked to a rectifier and electrodes to apply an electric field to primary neurons to attempt cell alignment induction, which has been shown to be conducive to nerve repair. ImageJ results reported that 88% of cells exhibited alignment upon electrical stimulation (10 V/mm, 1 Hz) of 30 minutes per day for 5 days as opposed to disordered growth exhibited in the nerve cells with no electrical stimuli [12, 38, 80, 120, 122].

3.1.3 Microsupercapacitors

A flexible, bioresorbable microsupercapacitor, a type of electrochemical supercapacitor for storage and delivery of energy, was built onto a glass substrate using metal electrodes (W, Fe, or Mo) and a NaCl agarose hydrogel for the electrolyte with PLGA or polyanhydride encapsulating layers, respectively. The device was flexible and able to adhere to a finger nail demonstraing conformability to body tissues. Tungsten had the highest mechanical and equivalent series resistance (ESR) durability compared to Fe and Mo at 1000 cycles and 500 cycles, respectively. The device had stable capacitance for ~6 hours with PLGA packaging (10–15 μm) and extended up to 36 hours (10×10^{-3} M PBS, pH 7.4, 37 °C) with the addition of thicker, double-sided layers of a hydrophobic polyanhydride packaging material (150 μm) [73].

3.2 *Biosensing*

Standard clinical practice heavily relies on electronic implants for biosensing. Common applications being neurological diagnostics, traumatic brain injury monitoring, and care for cardiac events. Current conventional electronic sensors often lack flexibility, failing to conform to the curvilinear anatomy causing unnecessary mechanical stresses on surrounding tissues and necessitating device removal. Transient electronic biosensors can serve as a solution to existing mechanical restraints with no need for device removal [73]. Sensors outlined below have been designed for electrical sensing, pressure sensing, and biochemical sensing in applications ranging from wound healing, intracranial and organ pressure monitoring to pH, neurologic, and biomarker sensing. Furthermore, with advances in resorbable wireless communications and memory, computerized and stored data acquisition along with self-sufficient power supply is becoming possible to collect and store data from completely transient implantable devices for translation [14, 79, 118].

3.2.1 Electrophysiologic Monitoring

Lightweight, transient electrophysiological (EP) monitors have been developed from an organic electrochemical transistor (OECT) and multifunctional, elastic sensor using CMOS technology for uses in cardiovascular, muscle, and nerve monitoring. These transient systems are particularly advantageous over conventional electronics by minimizing e-waste, providing easy transport for short-term clinical needs, and avoiding sensor removal [35, 38, 41, 58, 89, 97]. The OECT device was composed of a solvent-cast PLGA substrate, gold source and drain, and PEDOT:PSS conductive components [11, 42]. Using this technology, an EKG capable of detecting 50 μV signals was fabricated with reported response times around 1.5 ms (0.1 M PBS at pH 7.4) with signal-to-noise ratios (S:N) comparable to standard Faradaic electrodes [11].

For high-resolution electrophysiological and postoperative monitoring Yu and colleagues developed three key bioresorbable devices for EKG recording, electrocorticography (ECoG) and electroencephalography (EEG) monitoring with operation times lasting between a few hours to 30 days for heart function tracking and spatiotemporal mapping of brain activity, respectively [11]. The neural electrode array was composed of a PLGA substrate layer, phosphorous-doped Si-NM electrodes and interconnects, and SiO_2 for insulation around interconnects and interlayer dielectrics [119]. Areas of highly doped Si served as terminal pads and external connections. In vivo comparison studies of the bioresorbable electrodes compared to control (commercial stainless steel microwave electrodes) embedded 0.5 mm from the cortical surface into the left hemisphere were completed up to 33 days with the bioresorbable device demonstrating a superior S:N compared to control after application of bicuculline methiodide to induce epileptic activity. A multiplexed neural electrode array was also fabricated for more efficient ECoG sensing [119]. The flexible device consisted of PLGA as the substrate, mono-Si-NM as the semiconductor and neural interface electrodes, and SiO_2 as an insulating layer. SiO_2 and Si_3N_4 further served as gate dielectrics, interlayer dielectrics, and encapsulating layers with sensors and an external readout joined by electrical interconnects consisting of Mo [49, 119]. Based on the accelerated dissolution tests, the whole multiplexing device is expected to completely resorb in 6 months [119]. The microscale ECoG performance of a 64-electrode array was successfully used to visualize epileptic-related peaks after topical application of picrotoxin followed by immediate placement of the device onto the left hemisphere of the cortex in anaesthetized rats [119]. In future studies for epilepsy diagnostic applications, it would be ideal to increase operational lifetime as locating seizure origin takes 1–3 months of monitoring on average in practice. Somatosensory evoked potential (SSEP) studies were completed in vivo using the multiplexing device exhibiting the ability to record low-amplitude-induced cortical activity from mechanically stimulated whiskers. This device could also be used for monitoring of skeletal muscle tissue and organ function as well as early detection of implant failures (i.e., pressure and flow after coiling, vascular grafting, and repair of CSF leaks) during postoperative timeframes to evade expensive and invasive means of monitoring and consequences of neglecting implant failure [119].

3.2.2 Environmental Sensing

A multifunctional, completely bioresorbable wireless sensor was tested in vivo by Kang and colleagues for applications including but not limited to traumatic brain injury, extremity compartment syndromes, tissue stimulation, biomolecular sensing and recording. The device carried potential for temperature, intracranial pressure (ICP), fluid flow, motion, and/or pH sensing upon minor design modifications [118]. Furthermore, incorporation of chemical sensing, energy harvesting, and radiofrequency communication attachments is also possible [58]. The final device (1 mm × 2 mm × 0.08 mm) incorporated nanoporous silicon or Mg foil as the substrate with a cavity etched coated with a PLGA membrane to create an air pocket [58]. The piezoresistive component was a serpentine-shaped Si-NM placed on the perimeter of the etched gap where maximum strain was induced for a given applied pressure and the passivation layer was composed of SiO_2 [58]. Mo or Mg interconnect wires were used as the bridge between PLGA and the Mo wires for wireless communication with PLGA insulation [58]. A pressure range of 0–70 mmHg was detectable by this sensor with reliable operation up to 72 hours in deionized (DI) water (37 °C) and artificial cerebrospinal fluid (aCSF) using a 120 μm-thick polyanhydride encapsulation layer. A temperature sensor without an etched air pocket avoided water penetration for up to 6 days [58]. No evidence of device rejection was observed up to an 8-week time point in rats [58]. A PLLA-based piezoelectric device was also constructed for intra-organ pressure monitoring [58]. In this device, PLLA serves as the piezoelectric surrounded by Mo or Mg electrodes and is coated with a PLA encapsulation layer. The device undergoes complete dissolution after approximately 2 months [17].

An organic electrochemical sensor was developed to successfully sense dopamine and vitamin C by using electrodes made of PEDOT:PSS and silk fibroin as a substrate, which degraded within 4 weeks. Other sensors have been developed for bacterial sensing using conductive graphene components on a silk substrate to detect bacteria on tooth enamel [17]. Additionally, a novel, bioresorbable hydration sensor consisting of phosphorous-doped Si electrodes, Mg contacts and interconnects, SiO_2 dielectric, and PLGA substrate was used to monitor changes in skin moisture through impedance measurements comparably to a commercial moisture meter and was expected to completely disappear within a few months (0.1 M PBS at 37 °C). Hydration sensing could be used to monitor wound healing as both excess and inadequate hydration have negative implications for the wound healing process [35, 89].

3.2.3 Elastic Sensors for Electrophysical, Chemical, and Mechanical Sensing

Devices that need to accommodate high strains from body motion and curvature have been developed to create CMOS inverters, pH, pressure, and electrophysiological (EP) sensors. Stretchable sensors developed for monitoring tendon healing after surgery to guide patient-specific treatment regimens were created by designing a scalable, dual-functioning pressure sensor and strain gauge (12 mm × 7 mm × 1.4 mm) using the elastomer, POMaC, as encapsulating layers due to its softness comparable to body tissue and low tensile modulus to prevent interference with the healing process (Fig. 3e) [41]. The elastomer, PGS, was used as the dielectric material for the pressure sensor capacitor and nonadhesive layer in the strain gauge. Mg served as the electrodes supported by a PLLA substrate [7]. The device was capable of sensing strain 0–15% for five loading-unloading cycles and 0–100 kPa of pressure for six cycles with negligible hysteresis and S:N of 2:1. Moreover, the device was able to manage stable capacitance for over 20,000 cycles using applied relevant strain ranges for tendons (5–10%) along with 15–45 kPa of applied pressure with sustained function after 3.5 weeks in vivo [7]. No significant differences were found in immunochemistry analysis compared to silicone controls after 8 weeks. Other stretchable pressure sensors used PGS as a dielectric between Fe-Mg electrode layers taking a few months to completely degrade for cardiovascular monitoring [7].

Hwang and colleagues specially designed implantable devices capable of stretching up to ~30% while sustaining functionality. The design included Si nanoribbons with serpentine and noncoplanar interconnects joining device islands on the elastomeric substrate composed of POC. Under mechanical stress, the specially designed interconnects deform by in-plane and out-of-plane buckling while leaving device islands unaffected [42]. POC was chosen as a substrate due to its biocompatibility, controllable mechanical properties, biodegradability, and exceptional elasticity [6, 14]. The CMOS inverters consisted of Mg electrodes and interconnects, Si-NMs for the semiconducting nanoribbons, and SiO_2 for the gate and interlayer dielectrics and encapsulation layer. In each device, the interconnections were encapsulated with thin layers of POC to better perform under applied stress, prevent delamination, and for protection. SiO_2 and SiN_x dielectric materials were suggested to prolong device performance without substantial sacrifice to device sensitivity [42].

The hyperelastic device design was used to create a stretchable pH sensor and EP sensor [42]. In the pH sensor, Si nanoribbons underwent functionalization with 3-aminopropyltriethoxysilane. Under acidic conditions, the functionalized nanoribbons become protonated, changing surface charge, which serve as an electrostatic gate to subsequently diminish or gather charge carriers in doped n-type or p-type Si nanoribbons ($\sim 10^{20}$ cm^{-3}) permitting detection of pH changes. The pH sensors had a sensitivity of 0.1 ± 0.01 µS/pH for phosphorous-doped Si and 0.3 ± 0.02 µS/pH in boron-doped Si. The pH sensor containing p-type Si (200 nm) exhibited stable detection at 5 days (aqueous solution, pH 7.4, 37 °C) [42]. The EP sensors were attached topically to the chest and forearm of a patient to obtain EKG and electromyogram (EMG) readings (Fig. 3f) [42]. The serpentine meshes were composed of Mg and SiO_2 layers. Mg electrodes were coupled to the skin with a thin layer of POC between the two. Output EKG and EMG readings were compared to control (conventional gel electrode) (Fig. 3g), and devices were able to completely dissolve in pH 10 aqueous solution at room temperature.

3.3 Therapeutics

Developing bioresorbable electronic systems permits improved means of localized controlled release of drug delivery and on-demand non-pharmacologic (i.e., thermal and electric) therapies with envisioned indications not limited to nerve regeneration, bacterial infections, cancer therapy, and coronary artery disease [59, 66, 71, 97, 104]. Noteworthy advances in transient electronic therapeutic devices include heat-mediated drug delivery, noninvasive postsurgical electrical stimulation, and multi-functional capabilities sometimes incorporating memory storage and self-sufficient sources of power as summarized below.

3.3.1 Heat-Stimulated Drug Release

To manage postoperative infection secondary to *Staphylococcus aureus (S. aureus)* and *Escherichia coli (E. coli)*, Tao and colleagues designed a resorbable therapeutic device capable of remote-controlled, heat-stimulated drug release. The device consisted of a silk substrate, Mg containing resistor linked to a Mg heater with a MgO interlayer dielectric in between protected by a silk encapsulation layer [104]. The drug-loaded model of the device used an ampicillin-silk matrix that coated the encapsulation layer [104]. A remote primary coil placed within 1 mm of the receiving coil activated the device through near-field coupling. A steady, local temperature of 42 °C or 49 °C could be produced from the device for two 10-minute treatment intervals [104]. Increased temperatures caused simultaneous crystallization of the silk fibroin and enhanced diffusion of ampicillin to augment its release into the surrounding milieu. Drug release could be controlled by tuning silk crystallinity and thickness of the encapsulation layer [104]. In vitro and in vivo experiments both demonstrated successful bacterial killing and visible surgical site healing with complete device dissolution within 15 days in vivo [104].

Multidrug release from an electrically active resorbable device was introduced by Lee and colleagues in 2015. Drug-containing lipids expanded to release three chemically dissimilar medications: parathyroid hormone (pTH), dextran, and doxorubicin, in response to remote-controlled heat stimuli [71]. The device components were comprised of Mo conductive pieces, and PLGA constituted the interconnects, dielectric layers and encapsulation with a novel multilayered lipid and cholesterol design to support embedded drugs with minute leakage over the course of months without heat stimuli. Inductively coupled coils caused temperature increases (41–45 °C) wirelessly by applying power from a waveform and radiofrequency generator to an external coil (1.0 W at 12.5–14 MHz). The regimen was able to be applied daily to release medication induced by temperature-related changes in lipid structure for up to 1 week [71]. Strategies employed to modify drug or drug dosages included substituting lipids to alter diffusivity of and attractive forces with loaded drugs, increasing the quantity of lipid layers, and tailoring lipid to drug ratios [71]. Moreover, biocompatibility was demonstrated with no significant increase in immune cells at the implantation site or weight loss in mice with implanted lipid containing devices compared to high-density polyethylene (HDPE) and the resorbable device without lipids, respectively [71].

3.3.2 Tissue Regeneration

Peripheral nerve injuries account for an astounding 3% of all trauma-related injuries. Peripheral nerve repair procedures account for 5,000,000 disability days in the USA, and only half of patients victim to peripheral nerve damage report satisfactory functional recovery after surgical intervention [87, 92]. Studies have demonstrated improved functional recovery while reducing recovery times with electrical stimulation at the surgical site. An implantable, bioresorbable, and self-operating electrical stimulating device was reported to provide direct electrical stimulation postoperatively in 1 hour doses for up to 6 days with superior effects to a negative control (no stimulation) in rats, and required no surgical removeal unlike its predecessors [32, 66]. The transient device (40 mm × 10 mm × 200 μm) combined a radiofrequency power harvester using a loop antenna composed of Mg, a dielectric interlayer of PLGA, a doped Si-NM radiofrequency diode with Mg electrodes, and a parallel plate capacitor made of a SiO_2 dielectric layer in between the conducting layers linked to an electrical interface formed from protruding electrodes of the energy harvester embedded in PLGA. Strips of Mg or Mo for electrical connections were wrapped around the nerve [66]. The electrical stimulus is delivered from the receiving antenna to the electrical connections to the nerves (Fig. 3a). Longer postsurgical stimulation regimens showed to be superior in increasing denervated muscle mass, reaction times, and force measurements (1 vs. 6 days of stimulation) [66].

With a few refinements, other treatments can be explored using this technology including muscle, organ, spinal cord, and cardiac tissue stimulation therapies [66].

3.3.3 Multifunctional Therapies

Multifunctional sensing and treatment devices have been a substantial advancement in bioresorbable electronics to provide highly sensitive and minimally invasive monitoring and targeted treatment options for patients, and in some instances, overcoming the drawback of external energy supplies necessary for previously developed resorbable electronics [97, 120].

Son and colleagues engineered a flexible, bioresorbable vascular stent (210 μm thickness, 5.5 mm diameter, ~10 mm length) capable of controllable drug release, scavenging of reactive oxygen species (ROS), flow and temperature sensing, resistive random-access memory (RRAM) data storage, and wireless data communication [97]. Components of this device included Mg, MgO, Mg alloys, Zn, ZnO, gold (Au), SiO_2, cerium dioxide (CeO_2), and PLA [97]. Cerium dioxide nanoparticles scavenged reactive oxygen species, while near-infrared (IR) responsive (800 nm) Au-nanorod (NR) core nanoparticles with mesoporous silica shells released rapamycin for restenosis prophylaxis. A temperature sensor associated with thermal regulation acted to reduce the risk of heat-induced necrosis and embolization. Conductance of Mg in PBS with PLA (120 mm) and MgO (300 nm) insulation was reported to be stable for up to 66 days, whereas without the PLA layer, the Mg containing electronic components degraded within 30 minutes.

By using a triboelectric generator, an autonomous sensing and responsive therapeutic device was developed by Zhang and colleagues with proof-of-concept in vivo demonstration of applicability in epilepsy and anti-infective treatments [120]. Up to 60 V of open-circuit voltage was able to be generated upon 2 Hz stimuli from a device composed of silk and Mg [120]. Stable operation of the triboelectric nanogenerators was demonstrated to vary from 10 minutes to 10 hours by tuning surface features and encapsulating layers [120]. For epilepsy management, the device monitored and detected epileptic signals (Fig. 3b) to induce thermally mediated drug release of phenobarbital from silk films (Fig. 3c) while concomitantly delivering electronic communication for seizure identification and alert notification to a mobile device (Fig. 3d) [120].

4 Summary and Outlook

Since the development of transient electronics in the early 2000s, great strides have been achieved in advanced applications of biodegradable electronic devices in an attempt to address unsatisfactory clinical sensing and treatment in acute scenarios. Current bioresorbable electronics are primarily constituted of thin layers of conducting, semiconducting, and insulating materials capable of degrading in the body over finite timeframes. Engineering designs have successfully incorporated active (i.e., transistors, diodes) and passive (i.e., resistors, inductors, capacitors) components to develop bioresorbable medical devices like pressure and temperature sensors for intracranial pressure monitoring, direct, self-powered electrically stimulating scaffolds for nerve regeneration, electrocardiograms and multifunctional stents for cardiovascular management, and remote-controlled drug release devices for targeted drug delivery. Benefits to using bioresorbable electronic implants include evasion of secondary surgeries for device removal, sensitive and direct tissue interfacing, and reduced risk of tissue damage due to device migration, infection, or long-term immune-mediated reactions. Furthermore, bioresorbable electronics degrade completely into biocompatible by-products reducing the amount of environmental and economic strains attributed to electronic waste and associated recycling costs.

Aside from the advances made and the promising outlook of transient electronic devices for improved patient care, there still remains inadequacies before implemenation into a clinical setting is feasible. Namely, lack of tightly controlled dissolution and device lifetimes, limited manufacturing sites for scalability, few multifunctional designs, and high-power, self-sufficient energy supplies inhibit implementation into present healthcare practice. Areas of research dedicated to encapsulation materials and novel transient device designs, materials processing, and wireless communication are critical to attaining desired device performances with practical translatability. Unexplored and combinations of existing device designs and materials can bring transient electronics closer toward reducing unnecessary adverse effects and waste in the clinical setting while considerably enhancing biosensing and patient-specific treatments not achievable by current non-resorbable electronic technologies.

References

1. Acar, H., et al. (2014). Study of physically transient insulating materials as a potential platform for transient electronics and bioelectronics. *Advanced Functional Materials, 24*(26), 4135–4143. https://doi.org/10.1002/adfm.201304186.
2. Adnan, S. M., et al. (2016). Water-soluble glass substrate as a platform for biodegradable solid-state devices. *IEEE Journal of the Electron Devices Society, 4*(6), 490–494. https://doi.org/10.1109/JEDS.2016.2606340.
3. Bai, W., et al. (2018). Flexible transient optical waveguides and surface-wave biosensors constructed from monocrystalline silicon. *Advanced Materials, 30*(32), 1–12. https://doi.org/10.1002/adma.201801584.
4. Balde, C. P., et al. (2017). The global e-waste monitor 2017. *United Nations University, IAS – SCYCLE, Bonn, Germany., 35*, 3397. https://doi.org/10.1016/j.proci.2014.05.148.
5. Bettinger, C. J., & Bao, Z. (2010). Organic thin-film transistors fabricated on resorbable biomaterial substrates. *Advanced Materials, 22*(5), 651–655. https://doi.org/10.1002/adma.200902322.
6. Boutry, C. M., et al. (2015). A sensitive and biodegradable pressure sensor array for cardiovascular monitoring. *Advanced Materials, 27*(43), 6954–6961. https://doi.org/10.1002/adma.201502535.

7. Boutry, C. M., et al. (2018). A stretchable and biodegradable strain and pressure sensor for orthopaedic application. *Nature Electronics*. Springer US, *1*(5), 314–321. https://doi.org/10.1038/s41928-018-0071-7.
8. Bowen, P. K., Drelich, J., & Goldman, J. (2013). Zinc exhibits ideal physiological corrosion behavior for bioabsorbable stents. *Advanced Materials, 25*(18), 2577–2582. https://doi.org/10.1002/adma.201300226.
9. Brenckle, M. A., et al. (2015). Modulated degradation of transient electronic devices through multilayer silk fibroin pockets. *ACS Applied Materials and Interfaces, 7*(36), 19870–19875. https://doi.org/10.1021/acsami.5b06059.
10. Callister Jr, W. D. (University of U.) and Rethwisch, D. G. (University of I). (2014). *Materials Science and Engineering: An Introduction*. 9th edn. Edited by D. Sayre, J. Metzger, and M. Price. Hoboken: Wiley.
11. Campana, A., et al. (2014). Electrocardiographic recording with conformable organic electrochemical transistor fabricated on resorbable bioscaffold. *Advanced Materials, 26*(23), 3874–3878. https://doi.org/10.1002/adma.201400263.
12. Chen, Y., et al. (2015). A touch-communication framework for drug delivery based on a transient microbot system. *IEEE Transactions on Nanobioscience, 14*(4), 397–408. https://doi.org/10.1109/TNB.2015.2395539.
13. Chen, Y., et al. (2016). Physical–chemical hybrid transiency: A fully transient li-ion battery based on insoluble active materials. *Journal of Polymer Science, Part B: Polymer Physics, 54*(20), 2021–2027. https://doi.org/10.1002/polb.24113.
14. Chen, Y., et al. (2018). Advances in materials for recent low-profile implantable bioelectronics. *Materials, 11*(4), 1–24. https://doi.org/10.3390/ma11040522.
15. Cheng, H. (2016). Inorganic dissolvable electronics: Materials and devices for biomedicine and environment. *Journal of Materials Research, 31*(17), 2549–2570. https://doi.org/10.1557/jmr.2016.289.
16. Cheng, H., & Vepachedu, V. (2016). Recent development of transient electronics. *Theoretical and Applied Mechanics Letters*. Elsevier Ltd, *6*(1), 21–31. https://doi.org/10.1016/j.taml.2015.11.012.
17. Curry, E. J., et al. (2018). Biodegradable piezoelectric force sensor. *Proceedings of the National Academy of Sciences, 115*(5), 909–914. https://doi.org/10.1073/pnas.1710874115.
18. Dagdeviren, C., et al. (2013). Transient, biocompatible electronics and energy harvesters based on ZnO. *Small, 9*(20), 3398–3404. https://doi.org/10.1002/smll.201300146.
19. Dey, J., et al. (2008). Development of biodegradable crosslinked urethane-doped polyester elastomers. *Biomaterials*. Elsevier Ltd, *29*(35), 4637–4649. https://doi.org/10.1016/j.biomaterials.2008.08.020.
20. Erdogan, S., Kaya, M., & Akata, I. (2017). Chitin extraction and chitosan production from cell wall of two mushroom species (Lactarius vellereus and Phyllophora ribis), *AIP Conference Proceedings*, 1809. https://doi.org/10.1063/1.4975427.
21. Fedoročková, A., & Raschman, P. (2008). Effects of pH and acid anions on the dissolution kinetics of MgO. *Chemical Engineering Journal, 143*(1–3), 265–272. https://doi.org/10.1016/j.cej.2008.04.029.
22. Feig, V. R., Tran, H., & Bao, Z. (2018). Biodegradable polymeric materials in degradable electronic devices. *ACS Central Science, 4*(3), 337–348. https://doi.org/10.1021/acscentsci.7b00595.
23. Forrest, S. R. (2004). The path to ubiquitous and low-cost organic electronic appliances on plastic. *Nature, 428*(6986), 911–918. https://doi.org/10.1038/nature02498.
24. Fraietta, J. A., & Kietrys, D. M. (2015). Factors affecting the immune system. In K. Helgeson & B. Shelly (Eds.), *Pathology: implications for the physical therapist* (4th ed., pp. 276–279). Saunders/Elsevier: St. Louis. Available at: https://books.google.com/books?id=1he0BQAAQBAJ&pg=PA277&lpg=PA277&dq=ZINC+has+been+identified+as+a+cofactor+for+over+70 +different+enzymes&source=bl&ots=fYFUyYEp6Q&sig=5MF32GPY50jyv4B8x5IYcgzn9fE&hl=en&sa=X&ved=2ahUKEwj0wcTqpuHdAhWLuFMKHTKuA78Q6AEwBHoECAUQ.

25. Fu, K., et al. (2015). Transient rechargeable batteries triggered by Cascade reactions. *Nano Letters, 15*(7), 4664–4671. https://doi.org/10.1021/acs.nanolett.5b01451.
26. Fu, K., et al. (2016). All-component transient lithium-ion batteries. *Advanced Energy Materials, 6*(10), 1–9. https://doi.org/10.1002/aenm.201502496.
27. Fu, K. K., et al. (2016). Transient electronics: Materials and devices. *Chemistry of Materials, 28*(11), 3527–3539. https://doi.org/10.1021/acs.chemmater.5b04931.
28. Gao, Y., Zhang, Y., et al. (2017). Moisture-triggered physically transient electronics. *Science Advances, 3*(9), 1–9. https://doi.org/10.1126/sciadv.1701222.
29. Gao, Y., Sim, K., et al. (2017). Thermally triggered mechanically destructive electronics based on electrospun poly(ϵ-caprolactone) nanofibrous polymer films. *Scientific Reports*. Springer US, *7*(1), 1–8. https://doi.org/10.1038/s41598-017-01026-6.
30. Gentile, P., et al. (2014). An overview of poly(lactic-co-glycolic) acid (PLGA)-based biomaterials for bone tissue engineering. *International Journal of Molecular Sciences, 15*, 3640–3659. https://doi.org/10.3390/ijms15033640.
31. Göpferich, A., & Tessmar, J. (2002). Polyanhydride degradation and erosion. *Advanced Drug Delivery Reviews, 54*(7), 911–931. https://doi.org/10.1016/S0169-409X(02)00051-0.
32. Gordon, T. (2016). Electrical stimulation to enhance axon regeneration after peripheral nerve injuries in animal models and humans. *Neurotherapeutics, 13*(2), 295–310. https://doi.org/10.1007/s13311-015-0415-1.
33. He, X., et al. (2016). Transient resistive switching devices made from egg albumen dielectrics and dissolvable electrodes. *ACS Applied Materials and Interfaces, 8*(17), 10954–10960. https://doi.org/10.1021/acsami.5b10414.
34. Huang, X., et al. (2018). A fully biodegradable battery for self-powered transient implants. *Small, 14*(28), 1–8. https://doi.org/10.1002/smll.201800994.
35. Huang, X. (2018). Materials and applications of bioresorbable electronics. *Journal of Semiconductors, 39*(1), 011003. https://doi.org/10.1088/1674-4926/39/1/011003.
36. Hwang, S.-W., et al. (2012). A physically transient form of silicon electronics. *Science, 337*(6102), 1640–1644. https://doi.org/10.1126/science.1142996.
37. Hwang, S. W., Kim, D. H., et al. (2013). Materials and fabrication processes for transient and bioresorbable high-performance electronics. *Advanced Functional Materials, 23*(33), 4087–4093. https://doi.org/10.1002/adfm.201300127.
38. Hwang, S. W., Huang, X., et al. (2013). Materials for bioresorbable radio frequency electronics. *Advanced Materials, 25*(26), 3526–3531. https://doi.org/10.1002/adma.201300920.
39. Hwang, S. W., Park, G., Cheng, H., et al. (2014). 25th anniversary article: Materials for high-performance biodegradable semiconductor devices. *Advanced Materials, 26*(13), 1992–2000. https://doi.org/10.1002/adma.201304821.
40. Hwang, S. W., Park, G., Edwards, C., et al. (2014). Dissolution chemistry and biocompatibility of single-crystalline silicon nanomembranes and associated materials for transient electronics. *ACS Nano, 8*(6), 5843–5851. https://doi.org/10.1021/nn500847g.
41. Hwang, S. W., Song, J. K., et al. (2014). High-performance biodegradable/transient electronics on biodegradable polymers. *Advanced Materials, 26*(23), 3905–3911. https://doi.org/10.1002/adma.201306050.
42. Hwang, S. W., et al. (2015). Biodegradable elastomers and silicon nanomembranes/nanoribbons for stretchable, transient electronics, and biosensors. *Nano Letters, 15*(5), 2801–2808. https://doi.org/10.1021/nl503997m.
43. Institute of Medicine (US) Panel on Micronutrients. (2001). *Iron*. Washington D.C.: National Academies Press (US).
44. Institute of Medicine (US) Panel on Micronutrients. (2001). Molybdenum. In *Dietary reference intakes for Vitamin A, Vitamin K, Arsenic, Boron, Chromium, Copper, Iodine, Iron, Manganese, Molybdenum, Nickel, Silicon, Vanadium, and Zinc* (p. 420). Washington, D.C.: National Academies Press (US).
45. Institute of Medicine (US) Panel on Micronutrients. (2001). Zinc. In *Dietary reference intakes for Vitamin A, Vitamin K, Arsenic, Boron, Chromium, Copper, Iodine, Iron,*

Manganese, Molybdenum, Nickel, Silicon, Vanadium, and Zinc (pp. 442–444). Washington, D.C.: National Academies Press (US).
46. Irimia-Vladu, M., et al. (2010). Biocompatible and biodegradable materials for organic field-effect transistors. *Advanced Functional Materials, 20*(23), 4069–4076. https://doi.org/10.1002/adfm.201001031.
47. Irimia-Vladu, M., Głowacki, E. D., et al. (2012). Green and biodegradable electronics. *Materials Today*. Elsevier Ltd, *15*(7–8), 340–346. https://doi.org/10.1016/S1369-7021(12)70139-6.
48. Irimia-Vladu, M., Głowacki, E. D., et al. (2012). Indigo – a natural pigment for high performance ambipolar organic field effect transistors and circuits. *Advanced Materials, 24*(3), 375–380. https://doi.org/10.1002/adma.201102619.
49. Irwin, J. D., & Kerns Jr., D. V. (1995). In A. Apt (Ed.), *Introduction to electrical engineering*. Upper Saddle River: Prentice-Hall.
50. Janotti, A., & Walle, C. G. Van De (2009). Fundamentals of zinc oxide as a semiconductor. *72*(12). https://doi.org/10.1088/0034-4885/72/12/126501.
51. Jia, X., et al. (2016). Toward biodegradable Mg-air bioelectric batteries composed of silk fibroin-polypyrrole film. *Advanced Functional Materials, 26*(9), 1454–1462. https://doi.org/10.1002/adfm.201503498.
52. Jia, X., et al. (2017). A biodegradable thin-film magnesium primary battery using silk fibroin–ionic liquid polymer electrolyte. *ACS Energy Letters, 2*(4), 831–836. https://doi.org/10.1021/acsenergylett.7b00012.
53. Jiang, H. L., & Zhu, K. J. (1999). Preparation, characterization and degradation characteristics of polyanhydrides containing poly(ethylene glycol). *Polymer International, 48*(1), 47–52. https://doi.org/10.1002/(SICI)1097-0126(199901)48:1<47::AID-PI107>3.0.CO;2-X.
54. Jin, S. H., et al. (2014). Solution-processed single-walled carbon nanotube field effect transistors and bootstrapped inverters for disintegratable, transient electronics. *Applied Physics Letters, 105*(1), 013506. https://doi.org/10.1063/1.4885761.
55. Kang, S. K., et al. (2014). Dissolution behaviors and applications of silicon oxides and nitrides in transient electronics. *Advanced Functional Materials, 24*(28), 4427–4434. https://doi.org/10.1002/adfm.201304293.
56. Kang, S. K., Hwang, S. W., et al. (2015). Biodegradable thin metal foils and spin-on glass materials for transient electronics. *Advanced Functional Materials, 25*(12), 1789–1797. https://doi.org/10.1002/adfm.201403469.
57. Kang, S. K., Park, G., et al. (2015). Dissolution chemistry and biocompatibility of silicon- and germanium-based semiconductors for transient electronics. *ACS Applied Materials and Interfaces, 7*(17), 9297–9305. https://doi.org/10.1021/acsami.5b02526.
58. Kang, S. K., et al. (2016). Bioresorbable silicon electronic sensors for the brain. *Nature*. Nature Publishing Group, *530*(7588), 71–76. https://doi.org/10.1038/nature16492.
59. Kang, S. K., et al. (2018). Advanced materials and devices for bioresorbable electronics. *Accounts of Chemical Research*. American Chemical Society, *51*(5), 988–998. https://doi.org/10.1021/acs.accounts.7b00548.
60. Karandikar, S., et al. (2017). Nanovaccines for oral delivery-formulation strategies and challenges. *Nanostructures for Oral Medicine*. Elsevier Inc. https://doi.org/10.1016/B978-0-323-47720-8.00011-0.
61. Kim, D., et al. (2008). Materials and noncoplanar mesh designs for integrated circuits with linear elastic responses to extreme mechanical deformations. *Proceedings of the National Academy of Sciences, 105*(48), 1–6.
62. Kim, D. H., et al. (2009). Silicon electronics on silk as a path to bioresorbable, implantable devices. *Applied Physics Letters, 95*(13), 93–96. https://doi.org/10.1063/1.3238552.
63. Kim, D. H., et al. (2010). Dissolvable films of silk fibroin for ultrathin conformal bio-integrated electronics. *Nature Materials, 9*(6), 1–7. https://doi.org/10.1038/nmat2745.
64. Kim, Y. J., et al. (2013). Self-deployable current sources fabricated from edible materials. *Journal of Materials Chemistry B, 1*(31), 3781–3788. https://doi.org/10.1039/c3tb20183j.

65. Ko, J., et al. (2017). Human hair keratin for biocompatible flexible and transient electronic devices. *ACS Applied Materials and Interfaces, 9*(49), 43004–43012. https://doi.org/10.1021/acsami.7b16330.
66. Koo, J., et al. (2018). Wireless bioresorbable electronic system enables sustained nonpharmacological neuroregenerative therapy. *Nature Medicine*. Springer US, *24*, 1830–1836. https://doi.org/10.1038/s41591-018-0196-2.
67. Kumar, A., Holuszko, M., & Espinosa, D. C. R. (2017). E-waste: An overview on generation, collection, legislation and recycling practices. *Resources, Conservation and Recycling*. Elsevier B.V., *122*, 32–42. https://doi.org/10.1016/j.resconrec.2017.01.018.
68. Kumar, P., et al. (2016). Melanin-based flexible supercapacitors. *Journal of Materials Chemistry C*. Royal Society of Chemistry, *4*(40), 9516–9525. https://doi.org/10.1039/c6tc03739a.
69. Labet, M., & Thielemans, W. (2009). Synthesis of polycaprolactone: A review. *Chemical Society Reviews, 38*(12), 3484–3504. https://doi.org/10.1039/b820162p.
70. Landi, G., et al. (2017). Differences between graphene and graphene oxide in gelatin based systems for transient biodegradable energy storage applications. *Nanotechnology*. IOP Publishing, *28*(5), 054005. https://doi.org/10.1088/1361-6528/28/5/054005.
71. Lee, C. H., Kim, H., et al. (2015). Biological lipid membranes for on-demand, wireless drug delivery from thin, bioresorbable electronic implants. *NPG Asia Materials*. Nature Publishing Group, *7*(11), e227–e229. https://doi.org/10.1038/am.2015.114.
72. Lee, C. H., Jeong, J. W., et al. (2015). Materials and wireless microfluidic systems for electronics capable of chemical dissolution on demand. *Advanced Functional Materials, 25*(9), 1338–1343. https://doi.org/10.1002/adfm.201403573.
73. Lee, G., et al. (2017). Fully biodegradable microsupercapacitor for power storage in transient electronics. *Advanced Energy Materials, 7*(18), 1–12. https://doi.org/10.1002/aenm.201700157.
74. Lee, Y. K., Yu, K. J., Song, E., Farimani, A. B., et al. (2017). Dissolution of monocrystalline silicon nanomembranes and their use as encapsulation layers and electrical interfaces in water-soluble electronics. https://doi.org/10.1021/acsnano.7b06697.
75. Lee, Y. K., Yu, K. J., Song, E., Barati Farimani, A., et al. (2017). Dissolution of monocrystalline silicon nanomembranes and their use as encapsulation layers and electrical interfaces in water-soluble electronics. *ACS Nano, 11*(12), 12562–12572. https://doi.org/10.1021/acsnano.7b06697.
76. Lei, T., et al. (2017). Biocompatible and totally disintegrable semiconducting polymer for ultrathin and ultralightweight transient electronics. *Proceedings of the National Academy of Sciences, 114*(20), 5107–5112. https://doi.org/10.1073/pnas.1701478114.
77. Li, R., et al. (2013). An analytical model of reactive diffusion for transient electronics. *Advanced Functional Materials, 23*(24), 3106–3114. https://doi.org/10.1002/adfm.201203088.
78. Li, R., et al. (2017). Recent progress on biodegradable materials and transient electronics. *Bioactive Materials*. Elsevier Ltd, *3*(3), 322–333. https://doi.org/10.1016/j.bioactmat.2017.12.001.
79. Li, R., Wang, L., & Yin, L. (2018). Materials and devices for biodegradable and soft biomedical electronics. *Materials, 11*(2108), 1–23. https://doi.org/10.3390/ma11112108.
80. Liang, Q., et al. (2017). Recyclable and green triboelectric nanogenerator. *Advanced Materials, 29*(5). https://doi.org/10.1002/adma.201604961.
81. Lu, L., et al. (2018). Biodegradable monocrystalline silicon photovoltaic microcells as power supplies for transient biomedical implants. *Advanced Energy Materials, 8*(16), 1–8. https://doi.org/10.1002/aenm.201703035.
82. Luo, M., et al. (2014). A microfabricated wireless RF pressure sensor made completely of biodegradable materials. *Journal of Microelectromechanical Systems, 23*(1), 4–13. https://doi.org/10.1109/JMEMS.2013.2290111.
83. Makadia, H. K., & Siegel, S. J. (2011). Poly Lactic-co-Glycolic Acid (PLGA) as biodegradable controlled drug delivery carrier. *Polymers, 3*(763), 1377–1397. https://doi.org/10.3390/polym3031377.

84. Mannoor, M. S., et al. (2012). Graphene-based wireless bacteria detection on tooth enamel. *Nature Communications*. Nature Publishing Group, *3*, 763–768. https://doi.org/10.1038/ncomms1767.
85. Markoulaki, S., et al. (2009). Development of aliphatic biodegradable photoluminescent polymers. *Proceedings of the National Academy of Sciences, 106*(28), 11818–11819. https://doi.org/10.1073/pnas.0906359106.
86. Meitl, M. A., et al. (2006). Transfer printing by kinetic control of adhesion to an elastomeric stamp. *Nature Materials, 5*(1), 33–38. https://doi.org/10.1038/nmat1532.
87. Mobini, S., et al. (2017). Recent advances in strategies for peripheral nerve tissue engineering. *Current Opinion in Biomedical Engineering, 4*, 134–142. https://doi.org/10.1016/j.cobme.2017.10.010.
88. Najafabadi, A. H., et al. (2014). Biodegradable nanofibrous polymeric substrates for generating elastic and flexible electronics. *Advanced Materials, 26*(33), 5823–5830. https://doi.org/10.1002/adma.201401537.
89. Pal, R. K., et al. (2016). Conducting polymer-silk biocomposites for flexible and biodegradable electrochemical sensors. *Biosensors and Bioelectronics*. Elsevier, *81*, 294–302. https://doi.org/10.1016/j.bios.2016.03.010.
90. Park, S., et al. (2013). Inorganic/organic multilayer passivation incorporating alternating stacks of organic/inorganic multilayers for long-term air-stable organic light-emitting diodes. *Organic Electronics: physics, materials, applications*. Elsevier B.V., *14*(12), 3385–3391. https://doi.org/10.1016/j.orgel.2013.09.045.
91. Patrick, E., et al. (2011). Corrosion of tungsten microelectrodes used in neural recording applications. *Journal of Neuroscience Methods*. Elsevier B.V., *198*(2), 158–171. https://doi.org/10.1016/j.jneumeth.2011.03.012.
92. Pfister, B. J., et al. (2011). Biomedical engineering strategies for peripheral nerve repair: surgical applications, state of the art, and future challenges. *Critical Reviews in Biomedical Engineering, 39*(2), 81–124. doi: 2809b9b432c80c2c,0fb500fc3eef5342 [pii].
93. Pomerantseva, I., et al. (2009). Degradation behavior of poly(glycerol sebacate). *Journal of Biomedical Materials Research Part A, 91*(4), 1038–1047. https://doi.org/10.1002/jbm.a.32327.
94. Rahman, H. U., et al. (2013). Fabrication and characterization of PECVD silicon nitride for RF MEMS applications. *Microsystem Technologies, 19*(1), 131–136. https://doi.org/10.1007/s00542-012-1522-0.
95. Sabir, M. I., Xu, X., & Li, L. (2009). A review on biodegradable polymeric materials for bone tissue engineering applications. *Journal of Materials Science, 44*(21), 5713–5724. https://doi.org/10.1007/s10853-009-3770-7.
96. Shan, D., et al. (2018). Development of citrate-based dual-imaging enabled biodegradable electroactive polymers. *Advanced Functional Materials, 28*(34), 1801787. https://doi.org/10.1002/adfm.201801787.
97. Son, D., et al. (2015). Bioresorbable electronic stent integrated with therapeutic nanoparticles for endovascular diseases. *ACS Nano, 9*(6), 5937–5946. https://doi.org/10.1021/acsnano.5b00651.
98. Song, G., & Song, S. (2007). A possible biodegradable magnesium implant material. *Advanced Engineering Materials, 9*(4), 298–302. https://doi.org/10.1002/adem.200600252.
99. Stolica, N. (1985). Molybdenum. In A. J. Bard, R. Parsons, & J. Jordan (Eds.), *Standard potentials in aqueous solution* (pp. 462–483). New York: Marcel Dekker, Inc. Available at: https://books.google.com/books?id=XoZHDwAAQBAJ.
100. Su, L., et al. (2014). Study on the antimicrobial properties of citrate-based biodegradable study on the antimicrobial properties of citrate-based biodegradable polymers. *Frontiers in Bioengineering and Biotechnology, 2*(23). https://doi.org/10.3389/fbioe.2014.00023.
101. Sun, S., et al. (2016). The role of pretreatment in improving the enzymatic hydrolysis of lignocellulosic materials. *Bioresource Technology*. Elsevier Ltd, *199*, 49–58. https://doi.org/10.1016/j.biortech.2015.08.061.

102. Sun, Y., & Rogers, J. A. (2007). Inorganic semiconductors for flexible electronics. *Advanced Materials, 19*(15), 1897–1916. https://doi.org/10.1002/adma.200602223.
103. Tanskanen, P. (2013). Management and recycling of electronic waste. *Acta Materialia*. Acta Materialia Inc., *61*(3), 1001–1011. https://doi.org/10.1016/j.actamat.2012.11.005.
104. Tao, H., et al. (2014). Silk-based resorbable electronic devices for remotely controlled therapy and in vivo infection abatement. *Proceedings of the National Academy of Sciences, 111*(49), 17385–17389. https://doi.org/10.1073/pnas.1407743111.
105. Tao, H., Kaplan, D. L., & Omenetto, F. G. (2012). Silk materials - a road to sustainable high technology. *Advanced Materials, 24*(21), 2824–2837. https://doi.org/10.1002/adma.201104477.
106. Thimbleby, H. (2013). Technology and the future of healthcare. *Journal of Public Health Research, 2*(28), 160–167. https://doi.org/10.4081/arc.2013.e14.
107. Trolier-McKinstry, S. (Pennsylvania S. U.) and Newnham, R. (Pennsylvania S. U.) (2018) *Materials engineering: Bonding, structure, and structure-property relationships*. 1st edn. New York: Cambridge University Press.
108. Tsang, M., et al. (2015). Biodegradable magnesium/iron batteries with polycaprolactone encapsulation: A microfabricated power source for transient implantable devices. *Microsystems & Nanoengineering*. Taylor & Francis, 15024. https://doi.org/10.1038/micronano.2015.24.
109. Turner, R. E. (2006). Nutrition during pregnancy. In M. E. Shils et al. (Eds.), *Modern nutrition in health and disease* (10th ed., p. 778). Philadelphia: Lippincott Williams & Wilkins. Available at: https://books.google.com/books?id=S5oCjZZZ1ggC&printsec=frontcover&vq=molybdenum&source=gbs_ge_summary_r&cad=0#v=onepage&q=molybdenum&f=false.
110. Uwitonze, A. M., & Razzaque, M. S. (2018). Role of magnesium in vitamin D activation and function. *The Journal of the American Osteopathic Association, 118*(3), 181. https://doi.org/10.7556/jaoa.2018.037.
111. Vanýsek, P. (2012). Electrochemical series. *Handbook of Chemistry and Physics, 93*, 5–80.
112. Wang, X., et al. (2016). Food-materials-based edible supercapacitors. *Advanced Materials Technologies, 1*(3), 1600059. https://doi.org/10.1002/admt.201600059.
113. Wang, Y., et al. (2002). A tough biodegradable elastomer. *Nature Biotechnology, 20*(6), 602–606. https://doi.org/10.1038/nbt0602-602.
114. Witte, F., & Eliezer, A. (2012). Biodegradable metals. In N. Eliaz (Eds.), Degradation of Implant Materials (pp. 93-109). New York: Springer. Available at: https://www.google.com/books/edition/Degradation_of_Implant_Materials/NGydrqQb_CcC?hl=en&gbpv=0.
115. Yang, J., Webb, A. R., & Ameer, G. A. (2004). Novel citric acid-based biodegradable elastomers for tissue engineering. *Advanced Materials, 16*(6), 511–516. https://doi.org/10.1002/adma.200306264.
116. Yin, L., Cheng, H., et al. (2014). Dissolvable metals for transient electronics. *Advanced Functional Materials, 24*(5), 645–658. https://doi.org/10.1002/adfm.201301847.
117. Yin, L., Huang, X., et al. (2014). Materials, designs, and operational characteristics for fully biodegradable primary batteries. *Advanced Materials, 26*(23), 3879–3884. https://doi.org/10.1002/adma.201306304.
118. Yu, K. J., et al. (2016). Bioresorbable silicon electronics for transient spatiotemporal mapping of electrical activity from the cerebral cortex. *Nature Materials, 15*(7), 782–791. https://doi.org/10.1038/nmat4624.
119. Yu, X., et al. (2018). Materials, processes, and facile manufacturing for Bioresorbable electronics: A review. *Advanced Materials, 30*(28), 1–27. https://doi.org/10.1002/adma.201707624.
120. Zhang, Y., et al. (2018). Self-powered multifunctional transient bioelectronics. *Small, 14*(35), 1802050. https://doi.org/10.1002/smll.201802050.
121. Zhang, Z., Tsang, M., & Chen, I. W. (2016). Biodegradable resistive switching memory based on magnesium difluoride. *Nanoscale*. Royal Society of Chemistry, *8*(32), 15048–15055. https://doi.org/10.1039/c6nr03913h.

122. Zheng, Q., et al. (2016). Biodegradable triboelectric nanogenerator as a life-time designed implantable power source. *Science Advances, 2*(3), 1–9. https://doi.org/10.1126/sciadv.1501478.
123. Duncanson, A., & Stevenson, R. W. H. (1958). Some properties of magnesium fluoride crystallized from the melt. *Proceedings of the Physical Society, 72*(6), 1001–1006.
124. Horan, R. L., et al. (2005). In vitro degradation of silk fibroin. *Biomaterials, 26*(17), 3385–3393. https://doi.org/10.1016/j.biomaterials.2004.09.020.
125. Ma, C., et al. (2018). In vitro cytocompatibility evaluation of poly(octamethylene citrate) monomers toward their use in orthopedic regenerative engineering. *Bioactive Materials, 3*(1), 19–27. https://doi.org/10.1016/j.bioactmat.2018.01.002.
126. Torun Köse, G., Ber, S., Korkusuz, F. et al., (2003). Poly(3-hydroxybutyric acid-co-3-hydroxyvaleric acid) based tissue engineering matrices. *Journal of Materials Science: Materials in Medicine, 14*(2), 121–126. https://doi.org/10.1023/A:1022063628099.

Inorganic Dissolvable Bioelectronics

Huanyu Cheng

1 Introduction

Physical invariant performance has been a long-lasting pursuit as the hallmark of the modern electronics. However, certain opportunities emerge when part of or entire device system can disappear without a trace. For instance, implantable devices that are capable of fully resorbing in the human body would avoid the need for removal from a second surgical operation upon completion of their function. The application of dissolvable electronics also goes beyond the biomedical devices [1–3] to disposable environmental sensors [4–6] and physically destructive components for security [7]. In order to construct high-performance dissolvable devices, a realistic set of functional materials has been established. A typical device architecture configures several functional components connected by conducting interconnects on a substrate, followed by encapsulation or packaging. Insulating polymers are commonly used for the substrate and/or encapsulation layers, because of their soft nature and ease to interface with biological tissues. In comparison to conductive polymers that are intrinsically stretchable, biodegradable metals (e.g., Mg, Zn, Mo, W, etc.) in stretchable structures are of particular interest because of their low resistance and widespread use in the state-of-the-art devices. The high-performance device also involves the use of semiconducting (e.g., Si, Ge, SiGe, ZnO, etc.) and dielectric (e.g., Si_3N_4 and SiO_2) materials. It should be noted that the biodegradable bioelectronics could also be built with a set of organic materials as discussed in several review articles [6, 8, 9]. Because of the use of device-grade silicon, biodegradable active components (e.g., Si diodes [1, 10], n-/p-channel transistors [1, 5, 10, 11], photodetector [1], ring oscillators, and inverters [5, 10, 12]) can be easily demonstrated. Taken together with the biodegradable passive components such as resistor, inductor, and capacitor, various sensors (e.g., hydration sensor [5], pH sensor [13], strain and temperature sensor [1, 14], electrophysiological

H. Cheng (✉)
Department of Engineering Science and Mechanics, The Pennsylvania State University, University Park, PA, USA

Materials Research Institute, The Pennsylvania State University, University Park, PA, USA
e-mail: huanyu.cheng@psu.edu

H. Cao et al. (eds.), *Interfacing Bioelectronics and Biomedical Sensing*,
https://doi.org/10.1007/978-3-030-34467-2_3

sensor [13], pressure sensor (array) [15, 16], and flow sensor) are fabricated and characterized against their conventional counterparts. Integrating sensing components for diagnostics with therapeutic components into a system in several animal models furthers the utility of the technology toward biomedicine.

With a focus on inorganic dissolvable bioelectronics [17–20] in this book chapter, we will first briefly review the complete set of functional materials commonly used for inorganic dissolvable bioelectronics. Next, we will discuss various processing and manufacturing approaches for constructing the dissolvable bioelectronics. After introducing several strategies to the power supply as a key component for the bioelectronics, we will then provide application opportunities from functional transformation and active trigger to biomedical implants. The transient technology as an enabler for disposable sensors with environmentally benign end products and physically destructive electronics with application in data security will also be discussed in the conclusion and future perspective.

2 Materials

2.1 Semiconductors

As the most relevant semiconducting material for high-performance electronics in the modern integrated circuits industry, silicon has been the hallmark of the technology evolution over the past few decades, and it represents an excellent candidate in transient electronics. However, its use toward such application has not been explored until very recently, though the well-controlled and reproducible etching of device-grade silicon-crystal silicon has been extensively studied [21]. Recent advances of inorganic dissolvable bioelectronics nucleate around the realization that single-crystal silicon undergoes hydrolysis in physiologically relevant conditions (e.g., the body temperature of 37 °C and pH of ca. 7.4) (Fig. 1a) [1]. In comparison to the anisotropic etching of single-crystal silicon in solutions with high pH levels, the dissolution rate of silicon in near-neutral solutions (i.e., pH of ~7) becomes exceptionally small, primarily because of the significantly reduced amount of hydroxide ions that serve as a catalyst. The dissolution of silicon in near-neutral solutions follows the equilibrium of $Si + 4H_2O \rightarrow Si(OH)_4 + 2H_2$ [1, 22], where the orthosilicic acid $Si(OH)_4$ is stable (not stable in solutions with pH level over 12) and is assumed to diffuse away from the solid silicon surface.

The dissolution rate is calculated from the measurement of thickness change by immersing patterned silicon nanomembranes (created by reactive ion etching with sulfur hexafluoride gas) on a layer of thermal oxide in aqueous buffer solutions with a controlled pH level at a prescribed temperature. In the case of the anisotropic etching of silicon in the aqueous KOH solutions, the dissolution rate is shown to follow an empirical formula [21], which indicates a power law dependence on the KOH concentration and Arrhenius scaling on the temperature dependence. When the solution changes from base to near neutral, the Arrhenius scaling still holds for the temperature effect, and using a different power law exponent for $[OH^-]$ in the empirical formula indeed captures the experimental observation for the solution in a wide range of pH levels [2].

In addition to the pH level, the other anions are also found to have played an evident role on the dissolution of single-crystal silicon. In an effort to shed light on the effect of ion concentration, the patterned silicon is placed in a solution with different chlorides and phosphate concentrations (i.e., 0.05 M–1 M) at a fixed pH level of 7.5 [23]. This pH level is selected to ensure that the concentration of hydroxide ion is much smaller than those of chlorides and phosphates. The experimental study validated by density functional theory (DFT) simulations indicates that chlorides and phosphates can also form bonds with silicon to initiate the dissolution. The DFT simulations further reveal the bonding preference: Si-OH > $Si\text{-}HPO_4$ > Si-Cl (Fig. 1b). This insight helps explain the prior reports that show the direct correspondence of high dissolution rate and high ion concentration [2, 24]. While the protein (e.g., albumin) and silicic acid as the dissolution product are shown to slow down the dissolution of Si, cations such as Ca^{2+} and Mg^{2+} are shown to accelerate the dissolution because of their ability to deprotonate surface silanol groups and/or to enhance the reactivity of water and siloxane groups (Fig. 1c) [25]. The types (i.e., boron or phosphorous from high-temperature diffusion) and concentrations of doping in the intrinsically pure silicon are also found to reduce the dissolution rate of silicon [24], consistent with the anisotropic etching of silicon in base solutions [26].

The dissolution behavior of the other widely used semiconducting materials in the conventional electronics has also been examined, including polycrystalline silicon (poly-Si), amorphous silicon (a-Si), germanium (Ge), and an alloy of silicon and germanium (SiGe) [27]. The chemical reaction of poly-Si and a-Si follows the same one of that of the single-crystal silicon. The solution with dissolved oxygen reacts with Ge, and the reaction follows $Ge + O_2 \rightarrow H_2O + H_2GeO_3$. The dissolution measurement of the poly-Si and a-Si is conducted on the patterned array of materials on a layer of thermal oxide on a silicon wafer. The measurement of single-crystal SiGe [Si_8Ge_2 (100)] and Ge [Ge (100)] is achieved by patterning a titanium mask with an array of square openings and measuring the depth of the profile in the opening. Among this set of materials, a-Si and SiGe are found to be the fastest and slowest, respectively. Because of the wide direct bandgap and piezoelectric responses, ZnO is demonstrated as an active material, as well as energy harvesting and strain sensing devices [28]. In addition, amorphous indium-gallium-zinc oxide (a-IGZO) as the semiconductor has also been studied for a thin film transistor (TFT) technology [29] because of its optical transparency, high mobility, low-temperature deposition, and stable operation.

2.2 Conductors

As several metallic materials are essential for biological functions, a recommended daily intake ranging from 0.05 mm/day to 400 mm/day [30, 31] is prescribed for magnesium (Mg), zinc (Zn), iron (Fe), molybdenum (Mo), and many others. To

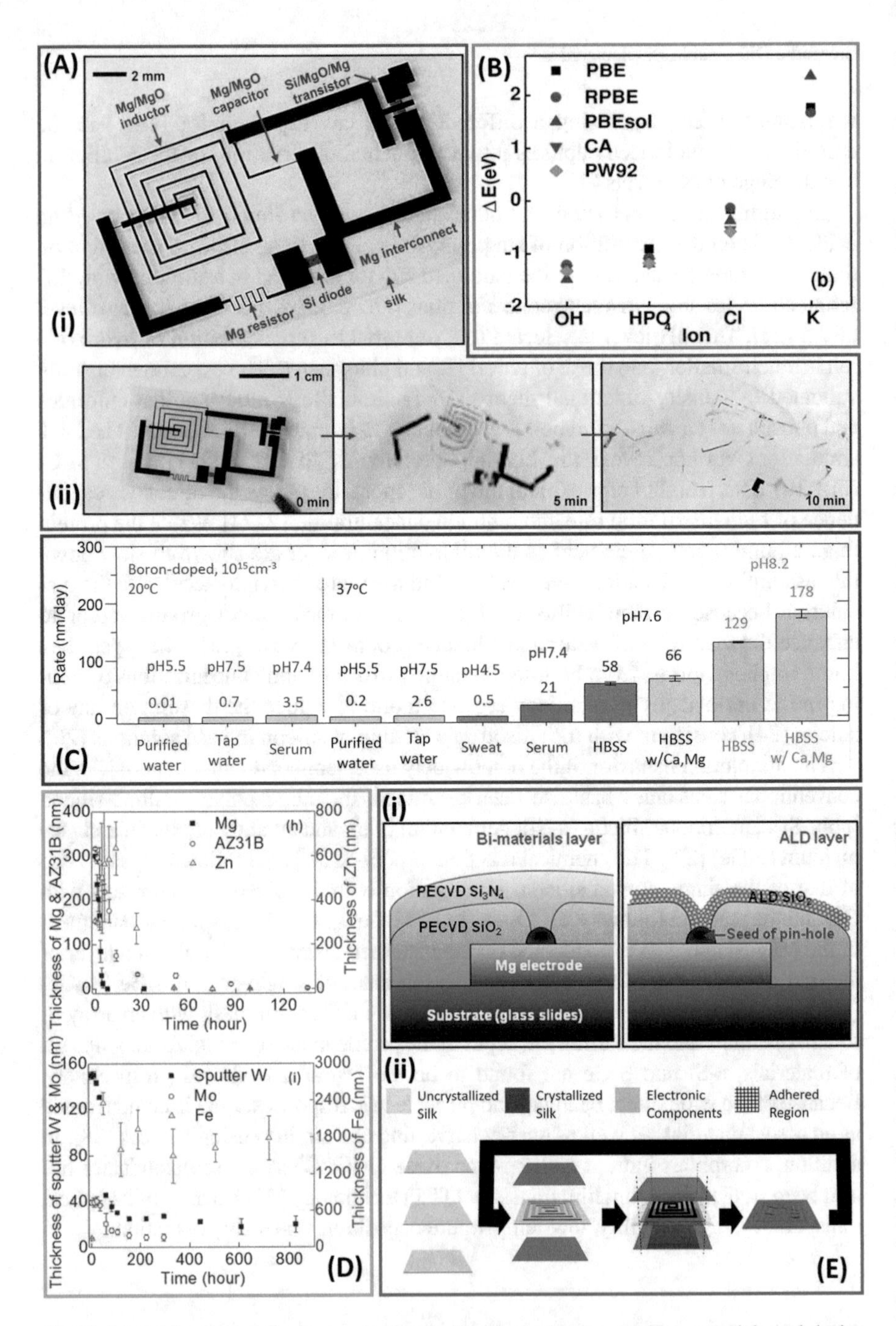

Fig. 1 Material selection for inorganic dissolvable bioelectronics. (**a**) Key materials and device structures of a physically transient form of silicon electronics. (**i**) Image of a device that includes inductors, capacitors, resistors, diodes, and transistors, with interconnects and interlayer dielectrics on a thin silk substrate. A top view of the device appears in the lower right inset. (**ii**) Sequence of images showing the dissolution of the device in the deionized (DI) water. (Reprinted with permission from [1]). (**b**) Comparison of the change in energy computed from density functional theory simulations for different ions. Bonding preference: OH > HPO4 > Cl. The K-Si bond is not

avoid going over the limit, a thin film form of these metallic materials should be used. Taken together with tungsten (W) that dissolves in physiological solutions with no obvious adverse effects [32], the dissolution behavior of these thin metal traces is studied and shown to be different from their bulk counterparts, because of the increasingly important role of grain size, pinholes, and many other effects (Fig. 1d) [11, 31]. The dissolution involves a fast degradation of the base metal and a slow dissolution of the residue oxide. It should be noted that the residue oxide might take a much longer timescale to dissolve (a week in DI water for Mg, AZ31B Mg alloy, and Zn with a thickness of 300 nm, or months for an initial 150-nm-thick W and 40-nm-thick Mo) or even insoluble such as iron oxides. It should also be noted that the application of metal foil goes from conducting traces to substrate with advantages of the low permeability of gases and liquids, the absence of swelling, and direct device fabrication [33].

The dissolution behavior of a porous metallic material can be captured by a reactive diffusion model. In addition to the dissolution that occurs at the surface of the thin film, the solution molecules also diffuse into its pores and react with the surrounding metal atoms. The design of thin trace allows the use of a one-dimensional (1-D) model as the thickness of the thin film is much smaller than its length and width. The concentration $w(y, t)$ of solution molecules (e.g., water) at a given location y as a function of time t follows the reactive diffusion eq. [34], $D\partial^2 w/\partial y^2 - kw = \partial w/\partial t$, $0 \leq y \leq h_0$, where h_0 is the initial thickness of the thin film, D is the diffusivity of solvent molecules in the thin film, and k is the reaction constant between solvent molecules and the thin film. Knowing the reaction chemistry between the solvent molecule and the metal atoms (i.e., the ratio between the two), integration of the spatiotemporal distribution of the solution molecules multiplied by the ratio over the thickness and time would yield the amount of the dissolved metal atoms and also the remaining thickness. The change in length and width of the metal trace is negligible when compared with that of its thickness. The electrical resistance R of the metal trace that is inversely proportional to its remaining thickness h can then be easily related to its initial resistance R_0 as $R = R_0 h_0/h$. The prediction of the electrical resistance from the model can quantitatively reproduce the trends observed in the experimental measurements.

Fig. 1 (continued) preferred over K-H. (Reprinted with permission from [23]). (**c**) Dissolution rates of Si in different aqueous solutions (HBSS: Hank's balanced salt solution). (Reprinted with permission from [25]). (**d**) Change in thickness of similar traces as a function of time during dissolution in DI water at room temperature for (top, initial thickness of 300 nm) Mg, AZ31B Mg alloy, and Zn and (bottom) Fe (initial thickness of 150 nm), sputter deposited W (initial thickness of 150 nm), and Mo (initial thickness of 40 nm). (Reprinted with permission from [11]). (**e**) Multilayered encapsulation strategies with transient materials. (**i**) Schematic illustrations of encapsulation methods for transient electronic devices, showing defects such as pinholes covered by a bilayer of SiO2/Si3N4 (left) a layer of ALD SiO2 (right). (Reprinted with permission from (Dissolution Behaviors and Applications of Silicon Oxides and Nitrides in Transient Electronics)). (**ii**) Silk pocket fabrication strategy that utilizes three uncrystallized silk films. Crystallization of the two outer layers renders them water-insoluble, while the inner device substrate layer can remain uncrystallized and water-soluble. Sealing the outer edges encapsulates the device in a protective pocket of silk fibroin. Repeating the process with an inner pocket as the device layer yields multilayered silk pockets. (Reprinted with permission from [46])

2.3 Insulators

Insulating materials are used in multiple places of the transient devices, including dielectric layers, substrates, and encapsulation layers. Both the substrate and encapsulation layers would interface with biological tissues, so biodegradable synthetic materials and those with natural origins have been explored [35]. Because of versatility in the synthesis method, a wide range of synthetic polymers has been developed for biodegradable substrates [5, 36], including diphenylalanine (FF) and polyfluorene (PF) [37], poly(vinyl alcohol) (PVA) [38], polycaprolactone (PCL) [36], polylactic acid (PLA) [39], polyglycolic acid (PGA) [40], and poly(lactic-co-glycolic acid) (PLGA) [39] that is a copolymer of PLA and PGA.

Representative examples of biodegradable materials with natural origin include silk [41], gelatin [42], shellac [8], rice paper, and biodegradable cellulose nanofibril [4]. Silk is widely used as an encapsulation layer because it doesn't elicit an immune response and its rate of dissolution can be well programmed. Gelatin and shellac have also been commonly used as capsules for drug delivery [42] and edible coating for food preservation [8], respectively. Biodegradable cellulose nanofibril that is completely derived from wood as paper can be relatively low-cost because of the available roll-to-roll industrial manufacturing capability. Its use as a substrate for eco-friendly electronics has been demonstrated in a fungal degradation test [4].

Without proper passivation and encapsulation, the material degradation would occur at an undesired time, and it compromises the device function. In an attempt to address such challenge, the two-stage dissolution is being used, where the device is expected to function stably with the protection of the encapsulation layer over the 1st stage and its rapid degradation is controlled by the dissolution of functional components over the 2nd stage. The functional timeframe depends on the material and its thickness of the encapsulation layer. The quantitative effect can be analytically studied by solving the aforementioned reactive diffusion model of a double-layered structure, where an encapsulation layer is placed on top of the functional material/component. For instance, the functional timeframe of an Mg conductor with a thickness of 300 nm can be effectively extended from 1 h to 3 h and 13 h, when a MgO encapsulation layer of 400 nm and 800 nm is used, respectively [1].

Though the thickness is an effective parameter to control the functional timeframe, considerations of physical space limitation and bending capability favor a thin geometry in the encapsulation layer. Therefore, the encapsulation materials that can serve as strong barriers against water are desirable. Widely used in the silicon-based high-performance electronics, silicon oxides (SiO_2) and nitrides (Si_3N_4) have been explored as dielectrics and encapsulations because of their capability to dissolve in the bio-solutions, i.e., $SiO_2 + 2H_2O \rightarrow Si(OH)_4$ [43] and $Si_3N_4 + 12H_2O \rightarrow 3Si(OH)_4 + 4NH_3$ [44]. Though the silicon oxides and nitrides are associated with different densities from different deposition methods, the model of reactive diffusion with the use of effective density is also found to be applicable. The low water diffusivity in the encapsulation layer such as silicon oxides and nitrides slows down the water permeation, but the pinholes and the other defects commonly found from several deposition processes provide the water leakage pathway. An effective method to mediate such an issue is to explore multilayered encapsulation layers for reducing the defect density in a cooperative manner. Demonstrated examples include alternating PECVD SiO_2 and Si_3N_4 layers (200/200/200/200/100/100 nm, total thickness of 1 μm) (Fig. 1e-i) [45] and alternating silk fibroin encapsulation layer with air pocket (Fig. 1e-ii) [46]. The composite layered structure can also be used to design composite insulating materials, where adding gelatin or sucrose to a PVA polymer matrix yields slower or faster transiency of the composite [47]. Because of the slow dissolution rate and single-crystal of Si nanomembranes, their use as a water barrier to encapsulate the biodegradable device (e.g., Mo electrode array for in vivo neural recordings) shows great promise, as submicron thickness easily ensures the device lifetime from days to years [25]. The same concept also applies to the ultrathin thermally grown silicon dioxide as biofluid barriers [48, 49]. Combining the thermally grown silicon dioxide with a thin layer of hafnium oxide (HfO_2) formed by atomic layer deposition [50] (and with coatings of parylene created by chemical vapor deposition [51]) or SiN_x formed by low-pressure chemical vapor deposition [52] in a multilayered structure further enhances the performance of the water and ion barrier independent of ion concentrations in the surrounding biofluids.

3 Manufacturing Processes

Various processing and manufacturing techniques for bioresorbable electronics are extensively discussed in a recent review article [53]. Therefore, we will only highlight a few representative key findings on the conducting materials and large-scale production. Patterning the conducting materials for interconnects and electrodes typically involves a deposition via physical vapor deposition techniques (e.g., e-beam evaporation or sputtering) through shadow mask [1, 11] on the target biodegradable substrates or transferring from conventional wafer substrate to the target [5] via the technique of transfer printing [54–56]. Because the resolution of the shadow mask and the glass temperature of the target substrate needs to be higher than the processing temperature, the former is limited to patterns with certain line width and the range of biodegradable substrates. Though the latter is versatile in terms of the wide range of biodegradable substrates, the use of lithographic processes complicates the fabrication.

As an alternative approach, printing of conductive pastes offers fast prototyping at a low cost. Initial exploration demonstrates moderate out-of-plane (but relatively small in-plane) conductivity in Zn and W pastes through the screen-

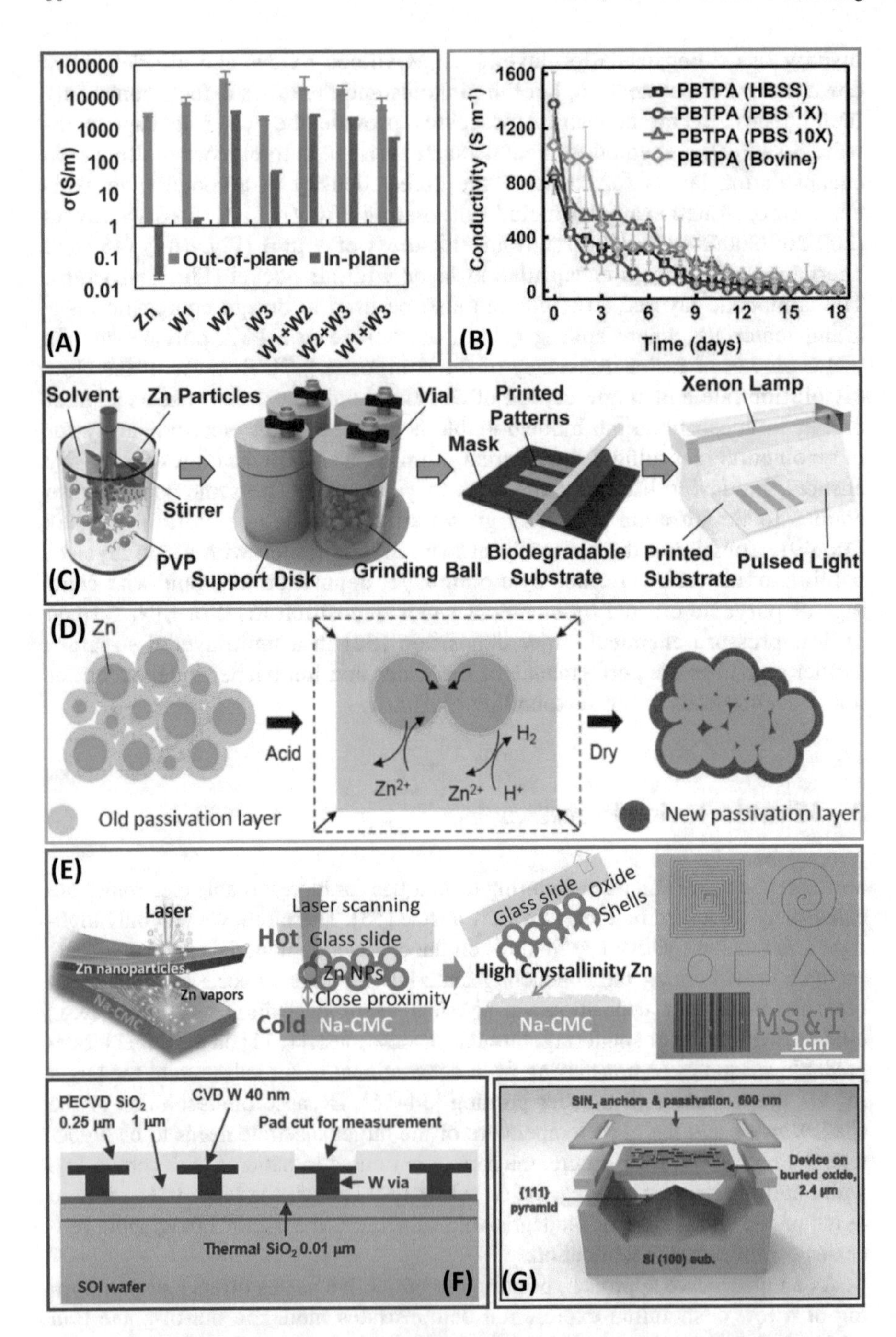

Fig. 2 Processing and manufacturing approaches. (**a**) Comparison of the out-of-plane and in-plane conductivities for all seven transient pastes. The transient pastes use either Zn (average diameter < 10 um) or W (average diameter of 0.6–1 um for W1, 4–6 um for W2, and 12 um for W3).

printing and room temperature curing on sodium carboxymethylcellulose (Na-CMC) substrates [57]. While the in-plane conductivity varies significantly depending on the ink formulation and curing time, the values for the representative Zn and W pastes are 0.024 S/m and 3.5×10^3 S/m (Fig. 2a) partly because of the difference in the contact area from their spherical and polyhedral shapes, respectively. Because of its high conductivity, resistance to thick oxide formation, and lower cost, Mo as the transient metal has also been explored as conductive inks with a conductivity of 1.4×10^3 S/m (Fig. 2b) [58]. To improve the in-plane conductivity, non-spherical Zn nanoparticles (NPs) are obtained from high-energy planetary ball milling to provide more surface contact area in between particles (Fig. 2c) [59]. Following a room temperature photonic sintering from a pulsed xenon lamp, the printed line from the resulting Zn NPs on a biodegradable substrate yield a high conductivity of 4.5×10^4 S/m. As the conductivity of the resulting Zn lines depends on the constituent of the paste for the screen-printing process and the post-processing, the paste with a ratio of Zn:PVP :glycerol:methanol = 7:0.007:2:1 by weight used in the process, followed by hot rolling (for additional void space removal from solvent evaporation) and photonic sintering, results in a conductivity of 6.0×10^4 S/m even without the use of non-spherical Zn NPs [60]. Different from the thermal processing in photonic sintering, the room temperature electrochemical sintering of screen-printed Zn microparticles can also yield highly conductive traces (conductivity of 3×10^5 S/m) (Fig. 2d) [61]. The surface passivation layer of ZnO is first dissolved in CH_3COOH/H_2O. Next, the self-exchange of Zn and Zn^{2+} at the Zn/H_2O interface leads to the Zn sintering, where the chemical specificity of metal particles and solutions is important.

The deposition of Zn NPs through screen-printing is limited by the mask and its resolution. In a different effort, conductive patterns of Zn NPs are directly created by aerosol jet printing, followed by photonic sintering to yield a conductivity of 2.2×10^4 S/m [62]. Though laser has also been used to sinter nanoparticles in printing electronics, its direct use on a polymeric substrate may easily cause the sub-

Fig. 2 (continued) (Reprinted with permission from [57]). (**b**) DC conductivity of Mo/PBTPA (1:4) as a function of immersion time in various solutions at 37 C. Blue, Hanks' balanced salt solution (HBSS); red, phosphate-buffered saline (PBS) 1× concentration; green, PBS 10× concentration; orange, bovine serum. (Reprinted with permission from [58]). (**c**) A schematic of the processes to acquire Zn nanoparticles (NPs) and sintered patterns using Zn NPs. (Reprinted with permission from [59]). (**d**) Proposed mechanism for the electrochemical sintering of Zn microparticles in CH_3COOH/H_2O. (Reprinted with permission from [61]). (**e**) Illustration of laser printing process and the photograph of printed Zn patterns in ambient condition: barcode and "Missouri S&T" logo on glass. (Reprinted with permission from [63]). (**f**) Schematic cross-sectional illustration of the various layers in a typical CMOS die with W interconnect metallization formed on an SOI wafer. (Reprinted with permission from [65]). (**g**) Schematic exploded view illustration of a unit cell in an array of devices after complete release. The etched areas are bounded by Si (111) planes, with freely suspended devices tethered by SiN_x anchors across the trenches. (Reprinted with permission from [66])

strate damage. However, it is possible to combine laser sintering with photonic sintering to result in an increase of conductivity to 3.5×10^4 S/m. Instead of applying laser for direct sintering of nanoparticles on polymeric substrates, a continuous-wave fiber laser driven by a galvanometer is used to irradiate a 2.5–5-μm-thick Zn NP suspension coated on a transparent glass slide with NP facing a Na-CMC substrate (Fig. 2e) [63]. Upon laser scanning, the Zn atoms escape through the cracks of the oxide shell and collide with background gas molecules before reaching the cold substrate for condensation. Using a small gap in between the Zn NPs and the substrate, a high crystalline Zn film can be printed on the substrate with quality similar to those obtained from vacuum-based thermal evaporation. Therefore, the resulting Zn traces from the evaporation-condensation process show a significantly improved conductivity of 1.1×10^6 S/m that is ca. 7% of its bulk value.

Creating a dissolvable printed circuit board (PCB) from Na-CMC (50 μm) also allows mounting of commercial off-the-shelf (COTS) components with dissolvable metal pastes to further enhance the capability of the transient devices [57]. In another effort to explore materials amenable to scale-up at the industrial scale, a polysaccharide circuit board with a nanocellulose thin-film on a pullulan support is demonstrated to allow printing of conductive lines and integration of COTS components such as an LED [64]. The pullulan backing layer can further be dissolved for a soft property and the polysaccharide circuit board also has a high water transmission rate, important for bio-integration. A different approach to large-scale production is to leverage the state-of-the-art Si CMOS foundry-compatible processes. The early effort starts with the removal of the 1st layer of Al by chemical-mechanical planarization (CMP) after electrical testing of transistors and ring oscillators fabricated on an 8-in. silicon on insulator (SOI) wafer in the usual way [65]. Next, films of W are formed through chemical vapor deposition with the same mask used for the Al and a tetrafluoromethane (CF_4)-based etch process (Fig. 2f). The chips are then transfer printed to a biodegradable substrate by eliminating the handle substrate of the SOI wafer by use of a temporary wafer. In the following work, anisotropic etching is used as a releasing strategy to transfer print arrays of devices fabricated on 6-inch wafers onto PLGA substrates (Fig. 2g). Applying encapsulation of PLGA layers with openings at contact pad regions to form a connection with interconnect leads results in a range of high-performance silicon devices sourced from a commercial foundry [66]. By using the technique of transfer printing, multiple biodegradable high-performance MOSFETs can be stack integrated vertically into 3D interconnected layouts, with layers separated by PLGA interlayer but connected by metal interconnects through via opening, on polymeric substrates, capable of functional transformation and electrical monitoring of the transient process [67].

4 Functional Components and Systems

4.1 Power Supply Components

As the power supply is the indispensable component for the device system, several strategies have been studied, including near-field radio frequency (RF) power transfer modules [10], mechanical energy harvesters [28], silicon-based solar cells [1, 27], and dissolvable batteries [42, 68–70]. The transient wireless RF power scavenger circuit consists of an RF antenna, an inductor, six capacitors, a resistor, and eight diodes (Fig. 3a) [10]. Designing the Mg antenna to operate at 950 MHz, the system can wirelessly harvest energy from an RF transmitter to turn on a LED at a distance of ca. 2 m. The use of silicon-based solar cells is not directly relevant for transient biomedical implants unless it operates at wavelengths with long penetration depths in biological tissues (i.e., red and near-infrared wavelengths) [71]. On a step further to the previous reports of solar cells based on amorphous silicon [27] or 3-um-thick single-crystal Si [1], an ultrathin solar cell array based on single-crystal Si can generate sufficient power to stably operate LEDs beneath a 4-mm-thick piece of porcine skin and fat for several days (Fig. 3b) [71].

Exploring dissolvable metals for the electrodes and biodegradable polymers for barrier layers and encapsulations, the fully biodegradable primary batteries have been demonstrated (Fig. 3c) [72]. The use of metallic cathodes reduces the operating voltage, but it is compensated by a series connection of batteries for the increased voltage. To demonstrate the idea, a Mo paste buried in the polyanhydride encasement connects four separated Mg-Mo cells in series. The porous thin polyanhydride film also serves as a top cover to retain the electrolyte but release the hydrogen gas. After the functional operation, the polyanhydride encasement first dissolves in the PBS at body temperature, followed by the slower dissolution of Mg and Mo (~ 11 days). Though the increase of the output current density is possible through enlarging the battery area, implantable devices would favor miniaturized cells with small footprint/size. To address this challenge, PCL is introduced as a packaging and functional material in the biodegradable magnesium/iron batteries, which significantly reduces the volume of the electrochemical cell and provides a stable performance at higher discharge rate (Fig. 3d) [73]. The energy density of PCL-coated Mg/Fe batteries in PBS is also observed to be two orders of magnitude higher than that of the Mg/Mo/PAH battery. The rechargeable modules have also been introduced to transient battery [74] when selecting lithium (Li) for the anode, vanadium oxide (V_2O_5) for the cathode because of its high capacity, water-soluble polyvinylpyrrolidone (PVP) for separators, and sodium alginate (Na-AG) for battery encasements. As the dissolution mechanism of this transient battery relies on the alkali solution formed from

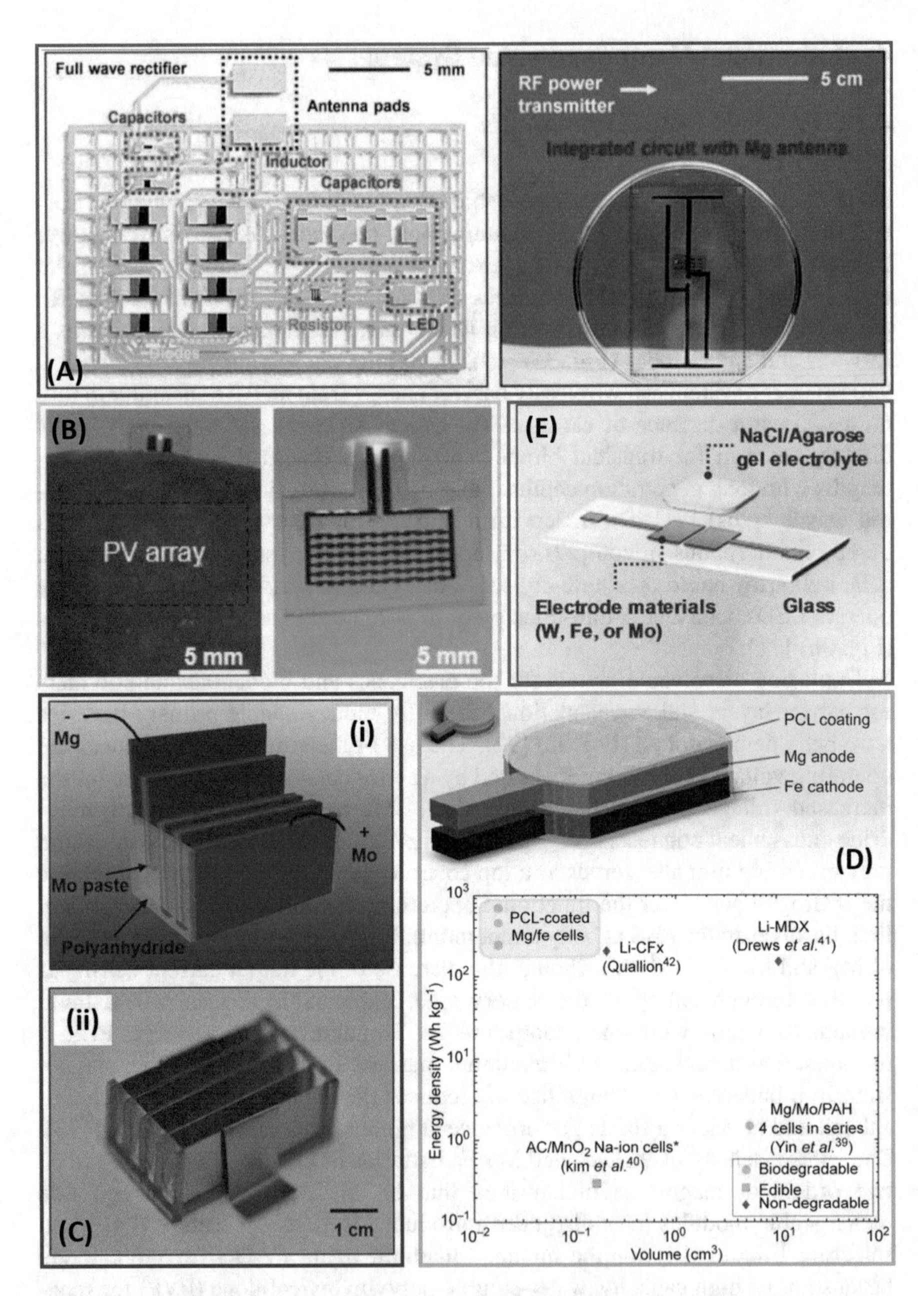

Fig. 3 Power supply components. (**a**) Transient RF power scavenging circuits and its integration with a transient antenna. The transient full-wave rectifying circuit consists of an array of diodes and capacitors, an inductor, a resistor, and antenna pads, fabricated with transient materials: Si NMs (semiconductors), Mg (electrodes), SiO2 (interlayer dielectrics), and silk (substrates). Image on the right shows a full-wave rectifying system powered wirelessly with an RF transmitter and an Mg receiving antenna for a working distance here is ~2 m. (Reprinted with permission from [10]).

reacting lithium (Li) anode with water, its application to biomedical implants is limited. Without proper encapsulation, the dissolution also occurs rapidly within seconds/minutes.

As an alternative approach to batteries, fully biodegradable micro-supercapacitors (MSCs) are built using water-soluble metals (e.g., W, Fe, and Mo) for electrodes and a hydrogel electrolyte (agarose gel) on a PLGA substrate (Fig. 3e) [75]. Even without the use of pseudocapacitive oxide layers, the metal electrodes in contact with the water-containing electrolyte undergo electrochemical oxidation from the repeated charge/discharge cycles to enhance the performance via pseudocapacitance for up to a few thousand cycles. The following decrease from the maximum may result from the reduction of the ion mobility because of hydrogel drying and the effectiveness of the current collection because of the increasing metal oxide. Applying a polyanhydride (PA) encapsulation layer increases the lifetime of the MSC.

4.2 *Functional Transformation and Active Control*

The function of basic components can also be transformed by utilizing a variety of types and thicknesses of dissolvable materials. With patterned encapsulation layer, the Mg layer in the location without encapsulation would quickly dissolve, yielding functional transformation of NAND gates to inverters and individual n-channel MOSFETs, respectively (Fig. 4a) [76]. A different yet simple approach can result from a switch between two components connected in parallel. When a resistor is in parallel connection with a diode, a low electrical bias flows mostly through the resistor, yielding the circuit as a resistor. But dissolving the resistor changes the circuit to a diode. In addition, dissolving part of the component could change the geometry and the circuit characteristics. In the demonstration of tunable antennas, dissolution-induced shortened length of the antenna progressively increases its resonance frequency (Fig. 4b) [76]. The same concept also applies to the inductor and many other components, where current flow within the pattern can be altered. Combining the idea of the external trigger such as a wireless microfluidic controlled

Fig. 3 (continued) (**b**) Optical images to show the use of a Si PV system to operate a blue LED during exposure to 200 mW/cm^2 NIR illumination. (Left) A functioning device under pig skin and fat. (Right) A functioning device after immersion in 1× PBS solution (pH 7.4, Mediatech Inc.) at 37 °C for 5 d. (Reprinted with permission from [71]). (**c**) (**i**) Configuration and (**ii**) optical images of a battery pack with four Mg-Mo cells connected in series. (Reprinted with permission from [72]). (**d**) Schematic of PCL-coated Mg/Fe batteries that harness physiological solution as the electrolyte and its comparison to reported biodegradable and edible energy sources in the energy density vs. volume plot (Li-CFx and Li-MDX are commercial primary batteries and ∗denotes data for batteries without packaging). (Reprinted with permission from [73]). (**e**) Schematic illustration of a planar supercapacitor with biodegradable metal thin-film (W, Fe, or Mo) electrodes and an NaCl/agarose gel electrolyte on a glass substrate. (Reprinted with permission from [73])

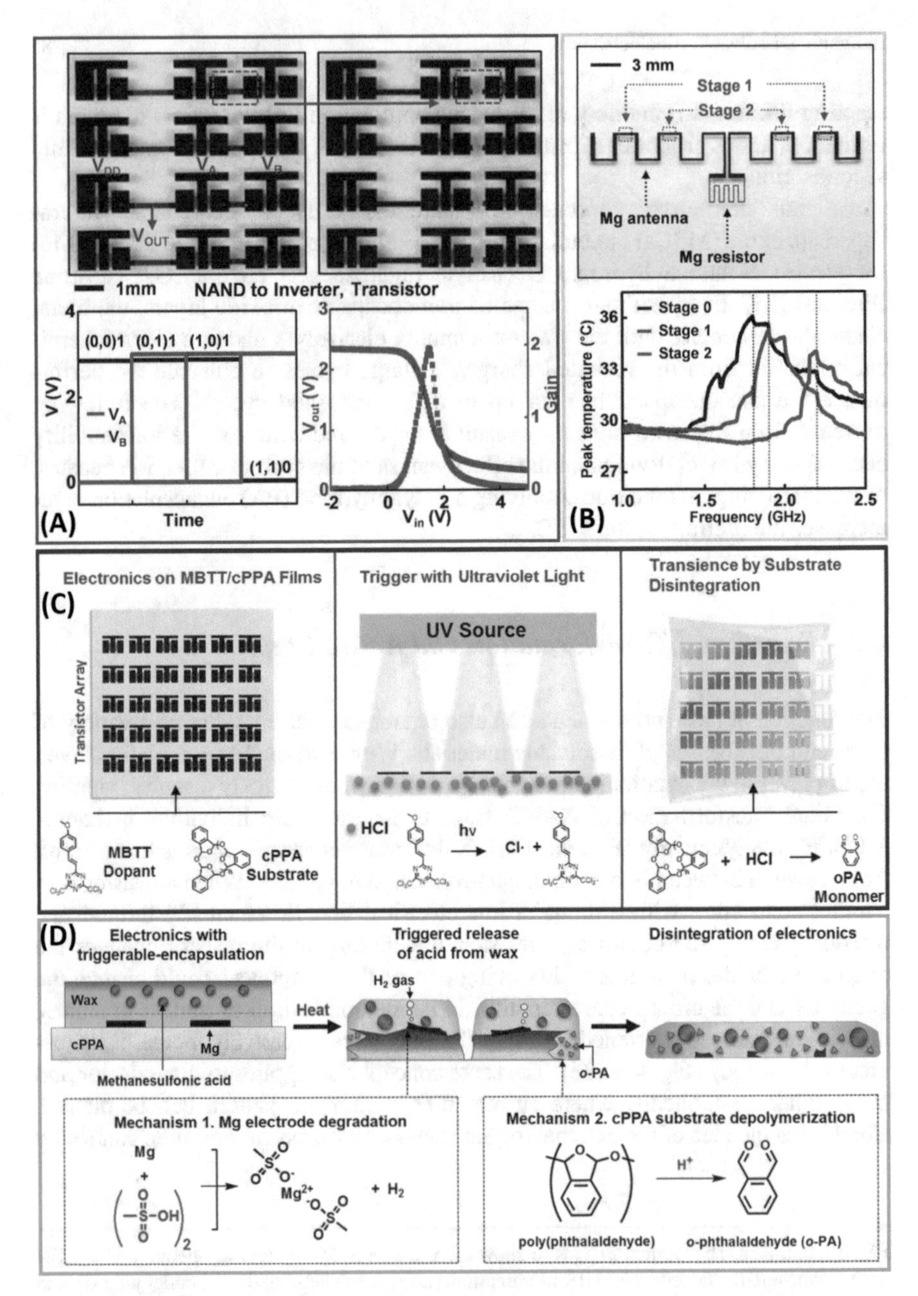

Fig. 4 Functional transformation and active trigger. (**a**) (Top) Images and (bottom) electrical characterization of transient electronic circuits before and after controlled transformation in function from a NAND gate (left) to an inverter and transistor (right) by dissolution of selected parts of the Mg interconnect structure. Electrical characterization includes output voltage responses of a NAND gate (left; V_A and V_B are the input voltages) and voltage transfer characteristics of an inverter (right). (**b**) Functional transformation in transient radio frequency (RF) 4devices. Image of a Mg antenna (thickness ca. 2.5 μm) with a Mg serpentine resistor (thickness ca. 2.5 μm) encapsulated with 400-nm and 800-nm-thick layers of MgO in the red (stage 1) and blue (stage 2) boxes, respectively. The other regions are encapsulated with layers of SiO_2 (ca. 2 μm) and

chemical etching component, the functional transformation such as tunable resonance frequency of an RF antenna or tunable gain of analog amplifiers and oscillation frequency of square waveform generators can be realized in a programmed and user-defined manner [77].

Despite the existence of various encapsulation materials and structures to define the functional lifetime, certain applications would benefit from an active control. One simple strategy is to exploit the metastable polymer for the substrate or encapsulation layer, which would provide a robust performance until a triggered depolymerization occurs to deactivate the device. The specific trigger can be swelling [78], chemical [79], humidity/moisture [74, 80], light [81, 82], or temperature [83]. By including a photo-acid generator (PAG) such as 2-(4-methoxystyryl)-4,6-bis(trichloromethyl)-1,3,5-triazine (MBTT) additive in a cyclic poly(phthalaldehyde) (cPPA), exposure of the metal-polymer to UV light (λ_{max} = 379 nm) yields hydrochloric acid (HCl) [84]. When used as a substrate, the depolymerization leads to the destruction of Si active devices such as N-MOSFET and PIN diodes (Fig. 4c) [81]. Moreover, when used as an encapsulation layer over a biodegradable resistor such as Mg, its partial degradation can remap the current flow for a functional change. The thermal trigger can be simply implemented by applying a wax coating with embedded acid microdroplets on the biodegradable device (e.g., an Mg electrode or an array of diodes on a cPPA substrate) (Fig. 4d) [83]. Exposure to heat melts the wax and releases the embedded acid to trigger the degradation of the Mg electrode and/or depolymerization of cPPA substrate. Coupling with an internal light emission such as integrated organic light-emitting diodes (OLEDs), the light trigger can even be controlled through the switch of the OLEDs [85]. Because of its easy coupling with light, magnetic, or RF, the demonstrated heat trigger [83] can be easily used to provide the other triggering modalities.

Fig. 4 (continued) MgO (ca. 800 nm). During different stages of functional transformation by dissolution in water (black, stage 0; red, stage 1; blue, stage 2), the maximum temperature of the resistive heater is evaluated with an IR camera during exposure to RF at different frequencies. The frequency of maximum energy-conversion efficiency (ECE) shifts toward higher values as the transformation proceeds from one stage to the next (stage 0, 1.8 GHz; stage 1, 1.9 GHz; stage 2, 2.2 GHz). (**a**, **b**: Reprinted with permission from [76]). (**c**) Exposing the MBTT/cPPA substrate (left) to UV generates HCl (middle) that causes the rapid depolymerization of the acid sensitive cPPA polymer. This leads to the destruction of the electronics on the substrate (right). In addition to the disintegration of the substrate polymer, generated HCl also degrades the Mg electrodes. MBBT 2-(4-methoxystyryl)- 4,6-bis(trichloromethyl)-1,3,5-triazine, cPPA cyclic poly(phthalaldehyde). (Reprinted with permission from [81]). (**d**) Heat-triggerable transient electronics coated with wax containing acid dispersions. Melting a wax coating releases MSA to rapidly destroy electronics by acidic degradation of Mg electrodes on glass. In addition, a more rapid destruction of the device is achieved by using a cPPA substrate due to acidic depolymerization of cPPA. (Reprinted with permission from [83])

4.3 Biomedical Implants

Biocompatibility and bioresorption demonstrated from both in vitro and in vivo investigations of Si nanomembranes [24, 86] and other transient materials (e.g., Mg/MgO on silk [24], W [86], Fe [87], poly-Si/a-Si/SiGe/Ge [27], and silicate spin-on-glass [33]) pave the way for the use of this set of functional materials toward implantable devices. When implanted, even simple sensors can provide significant information for the health conditions. When implementing the single-crystal Si as optical waveguides, the signals such as glucose concentration and oxygenation level can be detected (Fig. 5a) [88]. Because of its low propagation loss and ease of patterning, the single-crystal Si patterned in thin filamentary structures with high refractive index allows tight optical mode confinement at submicron precision scales for precise light delivery. The use of a patterned PLGA cladding layer only exposes the Si at the controlled locations for the evanescent fields to interact with the surrounding environment directly relevant to the local chemistry. The selection of the near-infrared (NIR) light (750–2500 nm) in the demonstration is based on its characteristic absorption by the biological materials.

Arrays of bioresorbable silicon electrodes configured in both passive and actively addressed formats with multiplexing capabilities are capable of recording in vivo electrophysiological signals from the cortical surface and the subgaleal space (Fig. 5b) [89]. Validated against the conventional clinical electrocorticography (ECoG) electrodes, the bioresorbable silicon electrodes can detect normal physiologic and epileptiform activity from both acute and chronic recordings, which is directly relevant to clinical problems (e.g., postoperation monitoring of brain activity, electrical monitoring of skeletal muscles or organ function). In addition, the bioresorbable silicon electronic sensors allow direct, continuous monitoring of intracra-

Fig. 5 (continued) circulator to a light source (1050 nm wavelength for Fiber 1, and 1200 nm wavelength for Fiber 2) and a power meter for delivering and detecting light, respectively. (**iii**) Image of a transient biosensor implanted into the subcutaneous region near the thoracic spine of a mouse. Inset: image of red light (wavelength: 660 nm) passing into the biosensor. (**iv**) In vivo measurements of blood oxygen saturation as a function of time during changes in the concentration of oxygen (labeled in blue) in the surrounding environment. Results measured by the transient biosensor and commercial oximeter are labeled in black and green, respectively. (Reprinted with permission from [88]). (**b**) Photograph of a four-channel bioresorbable electrode array placed on the cortical surface of the left hemisphere of a rat. Sleep spindles recoded by a bioresorbable electrode is compared against a nearby commercial stainless steel microwire electrode as a control placed at a depth of 0.5 mm from the cortical surface. (Reprinted with permission from [89]). (**c**) (**i**) Image of bioresorbable pressure and temperature sensors integrated with dissolvable metal interconnects (sputtered molybdenum, Mo, 2 μm thick) and wires (Mo, 10 μm thick). The inset optical image shows the serpentine Si-NM sensing structures, where the left Si-NM that is not above the air cavity responds only to temperature and the right one at the edge of the air cavity responds primarily to pressure. (**ii**) Diagram of a bioresorbable sensing system in the intracranial space of a rat, with electrical interconnects to an external wireless data-transmission unit for long-range operation. (**iii**) Measurements of ICP over 3 days reveal accurate, consistent responses (red) from bioresorbable devices encapsulated with biodegradable polyanhydride, when compared with the data from a commercial sensor (blue). (Reprinted with permission from [90])

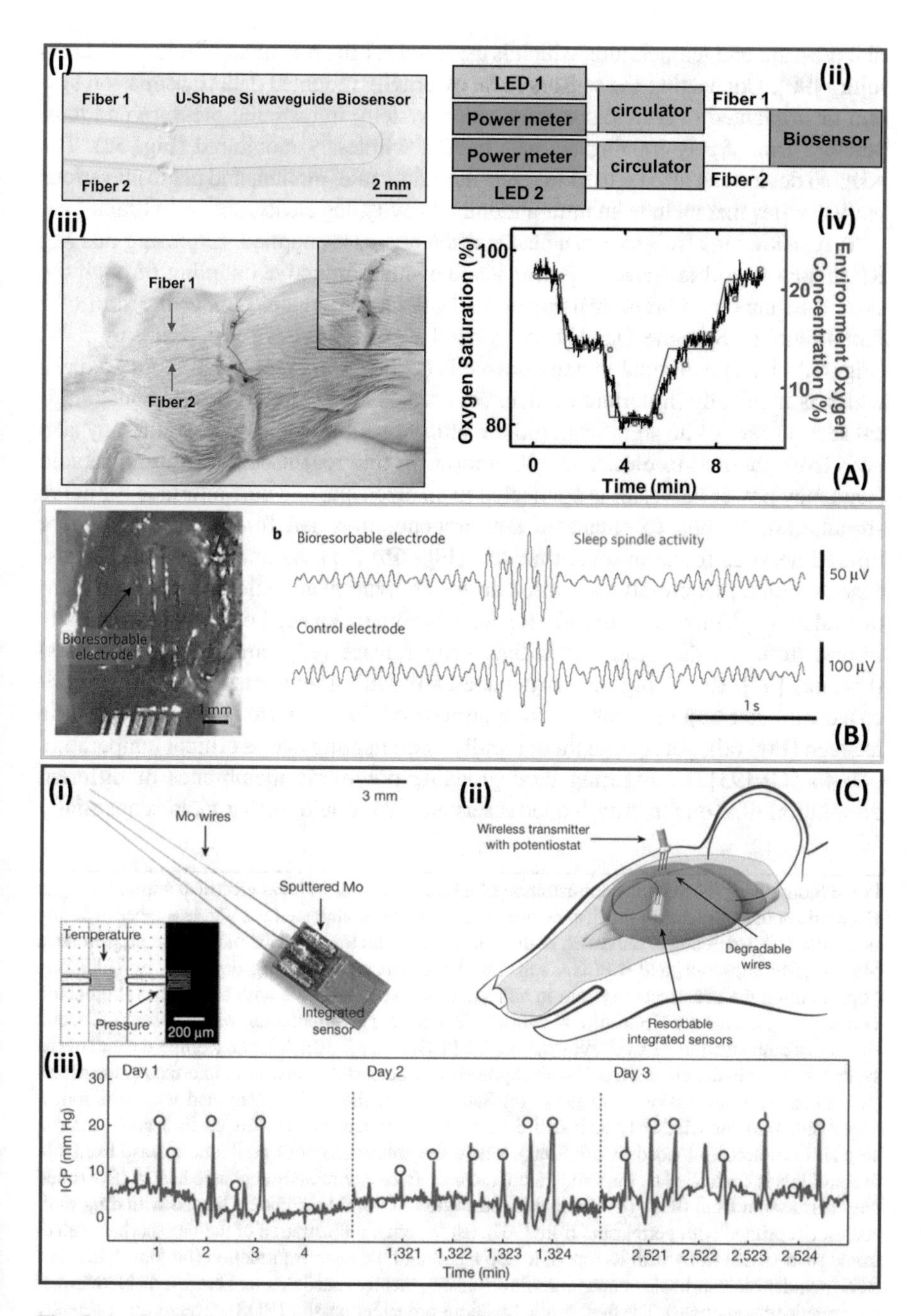

Fig. 5 Implantable sensors. (**a**) Monitoring of blood oxygen saturation in a live animal model using a transient Si waveguide-based biosensor. (**i**) Image of a transient biosensor. Two bioresorbable fibers connect to a U-shaped Si waveguide bonded to a film of PLGA with its top surface exposed. (**ii**) Schematic illustration of the operating principles: each of the two bioresorbable fibers (labeled Fiber 1 and Fiber 2), probing externally from the transient biosensor, connects via a fiber

nial pressure and temperature, which is essential for the treatment of traumatic brain injury [90]. Connecting the sensors to an externally mounted data transmission system or implanted near-field-communication system, intracranial pressure and temperature from freely moving animals can be wirelessly monitored (Fig. 5c). The adapted device can also be used to sense fluid flow rate, motion, and pH from various body cavities that include an intra-abdominal cavity, leg cavity, and deep brain.

A resistor can also serve as a heater when power is supplied. Exploring one Mg RF inductive coil to transmit power via near-field inductive coupling through the skin, a Si nanomembrane resistor in a silk pocket implanted under the skin of a female albino Sprague-Dawley rat is used to heat up the local tissue by 5 °C (Fig. 6a) [1]. The thermal therapy device is designed to operate for about 15 days, which is medically important to maintain asepsis at the wound site. The histological section reveals no significant inflammatory reactions. The mode of therapy also goes from thermal to electrical stimulation. In one recent demonstration, a radio frequency power harvester is connected to an electrode and cuff interface for nerve stimulation, leading to enhanced neuroregeneration and functional recovery of injured nervous tissue in rodent models (Fig. 6b) [91]. As another means for disease treatment, wirelessly controlled drug stabilization and release have been demonstrated by placing an antibiotic-loaded silk film on an Mg heater, where a remote trigger turns on the heater to initiate drug release (e.g., ampicillin molecules) (Fig. 6c) [92]. Replacing the drug-loaded silk with a temperature-sensitive lipid-based film can trap the drug for long periods of time in vitro (months) with little leakage (Fig. 6d), yet release them rapidly upon heating over a critical temperature (41–43 °C) [93]. Configuring biodegradable polymeric membranes of different dissolution time on the drug-loaded reservoirs [94] could further yield sequentially

Fig. 6 (continued) (**b**) Schematic illustration of a bioresorbable, wireless electrical stimulator as an electronic neuroregenerative medical device. The electronic component is a wireless receiver acting as a radio frequency power harvester, built with an Mg inductor, a Si NM radio frequency diode, a Mg/SiO_2/Mg capacitor, and a PLGA substrate interconnected with Mg deposited by sputtering (top). Folding the constructed system in half yields a compact device with a double-coil inductor. The bottom presents the electrode and cuff interface for nerve stimulation, which consists of metal electrodes embedded in a PLGA substrate with a PLGA encapsulation layer. Rolling the end of the system into a cylinder creates a cuff with exposed electrodes at the ends as an interface to the nerve. (Reprinted with permission from [91]). (**c**) Schematic of the device integrated with antibiotics-doped silk film for wirelessly activated drug release (orange dots embedded in a green matrix: ampicillin molecules loaded in silk films), where the enhancement of antibiotic release from silk fibroin films is controlled by the temperature increase from a wirelessly activated heater. (Reprinted with permission from [92]). (**d**) Schematic illustration of a lipid membrane loaded with drug molecules. (Reprinted with permission from [93]). (**e**) Schematic illustration of the bioresorbable electronic stent (BES) (left) with its top view (top right) and the layer information (bottom right). The BES includes bioresorbable temperature/flow sensors, memory modules, and bioresorbable/bioinert therapeutic nanoparticles. The therapeutic functions are either passive (ROS scavenging) or actively actuated (hyperthermia-based drug release) by NIR exposure. (Reprinted with permission from [97]). (**f**) Optical image of capacitive biodegradable, stretchable electrophysiological sensors with magnified view of the mesh electrode structure in the inset and exploded view schematic illustration of a biodegradable, stretchable pH sensors based on doped Si NRs with a photograph in the inset. (Reprinted with permission from [13])

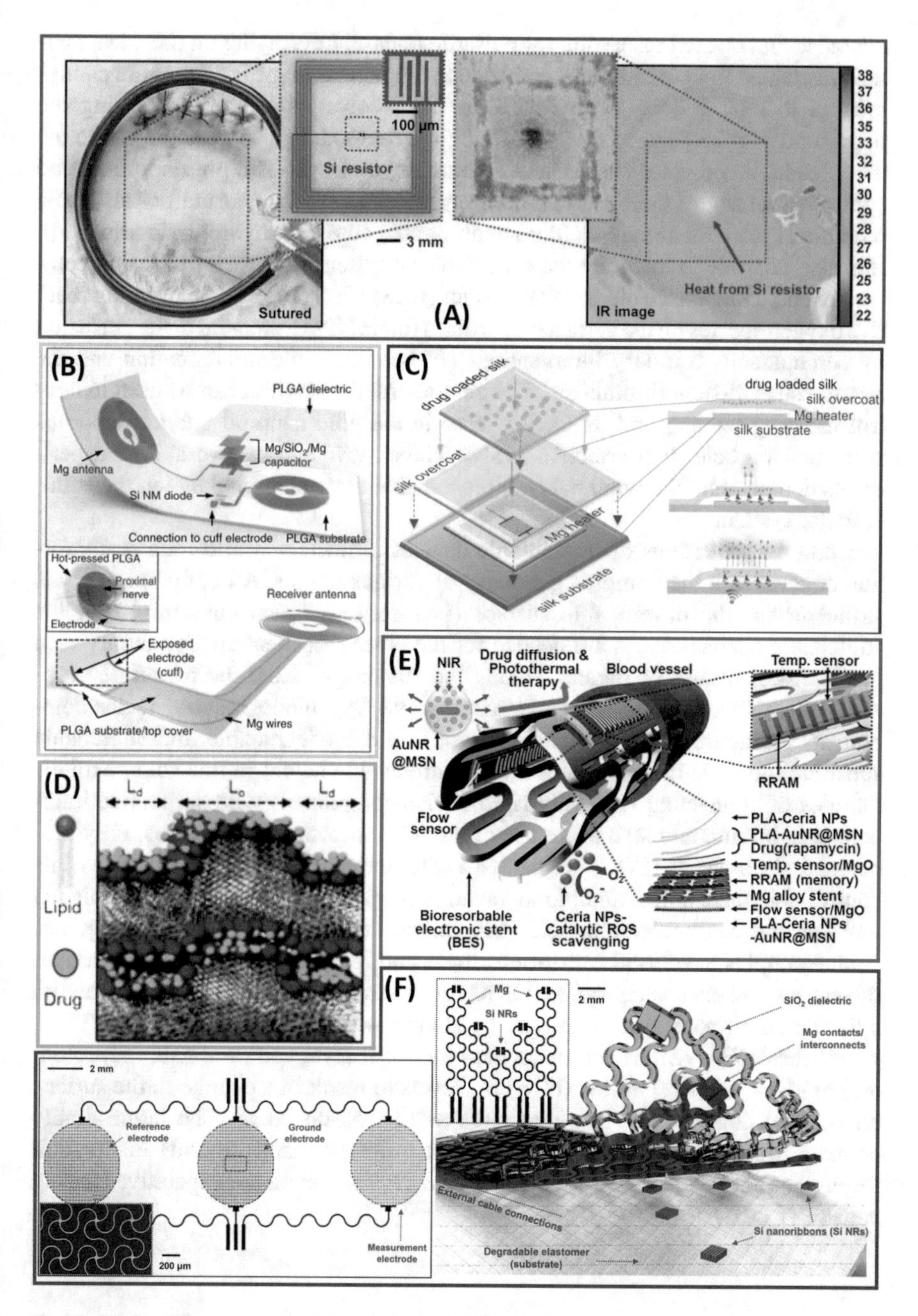

Fig. 6 Implantable actuators and drug delivery devices. (**a**) (Left) In vivo evaluations of a transient bioresorbable device for thermal therapy, where a primary coil is placed next to a sutured implant site of a transient thermal therapy device (image shown in the inset). (Right) Wirelessly powering the device through the skin results in a hot spot (5 °C above background) at the expected location captured by the thermal image (a magnified view in the inset). (Reprinted with permission from [1]).

degraded and opened reservoirs for pulsatile release. Leveraging on this idea, thermally actuated lipid membranes embedded with multiple types of drugs are configured in a 2 by 2 array with four different receiver coils [93], providing the opportunity for sequenced release of multiple drugs. It is also possible to use the biodegradable polymer-based interpenetrating networks for physisorption and controlled release of therapeutic proteins or vaccines [95]. Opportunities also exist with novel material structures (e.g., hydrophobic core with hydrophilic arm [96]). It is also possible to integrate the drug delivery system with various sensing components on a bioresorbable electronic stent (BES) (Fig. 6e) [97]. Scavenging reactive oxygen species by the ceria nanoparticles (ceria NPs) generated in the perfusion by percutaneous coronary interventions (PEI) reduces the inflammation and the risk to cause in-stent thrombosis. The photothermal activation can be used to control the drug loading and its local release in the gold nanorod core/mesoporous silica nanoparticle shell (AuNR@MSN). The ex vivo and in vivo animal experiments demonstrate the potential of bioresorbable electronic implants in the endovascular system.

Certain applications of implantable devices ultimately would require conformal contact with the complex geometry of various tissues. A flexible property is sufficient for the developable surface (i.e., zero Gaussian curvature), but the stretchable characteristics are needed for the other cases that are commonly seen in biological tissues or organs. Taking the advantages from the recent development of stretchable materials and structures [98, 99], biodegradable device components configured in a stretchable layout on a biodegradable and stretchable substrate illustrate the concept [13]. The stretchable layout in this study exploits the idea of connecting relatively rigid device components with serpentine interconnects, but it could also make use of the other stretchable structures. Poly(1,8-octanediol-co-citrate (POC) is selected as the stretchable elastomer because of its controllable mechanical and biodegradation properties [100]. The stretchable pH monitor and electrophysiological (EP) sensor demonstrate the utility (Fig. 6f). Both examples have used Mg for electrode and interconnect, SiO_2 for interlayer dielectrics and encapsulation, and POC for the substrate. The former is built with silicon nanoribbons functionalized by 3-aminopropyltriethoxysilane (APTES), where the $-NH_2$ ($-SiOH$) group in the functional layer can protonate (deprotonate) to $-NH_3^+$ ($-SiO^-$) at low (high) pH levels to result in a change in the surface charge and conductance of phosphorus and boron-doped Si. The latter simply couples Mg electrodes (i.e., measurement, reference, and ground) to the skin through a thin layer of POC to measure the EP signals via the capacitive sensing capability [101].

5 Conclusion and Future Perspective

The concept of inorganic dissolvable bioelectronics involves various types of biodegradable materials and components, manufacturing processes and approaches for device fabrication, and several demonstrated applications. As the corresponding materials are the building blocks for the dissolvable bioelectronics, the extended capabilities provided by new materials will help solve the challenges in the existing design. For instance, neither the inorganic coatings nor common biodegradable polymers are ideal for the encapsulating materials, because of the mechanical fragility and high-density defects in the former, and the water permeability and swelling from water uptake in the later. An alternative set of materials for the encapsulating material is evaluated in a recent study (Fig. 7a) [102], including soy, myrtle, and candelilla wax materials derived from soybeans (via soybean oil), myrica cerifera (myrtle), and candelilla shrubs, respectively, because of their environmental and biological degradation. Incorporating micro/nanoparticles of tungsten in the wax matrix also yields conductive composites that are hydrophobic and biodegradable.

The set of biodegradable materials can also be combined with the other environmentally friendly materials to create disposable sensors with environmental benign end products. Combining solution-processed single-walled carbon nanotubes (SWNTs) with Mo for electrodes/interconnects and SiO_x/SiN_x for gate/interlayer dielectrics on a PVA substrate yields transistors and bootstrapped inverters with transient behavior [103]. Consisting of water-soluble chlorophyll, biocompatible graphene, Mg electrodes, and PVA substrate, the flexible and transient photodetector yields zero waste with a minimum impact on the environment (Fig. 7b) [104]. In addition, the set of functional inorganic materials can be combined with the biodegradable organic materials (e.g., semiconducting polymers) with versatile synthesis strategies to create opportunities that complement the current inorganic electronics (Fig. 7c) [105]. Certain applications would also benefit from the transient behavior where the fluidic or moisture environment is absent. By using the material (cyclododecane, menthol, camphor, perfluorododecane, and hexamethylcyclotrisiloxane) that sublimes in the ambient condition for a substrate, encapsulation layer, or dielectric layer, the transient devices with a dry dissolution mode (Fig. 7d) avoid the need for a hermetical sealing against liquid reactants and variability in the dissolution rate from the uncontrolled liquid flow [106].

The capability to physically destroy components or electronics on demand represents an alternative approach to data encryption and access passwords in data security, as the destruction of the hardware is unrecoverable. The wireless microfluidic components discussed for the functional transformation on-demand can easily be explored to permanently remove the key constituent materials in the device system [7]. In the demonstration, an In/Ag solder paste connects an unpackaged near-

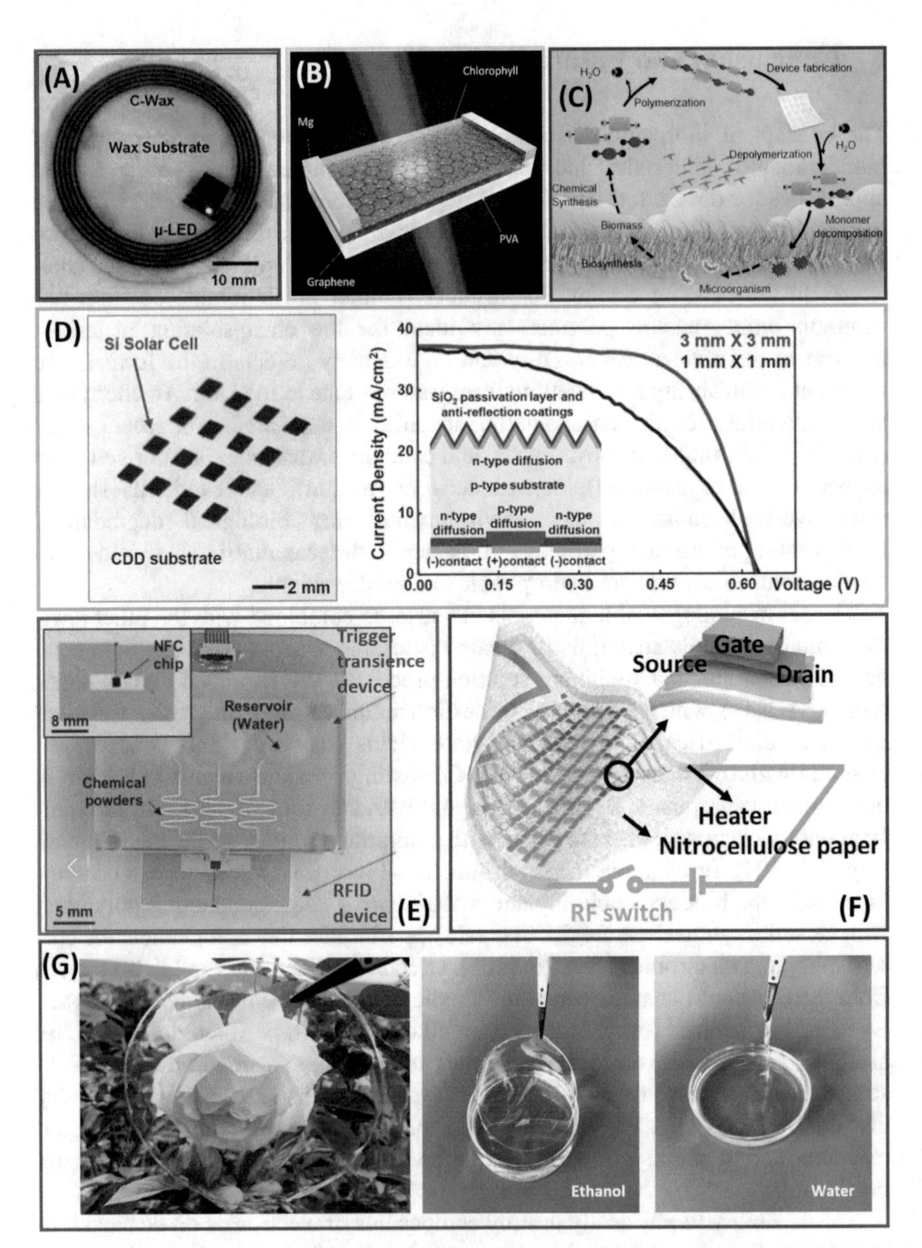

Fig. 7 Representative future directions. (**a**) Integration of insulating and conductive wax (C-wax) into a biodegradable electronic device, where the C-wax is used as the RF inductive coil for a wireless power delivery (the start and end points of the coil are electrically joined by a Mo wire). (Reprinted with permission from [102]). (**b**) Schematic diagram of the transient photodetector, where the CVD-grown graphene transferred on a PVA substrate is followed by deposition of Mg

field communication (NFC) chip to a Cu inductive coil. Directing the outlet of the KHF etchant generated from the microfluidic channel to the NFC chip would destroy the chip and the chip-coil connection for function abruption (Fig. 7e). In a different effort, the complete destruction of electronic devices is demonstrated by stamping a silver resistive heater on the backside of the highly flammable nitrocellulose paper with electronics (Fig. 7f) [107]. Upon the RF heat trigger from the embedded resistive heater, the physical degradation of the nitrocellulose paper with CNT transistors built on top occurs within a few seconds leading to unrecognizable residue (no ash).

The new materials may also be associated with unique properties suitable for a novel manufacturing approach. For instance, the water-soluble and biodegradable galactomannan, an earth-abundant reproducible polysaccharide derived from the seeds of a fast-growing mimosoid tree via a low-cost yet simple extraction and purification method, is stable in an organic solvent and has a suitable viscosity for printing (Fig. 7g) [108]. The interesting properties enable the use of the solvent-based processes that may be combined with recent advances in additive manufacturing for a novel manufacturing method. The environmentally benign end products could even be used for alkaline soil amendments. Taken together with novel device design, the new materials and their manufacturing processes further open up opportunities for the transient technology to address the applications that are previously challenging.

Fig. 7 (continued) and spincoating of chlorophyll. (Reprinted with permission from [104]). (**c**) Polymer synthesis, fabrication, and decomposition cycle for transient polymer electronics. Solid lines indicate the demonstrated processes and dashed lines are the envisioned processes. (Reprinted with permission from [105]). (**d**) Photographs of an array of solar cells on a cyclododecane (CDD) substrate for the demonstration of dry transient behavior. The right presents current-voltage curves and schematic illustration of an individual cell. The cells have sizes of 1 mm × 1 mm or 3 mm × 3 mm, laser cut from a commercially available silicon solar cell module. (Reprinted with permission from [106]). (**e**) Optical image (left) and exploded view schematic illustration (right) of the microfluidic system integrated with the RFID device with essential function provided by an unpackaged NFC chip (magnified view shown in the inset). (Reprinted with permission from [7]). (**f**) Schematic illustration of transient electronic devices based on the nitrocellulose paper that integrates top-gate CNT transistors (magnified view of an individual transistor shown in the inset) and an embedded resistive heater on the backside as the heat trigger from an RF switch. (Reprinted with permission from [107]). (**g**) Digital image of the galactomannan film with high optical transparency. The galactomannan film maintains its shape after immersing in ethanol for 24 h but partially dissolves and collapses after immersing in water for 5 min. (Reprinted with permission from [108])

References

1. Hwang, S.-W., et al. (2012). A physically transient form of silicon electronics. *Science, 337*(6102), 1640–1644.
2. Hwang, S. W., et al. (2014). 25th anniversary article: Materials for high-performance biodegradable semiconductor devices. *Advanced Materials, 26*(13), 1992–2000.
3. Darouiche, R. O. (2004). Treatment of infections associated with surgical implants. *New England Journal of Medicine, 350*(14), 1422–1429.
4. Jung, Y. H., et al. (2015). High-performance green flexible electronics based on biodegradable cellulose nanofibril paper. *Nature Communications, 6*.
5. Hwang, S. W., et al. (2014). High-performance biodegradable/transient electronics on biodegradable polymers. *Advanced Materials, 26*(23), 3905–3911.
6. Irimia-Vladu, M., et al. (2012). Green and biodegradable electronics. *Materials Today, 15*(7), 340–346.
7. Lee, C. H., et al. (2015). Materials and wireless microfluidic systems for electronics capable of chemical dissolution on demand. *Advanced Functional Materials, 25*(9), 1338–1343.
8. Irimia-Vladu, M. (2014). "Green" electronics: Biodegradable and biocompatible materials and devices for sustainable future. *Chemical Society Reviews, 43*(2), 588–610.
9. Tan, M. J., et al. (2016). Biodegradable electronics: Cornerstone for sustainable electronics and transient applications. *Journal of Materials Chemistry C, 4*(24), 5531–5558.
10. Hwang, S. W., et al. (2013). Materials for bioresorbable radio frequency electronics. *Advanced Materials, 25*(26), 3526–3531.
11. Yin, L., et al. (2014). Dissolvable metals for transient electronics. *Advanced Functional Materials, 24*(5), 645–658.
12. Hwang, S. W., et al. (2013). Materials and fabrication processes for transient and bioresorbable high-performance electronics. *Advanced Functional Materials, 23*(33), 4087–4093.
13. Hwang, S. W., et al. (2015). Biodegradable elastomers and silicon nanomembranes/nanoribbons for stretchable, transient electronics, and biosensors. *Nano Letters, 15*(5), 2801–2808.
14. Jamshidi, R., et al. (2015). Transient bioelectronics: Electronic properties of silver microparticle-based circuits on polymeric substrates subjected to mechanical load. *Journal of Polymer Science Part B: Polymer Physics, 53*(22), 1603–1610.
15. Boutry, C. M., et al. (2015). A sensitive and biodegradable pressure sensor array for cardiovascular monitoring. *Advanced Materials, 27*(43), 6954–6961.
16. Luo, M., et al. (2014). A microfabricated wireless RF pressure sensor made completely of biodegradable materials. *Journal of Microelectromechanical Systems, 23*(1), 4–13.
17. Kang, S. K., et al. (2018). Advanced materials and devices for bioresorbable electronics. *Accounts of Chemical Research, 51*(5), 988–998.
18. Cheng, H., & Yi, N. (2017). Dissolvable tattoo sensors: From science fiction to a viable technology. *Physica Scripta, 92*(1), 013001.
19. Cheng, H. (2016). Inorganic dissolvable electronics: Materials and devices for biomedicine and environment. *Journal of Materials Research, 31*(17), 2549–2570.
20. Fu, K. K., et al. (2016). Transient electronics: Materials and devices. *Chemistry of Materials, 28*(11), 3527–3539.
21. Seidel, H., et al. (1990). Anisotropic etching of crystalline silicon in alkaline-solutions. 1. Orientation dependence and behavior of passivation layers. *Journal of the Electrochemical Society, 137*(11), 3612–3626.
22. Sawada, Y., Tsujino, K., & Matsumura, M. (2006). Hydrogen evolution from atomically flat Si (111) surfaces exposed to 40% NH4F, oxygen-free water, or wet gas. *Journal of the Electrochemical Society, 153*(12), C854–C857.
23. Yin, L., et al. (2015). Mechanisms for hydrolysis of silicon nanomembranes as used in bioresorbable electronics. *Advanced Materials, 27*(11), 1857–1864.

24. Hwang, S. W., et al. (2014). Dissolution chemistry and biocompatibility of single-crystalline silicon nanomembranes and associated materials for transient electronics. *ACS Nano, 8*(6), 5843–5851.
25. Lee, Y. K., et al. (2017). Dissolution of monocrystalline silicon nanomembranes and their use as encapsulation layers and electrical interfaces in water-soluble electronics. *ACS Nano, 11*(12), 12562–12572.
26. Seidel, H., et al. (1990). Anisotropic etching of crystalline silicon in alkaline-solutions. 2. Influence of dopants. *Journal of the Electrochemical Society, 137*(11), 3626–3632.
27. Kang, S. K., et al. (2015). Dissolution chemistry and biocompatibility of silicon- and germanium-based semiconductors for transient electronics. *ACS Applied Materials & Interfaces, 7*(17), 9297–9305.
28. Dagdeviren, C., et al. (2013). Transient, biocompatible electronics and energy harvesters based on ZnO. *Small, 9*(20), 3398–3404.
29. Jin, S. H., et al. (2015). Water-soluble thin film transistors and circuits based on amorphous indium–gallium–zinc oxide. *ACS Applied Materials & Interfaces, 7*(15), 8268–8274.
30. Trumbo, P., et al. (2001). Dietary reference intakes: Vitamin A, vitamin K, arsenic, boron, chromium, copper, iodine, iron, manganese, molybdenum, nickel, silicon, vanadium, and zinc. *Journal of the American Dietetic Association, 101*(3), 294–301.
31. Zheng, Y., Gu, X., & Witte, F. (2014). Biodegradable metals. *Materials Science and Engineering: R: Reports, 77*, 1–34.
32. Anik, M., & Osseo-Asare, K. (2002). Effect of pH on the anodic behavior of tungsten. *Journal of the Electrochemical Society, 149*(6), B224–B233.
33. Kang, S. K., et al. (2015). Biodegradable thin metal foils and spin-on glass materials for transient electronics. *Advanced Functional Materials, 25*(12), 1789–1797.
34. Danckwerts, P. V. (1950). Absorption by simultaneous diffusion and chemical reaction. *Transactions of the Faraday Society, 46*(4–5), 300–304.
35. Nair, L. S., & Laurencin, C. T. (2007). Biodegradable polymers as biomaterials. *Progress in Polymer Science, 32*(8–9), 762–798.
36. Tian, H. Y., et al. (2012). Biodegradable synthetic polymers: Preparation, functionalization and biomedical application. *Progress in Polymer Science, 37*(2), 237–280.
37. Khanra, S., et al. (2015). Self-assembled peptide–Polyfluorene nanocomposites for biodegradable organic electronics. *Advanced Materials Interfaces, 2*(14), 1500265.
38. Bettinger, C. J., & Bao, Z. (2010). Organic thin-film transistors fabricated on resorbable biomaterial substrates. *Advanced Materials, 22*(5), 651–655.
39. Anderson, J. M., & Shive, M. S. (2012). Biodegradation and biocompatibility of PLA and PLGA microspheres. *Advanced Drug Delivery Reviews, 64*, 72–82.
40. Middleton, J. C., & Tipton, A. J. (2000). Synthetic biodegradable polymers as orthopedic devices. *Biomaterials, 21*(23), 2335–2346.
41. Kim, D.-H., et al. (2009). Silicon electronics on silk as a path to bioresorbable, implantable devices. *Applied Physics Letters, 95*(13), 133701.
42. Kim, Y. J., et al. (2013). Self-deployable current sources fabricated from edible materials. *Journal of Materials Chemistry B, 1*(31), 3781–3788.
43. Brady, P. V., & Walther, J. V. (1990). Kinetics of quartz dissolution at low-temperatures. *Chemical Geology, 82*(3–4), 253–264.
44. Bergstrom, L., & Bostedt, E. (1990). Surface-chemistry of silicon-nitride powders – Electrokinetic behavior and esca studies. *Colloids and Surfaces, 49*(3–4), 183–197.
45. Dameron, A. A., et al. (2008). Gas diffusion barriers on polymers using multilayers fabricated by Al2O3 and rapid SiO2 atomic layer deposition. *Journal of Physical Chemistry C, 112*(12), 4573–4580.
46. Brenckle, M. A., et al. (2015). Modulated degradation of transient electronic devices through multilayer silk fibroin pockets. *ACS Applied Materials & Interfaces, 7*(36), 19870–19875.

47. Acar, H., et al. (2014). Study of physically transient insulating materials as a potential platform for transient electronics and bioelectronics. *Advanced Functional Materials, 24*(26), 4135–4143.
48. Fang, H., et al. (2016). Ultrathin, transferred layers of thermally grown silicon dioxide as biofluid barriers for biointegrated flexible electronic systems. *Proceedings of the National Academy of Sciences, 113*(42), 11682–11687.
49. Lee, Y. K., et al. (2017). Kinetics and chemistry of hydrolysis of ultrathin, thermally grown layers of silicon oxide as biofluid barriers in flexible electronic systems. *ACS Applied Materials & Interfaces, 9*(49), 42633–42638.
50. Song, E., et al. (2018). Transferred, ultrathin oxide bilayers as biofluid barriers for flexible electronic implants. *Advanced Functional Materials, 28*(12), 1702284.
51. Song, E., et al. (2018). Ultrathin trilayer assemblies as long-lived barriers against water and ion penetration in flexible bioelectronic systems. *ACS Nano, 12*(10), 10317–10326.
52. Song, E., et al. (2017). Thin, transferred layers of silicon dioxide and silicon nitride as water and IoN barriers for implantable flexible electronic systems. *Advanced Electronic Materials, 3*(8), 1700077.
53. Yu, X., et al. (2018). Materials, processes, and facile manufacturing for bioresorbable electronics: A review. *Advanced Materials, 30*(28), 1707624.
54. Kim, D. H., et al. (2008). Materials and noncoplanar mesh designs for integrated circuits with linear elastic responses to extreme mechanical deformations. *Proceedings of the National Academy of Sciences of the United States of America, 105*(48), 18675–18680.
55. Kim, D. H., et al. (2008). Stretchable and foldable silicon integrated circuits. *Science, 320*(5875), 507–511.
56. Ying, M., et al. (2012). Silicon nanomembranes for fingertip electronics. *Nanotechnology, 23*(34), 344004.
57. Huang, X., et al. (2014). Biodegradable materials for multilayer transient printed circuit boards. *Advanced Materials, 26*(43), 7371–7377.
58. Lee, S., et al. (2018). Metal microparticle–polymer composites as printable, bio/ecoresorbable conductive inks. *Materials Today, 21*(3), 207–215.
59. Mahajan, B. K., et al. (2017). Mechanically milled irregular zinc nanoparticles for printable bioresorbable electronics. *Small, 13*(17), 8.
60. Li, J., et al. (2018). Processing techniques for bioresorbable nanoparticles in fabricating flexible conductive interconnects. *Materials (Basel), 11*(7), 1102.
61. Lee, Y. K., et al. (2017). Room temperature electrochemical sintering of Zn microparticles and its use in printable conducting inks for bioresorbable electronics. *Advanced Materials, 29*(38), 1702665.
62. Mahajan, B. K., et al. (2018). Aerosol printing and photonic sintering of bioresorbable zinc nanoparticle ink for transient electronics manufacturing. *SCIENCE CHINA Information Sciences, 61*, 1–10.
63. Shou, W., et al. (2017). Low-cost manufacturing of bioresorbable conductors by evaporation–condensation-mediated laser printing and sintering of Zn nanoparticles. *Advanced Materials, 29*(26), 1700172.
64. Daniele, M. A., et al. (2015). Sweet substrate: A polysaccharide nanocomposite for conformal electronic decals. *Advanced Materials, 27*(9), 1600–1606.
65. Yin, L., et al. (2015). Materials and fabrication sequences for water soluble silicon integrated circuits at the 90 nm node. *Applied Physics Letters, 106*(1), 014105.
66. Chang, J. K., et al. (2017). Materials and processing approaches for foundry-compatible transient electronics. *Proceedings of the National Academy of Sciences of the United States of America, 114*(28), E5522–E5529.
67. Chang, J. K., et al. (2018). Biodegradable electronic systems in 3D, heterogeneously integrated formats. *Advanced Materials, 30*(11), 1704955.
68. Sammoura, F., Lee, K. B., & Lin, L. W. (2004). Water-activated disposable and long shelf life microbatteries. *Sensors and Actuators a-Physical, 111*(1), 79–86.

69. Tsang, M., et al. (2014). A MEMS-enabled biodegradable battery for powering transient implantable devices. 2014 IEEE 27th International Conference on Micro Electro Mechanical Systems (MEMS) IEEE, Piscataway, New Jersey, US.
70. Chen, Y., et al. (2016). Physical–chemical hybrid transiency: A fully transient li-ion battery based on insoluble active materials. *Journal of Polymer Science Part B: Polymer Physics, 54*(20), 2021–2027.
71. Lu, L., et al. (2018). Biodegradable monocrystalline silicon photovoltaic microcells as power supplies for transient biomedical implants. *Advanced Energy Materials, 8*(16), 1703035.
72. Yin, L., et al. (2014). Materials, designs, and operational characteristics for fully biodegradable primary batteries. *Advanced Materials, 26*(23), 3879–3884.
73. Tsang, M., et al. (2015). Biodegradable magnesium/iron batteries with polycaprolactone encapsulation: A microfabricated power source for transient implantable devices. *Microsystems & Nanoengineering, 1*, 15024.
74. Fu, K., et al. (2015). Transient rechargeable batteries triggered by cascade reactions. *Nano Letters, 15*(7), 4664–4671.
75. Lee, G., et al. (2017). Fully biodegradable microsupercapacitor for power storage in transient electronics. *Advanced Energy Materials, 7*(18), 1700157.
76. Hwang, S. W., et al. (2015). Materials for programmed, functional transformation in transient electronic systems. *Advanced Materials, 27*(1), 47–52.
77. Lee, C. H., et al. (2015). Wireless microfluidic systems for programmed, functional transformation of transient electronic devices. *Advanced Functional Materials, 25*(32), 5100–5106.
78. Cınar, S., et al. (2016). Study of mechanics of physically transient electronics: A step toward controlled transiency. *Journal of Polymer Science Part B: Polymer Physics, 54*(4), 517–524.
79. Seo, W., & Phillips, S. T. (2010). Patterned plastics that change physical structure in response to applied chemical signals. *Journal of the American Chemical Society, 132*(27), 9234–9235.
80. Gao, Y., et al. (2017). Moisture-triggered physically transient electronics. *Science Advances, 3*(9), e1701222.
81. Hernandez, H. L., et al. (2014). Triggered transience of metastable poly(phthalaldehyde) for transient electronics. *Advanced Materials, 26*(45), 7637–7642.
82. Zhong, S., et al. (2018). Enabling transient electronics with degradation on demand via light-responsive encapsulation of hydrogel/oxide bi-layer. *ACS Applied Materials & Interfaces, 10*(42), 36171–36176.
83. Park, C. W., et al. (2015). Thermally triggered degradation of transient electronic devices. *Advanced Materials, 27*(25), 3783–3788.
84. Pohlers, G., et al. (1997). Mechanistic studies of photoacid generation from substituted 4, 6-bis (trichloromethyl)-1, 3, 5-triazines. *Chemistry of Materials, 9*(6), 1353–1361.
85. Lee, K. M., et al. (2018). Phototriggered depolymerization of flexible poly(phthalaldehyde) substrates by integrated organic light-emitting diodes. *ACS Applied Materials & Interfaces, 10*(33), 28062–28068.
86. Chang, J. K., et al. (2018). Cytotoxicity and in vitro degradation kinetics of foundry-compatible semiconductor nanomembranes and electronic microcomponents. *ACS Nano, 12*(10), 9721–9732.
87. Mueller, P. P., et al. (2012). Histological and molecular evaluation of iron as degradable medical implant material in a murine animal model. *Journal of Biomedical Materials Research Part A, 100*(11), 2881–2889.
88. Bai, W., et al. (2018). Flexible transient optical waveguides and surface-wave biosensors constructed from monocrystalline silicon. *Advanced Materials, 30*(32), 1801584.
89. Yu, K. J., et al. (2016). Bioresorbable silicon electronics for transient spatiotemporal mapping of electrical activity from the cerebral cortex. *Nature Materials, 15*(7), 782–791.
90. Kang, S. K., et al. (2016). Bioresorbable silicon electronic sensors for the brain. *Nature, 530*(7588), 71–76.
91. Koo, J., et al. (2018). Wireless bioresorbable electronic system enables sustained nonpharmacological neuroregenerative therapy. *Nature Medicine, 24*, 1830–1836.

92. Tao, H., et al. (2014). Silk-based resorbable electronic devices for remotely controlled therapy and in vivo infection abatement. *Proceedings of the National Academy of Sciences of the United States of America, 111*(49), 17385–17389.
93. Lee, C. H., et al. (2015). Biological lipid membranes for on-demand, wireless drug delivery from thin, bioresorbable electronic implants. *NPG Asia Materials, 7*(11), e227.
94. Grayson, A. C. R., et al. (2003). Multi-pulse drug delivery from a resorbable polymeric microchip device. *Nature Materials, 2*(11), 767–772.
95. Acar, H., et al. (2016). Transient biocompatible polymeric platforms for long-term controlled release of therapeutic proteins and vaccines. *Materials, 9*(5), 321.
96. Miller, R. D., et al. (2015). Water soluble, biodegradable amphiphilic polymeric nanoparticles and the molecular environment of hydrophobic encapsulates: Consistency between simulation and experiment. *Polymer, 79*, 255–261.
97. Son, D., et al. (2015). Bioresorbable electronic stent integrated with therapeutic nanoparticles for endovascular diseases. *ACS Nano, 9*(6), 5937–5946.
98. Rogers, J. A., Someya, T., & Huang, Y. (2010). Materials and mechanics for stretchable electronics. *Science, 327*(5973), 1603–1607.
99. Kim, D. H., et al. (2011). Epidermal electronics. *Science, 333*(6044), 838–843.
100. Yang, J., Webb, A. R., & Ameer, G. A. (2004). Novel citric acid-based biodegradable elastomers for tissue engineering. *Advanced Materials, 16*(6), 511–516.
101. Jeong, J. W., et al. (2014). Capacitive epidermal electronics for electrically safe, long-term electrophysiological measurements. *Advanced Healthcare Materials, 3*(5), 642–648.
102. Won, S. M., et al. (2018). Natural wax for transient electronics. *Advanced Functional Materials, 28*(32), 1801819.
103. Jin, S. H., et al. (2014). Solution-processed single-walled carbon nanotube field effect transistors and bootstrapped inverters for disintegratable, transient electronics. *Applied Physics Letters, 105*(1), 013506.
104. Lin, S.-Y., et al. (2018). Transient and flexible photodetectors. *ACS Applied Nano Materials, 1*(9), 5092–5100.
105. Lei, T., et al. (2017). Biocompatible and totally disintegrable semiconducting polymer for ultrathin and ultralightweight transient electronics. *Proceedings of the National Academy of Sciences of the United States of America, 114*(20), 5107–5112.
106. Kim, B. H., et al. (2017). Dry transient electronic systems by use of materials that sublime. *Advanced Functional Materials, 27*(12), 1606008.
107. Yoon, J., et al. (2017). Flammable carbon nanotube transistors on a nitrocellulose paper substrate for transient electronics. *Nano Research, 10*(1), 87–96.
108. Yi, N., et al. (2018). Fully water-soluble, high-performance transient sensors on a versatile galactomannan substrate derived from the endosperm. *ACS Applied Materials & Interfaces, 10*(43), 36664–36674.

Wireless Powered Medical Implants via Radio Frequency

Parinaz Abiri, Alireza Yousefi, Hung Cao, Jung-Chih Chiao, and Tzung K. Hsiai

1 Introduction

While advancements in medical device development for the treatment and diagnosis of disease have grown drastically in recent years, the ability to power these devices has not been able to keep up with the growing trend in device miniaturization and adaptability. The need for large batteries, the danger and inconvenience of multiple surgeries for battery replacement, and the presence of long implanted wires that can cause life-threatening complications have been lingering limitations in the medical device world. Most prominently, lead-related complications, such as insulation breaks, conductor coil fractures, lead dislodgments, vascular occlusions, hemorrhaging, infection, and tissue perforation, have plagued implantable device for the past five decades [1–4]. Cardiac pacemakers and neuromodulation devices used for spinal cord stimulation, deep brain stimulation, and peripheral nerve stimulation all rely on the same problematic lead-based structure. The elimination of these problems

P. Abiri
Department of Bioengineering, University of California, Los Angeles, Los Angeles, CA, USA

A. Yousefi
Department of Electrical Engineering, University of California, Los Angeles, Los Angeles, CA, USA

H. Cao
Department of Electrical Engineering and Computer Science, University of California, Irvine, Irvine, CA, USA

J.-C. Chiao
Department of Electrical Engineering, Southern Methodist University, Dallas, TX, USA

T. K. Hsiai (✉)
Department of Bioengineering, University of California, Los Angeles, Los Angeles, CA, USA

Medical Engineering, California Institute of Technology, Pasadena, CA, USA
e-mail: thsiai@mednet.ucla.edu

H. Cao et al. (eds.), *Interfacing Bioelectronics and Biomedical Sensing*,
https://doi.org/10.1007/978-3-030-34467-2_4

has, therefore, become a necessary step in perfecting medical device technology to meet patient needs.

The challenge, however, lies in the ability to provide sufficient long-term power to a device within our bodies' limited anatomical confines. Numerous techniques have been introduced for wirelessly delivering power to a remote device [5]. These have included a variety of innovative power harvesting methods, such as optical energy [6], acoustic energy [7–11], mechanical energy [12–17], and biofuel harvesting [18–20]. While these sources of energy can be valuable in integration into some medical devices, they do present with limitations in size, motion, and power production. Specifically, due to the high-power demands in stimulation applications and limitations in anatomical real-estate, alternative methods have been sought.

Recent years have placed much focus on power harvesting through radio frequency (RF), with potential to produce sufficient energy for long-term implants. While this technology was developed over a century ago and has been used in practice for years in consumer electronics as well as theorized for charging of electric vehicles, its utilization in medical devices has been limited due certain inherent challenges. This chapter will focus on the different techniques of wireless power transfer through electromagnetic radiation, as well as their limitations and potential applications.

Wireless power transfer (WPT) through RF can ultimately be separated into three forms of power transmission and harvesting: near-field, far-field, and midfield.

1.1 Near-Field WPT

Inductive power transfer occurs as a result of a changing electrical current that induces the formation of an electromagnetic field around a transmitting antenna. Near-field acts near the source antenna, formally defined as less than one wavelength (λ) away from the antenna [21]. This region can be separated into two areas: (1) the reactive zone, ranging up to $\lambda/2\pi$ distance, and (2) the radiative zone, ranging from $\lambda/2\pi$ to λ distance [22]. In the reactive zone, in which we primarily operate in near-field applications, either the electric (E) or magnetic (H) field dominates. In the commonly used loop antenna, the magnetic field is dominant [22, 23].

This magnetic field is then able to induce a voltage in a nearby conductor, thus transferring power wirelessly from one conducting system to another (Fig. 1). When a transmitting coil (TX) and receiving coil (TX) interact in this phenomenon, they are said to be coupled by the magnetic field between them. The strength of the magnetic field from the transmitter antenna decreases rapidly in near-field, following the inverse-cube law over distance ($1/d^3$) [21]. This rapid decline has two implications: (1) Power is contained within only a few wavelengths away from the source, and any near-field effects are negligible further away; (2) the receiving coil must be within a small region to maintain power transfer efficiency from the transmitting to the receiving system. The power transfer efficiency in the near-field can be defined as [21]:

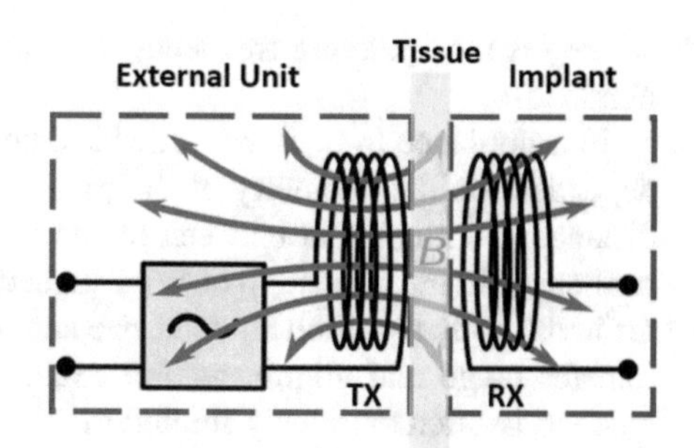

Fig. 1 Alternating current induces a magnetic field that generates a voltage in a nearby conductor

$$PTE = \eta = \frac{k^2 Q_1 Q_{2L}}{1 + k^2 Q_1 Q_{2L}} \cdot \frac{Q_{2L}}{Q_L}, \tag{1}$$

where k is the coupling coefficient, Q_1 is the quality factor of the transmitting coil, Q_2 is the quality factor of the receiving coil, $Q_{2L} = Q_2 Q_L / Q_2 + Q_L$, and $Q_L = R_{Load} / 2\pi f L_2$.

The coupling coefficient is dependent on coil geometries and relative position of the transmitting and receiving coils, as defined below [21]:

$$k = \frac{r_1^2 r_2^2}{\sqrt{r_1 r_2}\left(\sqrt{r_1^2 + D^2}\right)^3} \cos(\theta), \tag{2}$$

where r_1 is the radius of the transmitter, r_2 is the radius of the receiver, D is the distance between the transmitter and receiver, and θ is the misalignment angle.

Alternatively, the coupling coefficient can be defined by the transmitting and receiving coil inductances and the mutual inductance that is produced as a result of the two interacting conductors [21]:

$$k = \frac{M}{\sqrt{L_1 L_2}}, \tag{3}$$

where L_1 is the inductance of the transmitter, L_2 is the inductance of the receiver, and M is the mutual inductance.

As coil inductance is a function of conductor geometry, and mutual inductance is further dependent on the relative position of the interacting coils, these factors remain the primary variable that impact coupling coefficient. Most importantly, k is independent of the operating frequency. The quality factor, which represents the coil's ability to retain energy, is, however, heavily influenced by the transmission frequency [21]:

$$Q = \frac{2\pi f L}{R}, \tag{4}$$

where f is the operating frequency, L is antenna inductance, and R is the effective ohmic loss.

In accordance to these relationships, power transfer efficiency (PTE) is greatly dependent on the geometry of the transmitting and receiving coils as well as the distance and alignment between the coils. These dependencies cause some major challenges in the utilization of WPT in medical devices due to anatomical boundaries in the body that limit antenna size and introduce substantial variations in power transfer range and alignment. For example, in the case of neural implants, the receiver is often extremely limited in size; or, in the case of cardiac implants, the receiver faces the challenge of motion-induced variations in distance and alignment.

In addition to coil design, PTE is maximized in resonant inductive coupling by having the transmitter and receiver resonate at the same frequency. Since the conducting coils are inductive, resonance is achieved by having an LC tank circuit, consisting of the coil as the inductor and a corresponding capacitor. The relationship between these components is defined as follows [24]:

$$C = \frac{1}{L(2\pi f)^2} \tag{5}$$

where C is the capacitance of the resonant capacitor, L is the inductance of the coil, and f is the operating frequency.

While various forms of the LC tank circuit are utilized, typically, an inductive power transfer (IPT) system working in the near-field contains a transmitter system with a series LC tank (Fig. 2a) and a receiver system with a parallel LC tank (Fig. 2b). Power transfer efficiency (PTE) is highly impacted by the proper tuning of the LC tank [25].

Finally, tissue energy absorption is also a significant player in IPT systems. Depending on the dielectric constant, a substance absorbs different amount of energy. The level of tissue energy absorption depends on the tissue type (e.g., skin, fat, muscle, etc.) and the electric (E) field generated as a result of IPT. The strength of the E-field depends on the amount of power transferred and the frequency of transmission. Specific absorption rate (SAR) can, therefore, be defined as follows:

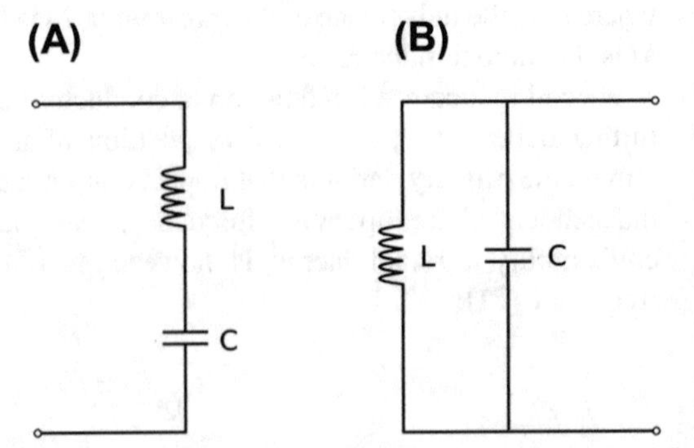

Fig. 2 (**a**) Series LC tank, (**b**) parallel LC tank

$$SAR = \frac{\sigma E^2}{\rho}, \tag{6}$$

where σ is the electrical conductivity of the tissue and ρ is the mass density of the tissue.

The Federal Communication Commission (FCC) has established SAR safety limits for humans. This has been established at 1.6 W/kg. An IPT system must follow these guidelines [26].

Several different near-field system architectures have been established to deliver power wirelessly to a stimulation system, each to be discussed in the following sections.

1.2 Batteryless Direct Stimulation

In the first architecture to be discussed, power is transferred sequentially as follows:

1. Power supply (Fig. 3a)
2. Transmitter unit with transmitter coil and circuitry (Fig. 3b)
3. Receiver unit with receiver coil and circuitry (Fig. 3c)
4. Control circuitry (Fig. 3d)
5. Stimulation electrodes (Fig. 3e)

This architecture defines the most direct method to transmit power from a remote power source. Power is delivered continuously from a transmitter to the control circuitry in the receiver, where rate and duration of stimulation are then controlled based on pre-defined settings or sensor input (Fig. 4).

This structure is utilized in the vast majority of systems implemented in literature. Heetderks et al. utilized this architecture in the analysis of power transfer efficiency for significantly large transmitter-to-receiver size ratios. Millimeter- and submillimeter-sized neural implant receivers (RX Ø = 0.4–1.5 mm, length = 3 mm) were coupled to large transmitters (TX Ø = 90–32 mm) at 2 MHz and 20 MHz operating frequency [27]. Given a 4 W power supply, 0.09–50 mW was received by the implant. Von Arx et al. also describe a fully integrated neuromuscular electrical stimulation system (FITNESS) operating at 4 MHz (Tx Ø = 80 mm, Rx = 2 × 8 mm^2) and 30 mm distance between the transmitter and receiver, capable of delivering

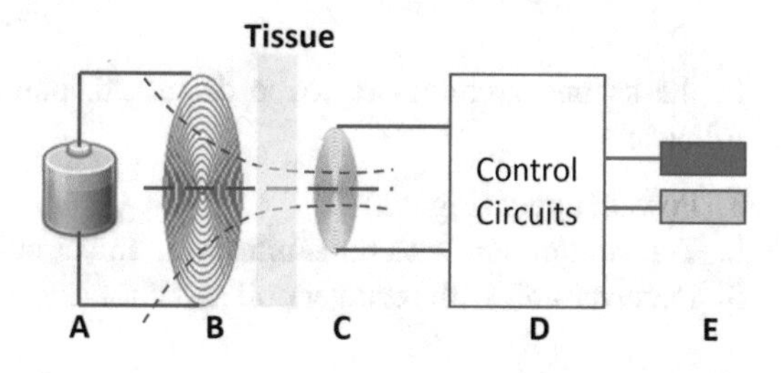

Fig. 3 Batteryless direct simulation system architecture

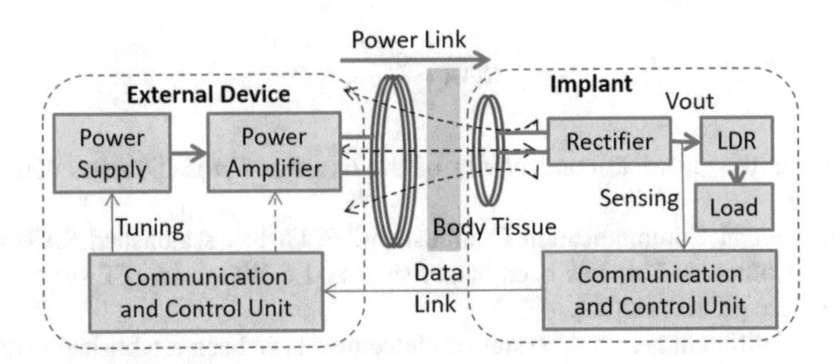

Fig. 4 Wireless power transfer system for IMDs

20 mW to the receiver [28]. Neagu et al. similarly focused their design on a small receiver (Rx Ø = 4.5 mm). With transmitter and receiver coil separations of 1 mm and 3.5 mm and operating frequency of 3 MHz, they obtained a maximum system output of 2 mW [29]. In all cases, it can be seen that power transfer efficiency suffers severely as a result of small receiver size, either limiting the received power or range of transmission or sacrificing on SAR regulatory compliance.

Larger receiver sizes can be more practical when needing higher power transfer efficiency but can also be limiting in implant location and device application. Ali et al. designed an inductive link with a larger receiver coil (TX Ø = 27 mm, RX Ø = 12.6 mm) for medical implants at 2.5 MHz frequency and achieved 40 mW output while simultaneously transferring data at a rate of 128 kbps [30]. Similarly, Li et al. designed a wirelessly powered system for medical devices (TX Ø = 25 mm, RX Ø = 9.5 mm) at 13.56 MHz and achieved 60 mW output and 92.5% efficiency at 3 mm coil separation [31]. Ghovanloo et al. and Parramon et al. were able to achieve longer range of power transfer at 10 mm and 15 mm, respectively [32, 33]. At 5 MHz and using larger receiver coils (TX Ø = 40 mm, RX Ø = 20 mm), Ghovanloo obtained 78% power transfer efficiency. Parramon increased operating frequency (10 MHz) and decreased coil sizes (TX Ø = 20 mm, RX Ø = 10 mm) to achieve 19 mW output. Monti et al. took advantage of the same system design at a significantly higher frequency of 434 MHz (TX Ø ≅ 68 mm, RX Ø ≅ 33 mm), obtaining 51 mW at 1 cm and 10 mW at 2 cm given a power supply of 1 W on the transmitter side [34].

1.3 Battery-Based Stimulation

In the second architecture to be discussed, power is transferred sequentially as follows:

1. Power supply (Fig. 5a)
2. Transmitter unit with transmitter coil and circuitry (Fig. 5b)
3. Receiver unit with receiver coil and circuitry (Fig. 5c)

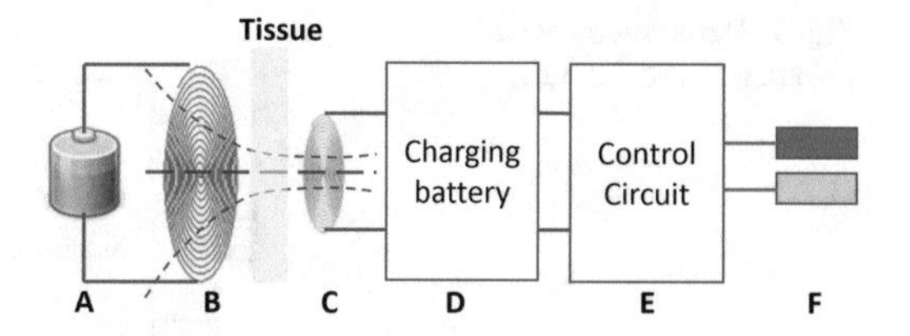

Fig. 5 Battery-based system architecture

4. Energy-storage unit (Fig. 5d)
5. Control circuitry (Fig. 5e)
6. Stimulation electrodes (Fig. 5f)

This architecture defines a high-reliability method to transmit power from a remote power source. Power can be transferred continuously or intermittently from a transmitter to an energy-storage unit (ESU), such as a small battery or capacitor, in the receiver. The ESU then delivers the power needed to feed the control unit as well as the stimulation pulse. Due to the intermediary presence of an ESU, the instantaneous power needed for stimulation does not need to be provided via the wireless connection, but rather from the ESU, thus allowing for more freedom in IPT system efficiency. This is important as it allows a lower efficiency system to be implemented, in which power is slowly accumulated to the necessary threshold and delivered when appropriate.

The battery-based architecture is valuable in cases in which there is sufficient space for an ESU and a low-efficiency system has to be tolerated despite the need for high instantaneous stimulation power. In Lee et al., a wireless cardiac defibrillator is designed using this architecture due to the high-power requirements for defibrillation. In their design, an IPT delivers power to a rechargeable battery, which then feeds into a control system and stimulator circuitry [35]. Lee et al. alternatively describe an IPT system designed to charge a series of capacitors preceding delivery to a stimulator. Given a 2.7 Vpp input, the system (TX Ø = 40 mm, RX Ø = 10 mm) with a 2 MHz operating frequency is able to charge pairs of 1 μF capacitors up to 2 V in 420 μs, achieving a high measured charging efficiency of 82% [36].

1.4 Remote-Controlled Stimulation

In the third architecture to be discussed, power is transferred sequentially as follows:

1. Power supply (Fig. 6a)
2. Control circuitry (Fig. 6b)
3. Transmitter unit with transmitter coil and circuitry (Fig. 6c)
4. Receiver unit with receiver coil and circuitry (Fig. 6d)
5. Stimulation electrodes (Fig. 6e)

This architecture significantly minimizes battery power consumption in a low-efficiency IPT system, such as one in which there is a large transmitter-to-receiver size ratio. The placement of the control circuitry in the transmitter avoids the need

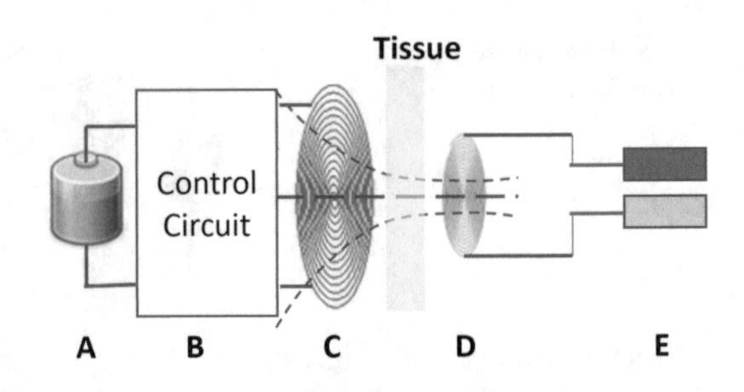

Fig. 6 Remote-controlled stimulation architecture

for continuous power transmission to the receiver. Thus, power is transmitted wirelessly in intermittent time intervals only at the time and during the period of stimulation. In this system, while the IPT system itself may have low power transfer efficiency, the overall end-to-end system efficiency is high. The degree of improvement in system efficiency is dependent on transmission frequency and duration. In Abiri et al., a miniature pacing system is described (TX Ø = 40 mm, RX = 3 × 15 mm) in which 1 mW of power is consumed to stimulate the heart at 2 V threshold with 500 ohm impedance electrodes (Fig. 7) [37]. This 1 mW average power consumption is a result of instantaneous power consumption of 1 W with pulse duration of 1 ms and power transmission frequency of 1 Hz.

1.5 Multi-coil Stimulation

In the fourth and final near-field architecture to be discussed, power is transferred sequentially as follows:

1. Power supply (Fig. 8a)
2. Transmitter unit with transmitter coil and circuitry (Fig. 8b)
3. Relay coil (Fig. 8c)
4. Receiver unit with receiver coil and circuitry (Fig. 8d)
5. Logic circuitry (Fig. 8e)
6. Stimulation electrodes (Fig. 8f)

In this architecture, the presence of one or more intermediary relay coils can significantly increase power transfer efficiency of the IPT system. This design allows for higher efficiency where there is anatomical leniency for the presence of additional relay coils.

Kurs et al. present a four-coil system defined by two primarily interacting coils called the "source" and "device" coils and their interacting loop coils: one that is part of the driving circuit connected to the "source" (defined as "A") and one that is part of the load connected to the "device" (defined as "B"). The system is designed such that coupling is primarily present between A and source, source and device,

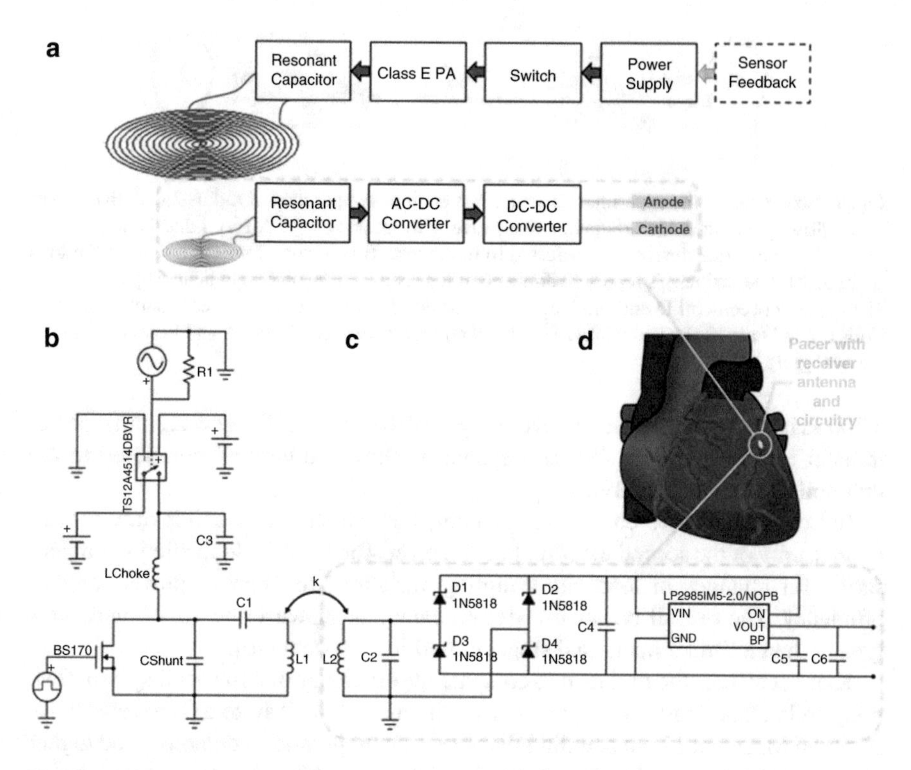

Fig. 7 System architecture. (**a**) A block diagram highlights the wireless pacing system with "remote-controlled" stimulation on the receiver side using intermittent power transfer from the transmitter side. (**b**) The transmitter circuitry consisted of the primary functional components of the pacemaker that remotely control and power the stimulator on the heart, including a switch, potential control input from a sensor, Class-E PA, and series resonant tank circuit. (**c**) The receiver circuitry consisted of only the components necessary to output a regulated voltage to the two electrodes, including a parallel resonant tank circuit, bridge rectifier, and voltage regulator. (**d**) A cartoon diagram depicts the pacer implant location into the anterior cardiac vein, which is made possible through miniaturization using a remote-controlled stimulation system [37]

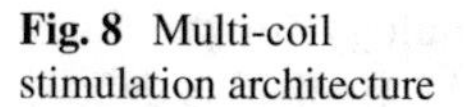

Fig. 8 Multi-coil stimulation architecture

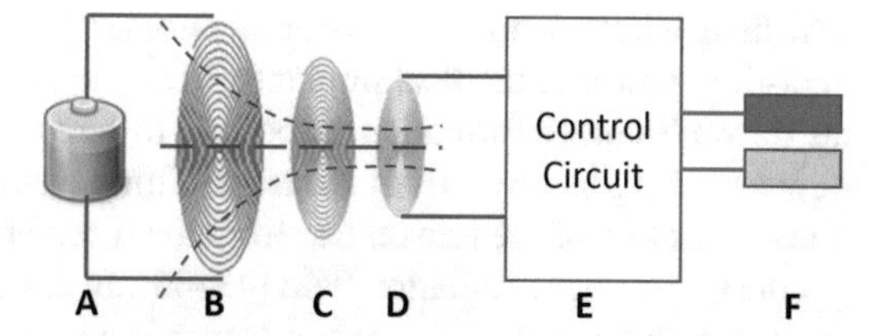

and device and B (Fig. 9). Their system is shown to transfer 60 W with 40% efficiency over 2 m distance [38].

Ramrakhyani et al. similarly describe a four-coil structure in which the transmitting system consists of a "driver" coil in proximity to a "primary" coil and the receiving system consists of a "secondary" coil in proximity to a "load" coil. Given

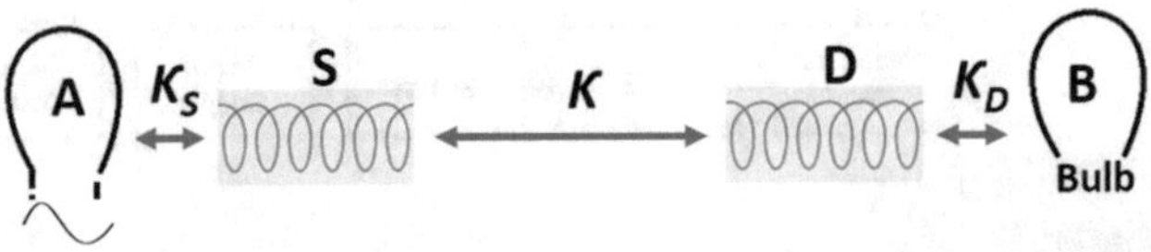

Fig. 9 Schematic of the experimental setup. A is a single copper loop of radius 25 cm that is part of the driving circuit, which outputs a sine wave with frequency 9.9 MHz. S and D are, respectively, the source and device coils referred to in the text. B is a loop of wire attached to the load (light bulb). The various k's represent direct couplings between the objects indicated by the arrows. The angle between coil D and the loop A is adjusted to ensure that their direct coupling is zero. Coils S and D are aligned coaxially. The direct couplings between B and A and between B and S are negligible

a 700 kHz operating frequency over ranges of 10–20 mm (Rx Ø = 22 mm), power transfer efficiency in the four-coil system is shown to be 80% compared to the two-coil system at 40% [39].

In Lee et al., a three-coil system is defined, in which a relay coil is only present in proximity of the secondary coil of the receiver. The local Rx loop allows compensation for changes in load and coupling variations to improve power transfer efficiency. The overall power transfer efficiency is shown to be 10.5% and 4.7% greater than a similar open- and single closed-loop system [40].

Kiani et al. take the initiative to compare the efficiency of a four-coil system (two relay coils), three-coil system (one relay coil), and two-coil system (zero relay coil). At 13.56 MHz and 12 cm coupling distance, their experiments demonstrated higher efficiency in a three-coil system, with 35%, 37%, and 15% efficiency in two-, three-, and four-coil inductive links, respectively [41].

Finally, the placement of multiple coils in a domino format in coaxial and noncoaxial structures was analyzed by Zhong et al., in which unequal spacing between the subsequent coils was found to be more efficient than equal spacing [42].

2 Far-Field WPT

Far-field wireless power transfer is another promising technology for providing real-time power to medical implants. In opposition to near-field, far-field is defined as transmission (information or power) to a receiving antenna that is more than a few wavelengths away from the transmitting antenna [22]. This distance results in the decoupling of the transmitter from the receiver; thus, the receiver has no impact on the transmitter's radiation field [43–45]. In comparison to near-field, the far-field approach often offers a longer transmission distance, smaller transmitter and receiver size, less sensitivity to movement, and the capability to deliver power to multiple devices. On the other hand, far-field also faces the challenge of meeting the FCC SAR limits, which is a major concern in long-term implantable devices.

These properties in far-field wireless power transfer arise from key physical characteristics. Firstly, the radiation field is uniform (assuming a uniform radiator) in shape with the electric (E) field moving orthogonal to the magnetic (M) field. Furthermore, the radiative power decreases in strength much more slowly compared to near-field (proportional to the distance cubed), where the power density (i.e., Poynting vector [43]) of electromagnetic radiation decreases as a function of the square of the distance:

$$S = \frac{P_t}{4\pi R^2}, \tag{7}$$

where S is the power density, P_t is the transmitted power, and R is the distance from the lossless isotropic (uniformly radiating) transmitter antenna.

Another advantage of the far-field approach lies in the ability to design the antenna such that it radiates more strongly in a specific direction. For example, the power beam can be focused using an array of antennas or dish antennas. The power density can thus be modified as follows:

$$S = \frac{P_t G}{4\pi R^2}, \tag{8}$$

where G is the antenna gain.

It is important to note that the far-field approach has lower efficiency in water-rich environments, such as biological tissues. The power transfer efficiency in far-field is described in [45] and can be decomposed into three power conversion efficiency components:

$$eff = \frac{P_{dc}^r}{P_{dc}^t} = \frac{P_{rf}^r}{P_{dc}^t} \times \frac{P_{rf}^r}{P_{rf}^t} \times \frac{P_{dc}^r}{P_{rf}^r}, \tag{9}$$

where P_{rf}^r / P_{dc}^t is the DC-to-RF (frequency up-conversion from DC to a designated high frequency) conversion efficiency at the energy transmitter, P_{rf}^r / P_{rf}^t is the RF-to-RF transmission efficiency (i.e., medium loss), and P_{dc}^r / P_{rf}^r is the RF-to-DC (conversion down-conversion from high frequency to DC) conversion efficiency at energy receiver (e.g., rectenna [46]). The major bottleneck in medical applications is the RF-to-RF transmission efficiency due to losses through the medium consisting primarily of water and live tissues. This efficiency can, however, be improved by directional transmission and beam-forming.

Due to the long-range radiative nature of far-field power harvesting, safety considerations are significant in its application to medical devices. While less commonly applied than near-field, the use of power harvesting using far-field electromagnetic waves in medical devices has been studied by a few investigators.

Bakogianni et al. examine the efficiency of a power link at various sub-1 GHz frequencies (433 MHz, 868 MHz, 915 MHz), noting optimal reliability at the highest frequency of 915 MHz at which −5 dBm power level is achieved at 40 cm distance and transmitter output of 30 dBm [47]. Liu et al. analyze the safety considering for an implantable rectenna (rectifying antenna) for far-field wireless power transfer using a planar inverted-F antenna in combination of a parasitic patch for improved power harvesting efficiency (Fig. 10). The proposed system (RX = 3 × 8 mm, patch = 4 × 8 mm) is shown to achieve a power level of −11 dBm at an operating frequency of 2.45 GHz and distance of 0.5 m [48].

Sun et al. present a batteryless wirelessly powered pacemaker that harvests energy at 9 GHz. A miniature receiver chip (RX = 4 mm × 1 mm) is thus presented with which a 1.3 V stimulation pulse is initiated at 2 cm distance between the transmitting and receiving units (Fig. 11) [49].

3 Midfield WPT

While the terms near-field and far-field are common jargons, the idea of midfield powering was only recently given prominence by Poon et al. at Stanford University, whose research has focused on increasing wireless power transfer efficiency to very

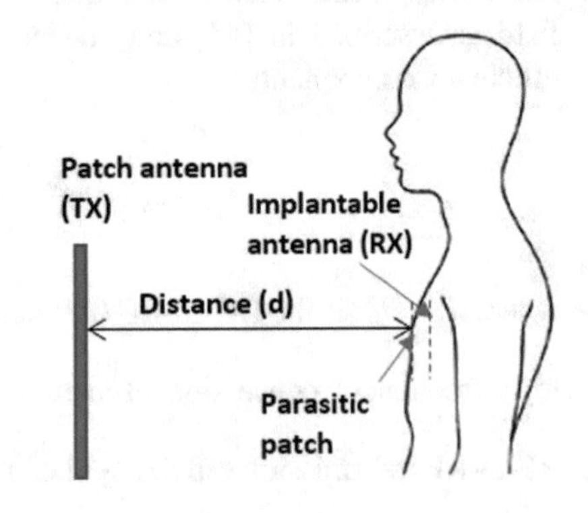

Fig. 10 Simulated environment of wireless power link with parasitic patch

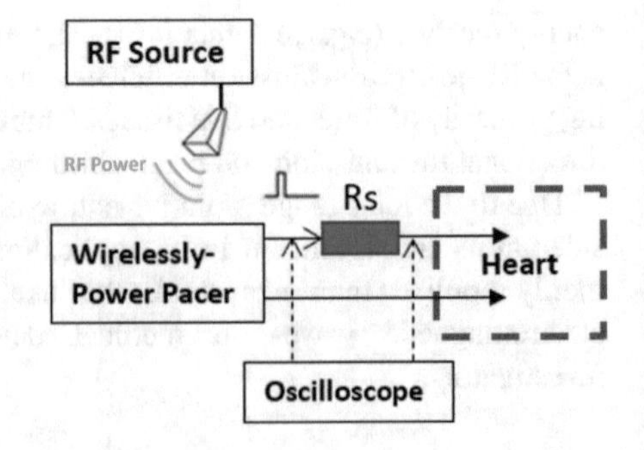

Fig. 11 Wireless test setup

small implants inside the body [50, 51]. In their numerical studies of the midfield approach, which utilizes a combination of inductive and radiative modes, they have shown that it can be an efficient way to transfer power to deep implants in the body despite millimeter-sized receiver antennas.

In their midfield approach, the human body is considered to be a multilayer tissue model, where each layer has specific EM wave propagation characteristic. By optimizing the physical realization of the source planar antenna, they attempt to achieve the highest efficiency for delivering power to a small receiver coil in the body. They rewrite the efficiency equation as the function of current density J_S on the source (planar structure) and dyadic Green's functions G_E and G_H (defined as $\boldsymbol{E}(\boldsymbol{r}) = i\omega\mu \int \bar{\boldsymbol{G}}_E(\boldsymbol{r}-\boldsymbol{r}')\boldsymbol{J}_S(\boldsymbol{r}')d\boldsymbol{r}'$ and $\boldsymbol{H}(\boldsymbol{r}) = \int \bar{\boldsymbol{G}}_H(\boldsymbol{r}-\boldsymbol{r}')\boldsymbol{J}_S(\boldsymbol{r}')d\boldsymbol{r}'$, where E and H are electric and magnetic fields). They then work toward reaching the maximum efficiency that is achievable by any arbitrary source surface current density J_S.

The source is then realized with a slot array structure, as shown in Fig. 12. The patterned metal plate (slot array) is placed near the skin at sub-wavelength distances away from the small receiver coil (Fig. 10). The ability to focus the field allows a much smaller antenna to harvest a relatively larger amount of energy. As shown in Ho et al., a miniature cardiac implant (RX Ø = 2 mm and 3.5 mm height) is able to receive 200 μW of power at 5 cm distance from the transmitter (TX = 60 mm × 60 mm) at 1.6 GHz operating frequency with 500 mW of input [50].

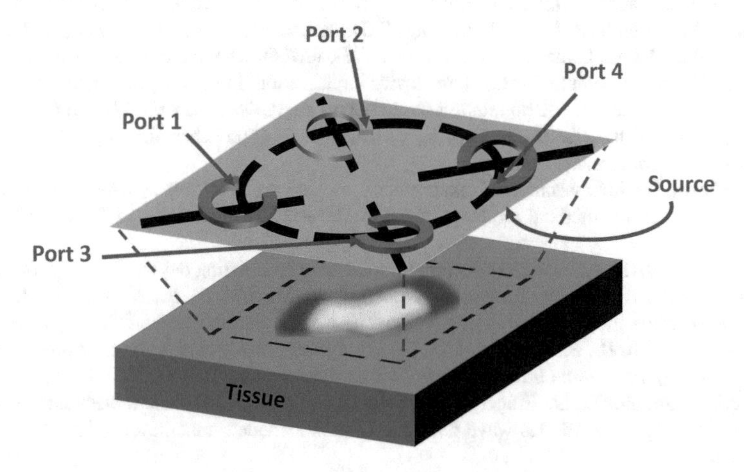

Fig. 12 Midfield energy transfer with a patterned metal plate. Schematic of the midfield source (dimensions 6 × 6 cm, operating frequency 1.6 GHz) and the magnetic field (time snapshot H y) on the skin surface

4 Future Directions and Conclusion

Wireless power transfer through RF serves as a promising endeavor in eliminating the need for implanted batteries in medical devices. While some challenges remain in achieving safety and efficacy of a long-term implantable system, novel system designs have presented with opportunities to eliminate inherent limitations of electromagnetic-based wireless power transfer systems. Device miniaturizations and increases in power transfer efficiency have allowed for new applications in the medical device field, including deep neural implants that necessitate significant size reductions, cardiac implants that must withstand variations from contractile motion, gastrointestinal implants that must withstand challenging environments, and more. In overcoming these challenges, wireless power transfer can meet many of today's unmet clinical needs.

References

1. Gul, E. E., & Kayrak, M. (2011). Common pacemaker problems: Lead and pocket complications. In: M. R. Das (Ed.), *Modern pacemakers – present and future*. Available from: https://www.intechopen.com/books/modern-pacemakers-present-and-future/common-pacemaker-problems-lead-and-pocket-complications.
2. Klug, D., Lacroix, D., Savoye, C., Goullard, L., Grandmougin, D., Hennequin, J. L., Kacet, S., & Lekieffre, J. (1997). Systemic infection related to endocarditis on pacemaker leads: Clinical presentation and management. *Circulation, 95*(8), 2098–2107.
3. Udo, E. O., Zuithoff, N. P. A., Van Hemel, N. M., De Cock, C. C., Hendriks, T., Doevendans, P. A., & Moons, K. G. M. (2012). Incidence and predictors of short- and long-term complications in pacemaker therapy: The FOLLOWPACE study. *Hear Rhythm., 9*(5), 728–735.
4. Lin, Y. S., Hung, S. P., Chen, P. R., Yang, C. H., Wo, H. T., Chang, P. C., Wang, C. C., Chou, C. C., Wen, M. S., Chung, C. M., & Chen, T. H. (2014). Risk factors influencing complications of cardiac implantable electronic device implantation: Infection, pneumothorax and heart perforation: A nationwide population-based cohort study. *Medicine, 93*(27), e213.
5. Ben, A. A., Kouki, A. B., & Cao, H. (2015). Power approaches for implantable medical devices. *Sensors (Switzerland).*
6. Murakawa, K., Kobayashi, M., Nakamura, O., & Kawata, S. (1999). A wireless near-infrared energy system for medical implants. *IEEE Engineering in Medicine and Biology Magazine, 18*, 70–72.
7. Penner, A. (2013). Acoustically powered implantable stimulating device. US8577460 B2.
8. Willis, N. P., Brisken, A. F., Cowan, M. W., Pare, M., Fowler, R., & Brennan, J. (2014). *Optimizing energy transmission in a leadless tissue stimulation system*. US8718773 B2.
9. Tran, B. C., Mi, B., & Harguth, R. S. (2013). *Systems and methods for controlling wireless signal transfers between ultrasound-enabled medical devices*. WO2009102640 A1.
10. Ozeri, S., Shmilovitz, D., Singer, S., & Wang, C. C. (2010). Ultrasonic transcutaneous energy transfer using a continuous wave 650 kHz Gaussian shaded transmitter. *Ultrasonics, 50*(7), 666–674.
11. Ozeri, S., & Shmilovitz, D. (2010). Ultrasonic transcutaneous energy transfer for powering implanted devices. *Ultrasonics, 50*(6), 556–566.
12. Hwang, G. T., Park, H., Lee, J. H., Oh, S., Park, K., II, Byun, M., Park, H., Ahn, G., Jeong, C. K., No, K., Kwon, H., Lee, S. G., Joung, B., & Lee, K. J. (2014). Self-powered cardiac pacemaker enabled by flexible single crystalline PMN-PT piezoelectric energy harvester. *Advanced Materials, 26*(28), 4880–4887.

13. Karami, M. A., & Inman, D. J. (2012). Powering pacemakers from heartbeat vibrations using linear and nonlinear energy harvesters. *Applied Physics Letters, 100*(4).
14. Glynne-Jones, P., Beeby, S. P., & White, N. M. (2001). Towards a piezoelectric vibration-powered microgenerator. *IEE Proceedings – Science, Measurement and Technology, 148*(2), 68.
15. Kim, H. S., Kim, J. H., & Kim, J. (2011). A review of piezoelectric energy harvesting based on vibration. *International Journal of Precision Engineering and Manufacturing, 12*(6), 1129–1141.
16. Miao, P., Holmes, A. S., Yeatman, E. M., & Green, T. C. (2003). Micro-machined variable capacitors for power generation. *Electrostatics, 178*, 53–58.
17. Tashiro, R., Kabei, N., Katayama, K., Tsuboi, F., & Tsuchiya, K. (2002). Development of an electrostatic generator for a cardiac pacemaker that harnesses the ventricular wall motion. *Journal of Artificial Organs, 5*(4), 239–245.
18. Bock, D. C., Marschilok, A. C., Takeuchi, K. J., & Takeuchi, E. S. (2012). Batteries used to power implantable biomedical devices. *Electrochimica Acta., 84*, 155–164.
19. Halliwell, C. M., Simon, E., Toh, C. S., Cass, A. E. G., & Bartlett, P. N. (2002). The design of dehydrogenase enzymes for use in a biofuel cell: The role of genetically introduced peptide tags in enzyme immobilization on electrodes. *Bioelectrochemistry, 55*(1–2), 21–23.
20. Simon, E., Halliwell, C. M., Toh, C. S., Cass, A. E. G., & Bartlett, P. N. (2002). Immobilisation of enzymes on poly(aniline)-poly(anion) composite films. Preparation of bioanodes for biofuel cell applications. *Bioelectrochemistry, 55*(1–2), 13–15.
21. Finkenzeller, K. (2010). *RFID handbook: Fundamentals and applications in contactless smart cards, radio frequency identification and near-field communication* (3rd ed.). Chippenham, Wiltshire: Wiley. 462 p.
22. Djuknic, G. M. (2002). Method of measuring a pattern of electromagnetic radiation. United States; US6657596B2.
23. Cao, H., Rao, S., Tang, S. J., Tibbals, H. F., Spechler, S., & Chiao, J. C. (2013). Batteryless implantable dual-sensor capsule for esophageal reflux monitoring. *Gastrointestinal Endoscopy, 77*, 649.
24. Chen, S. C. Q., & Thomas, V. (2001). Optimization of inductive RFID technology. *International Symposium on Electronics and the Environment, 2*, 82–87.
25. Kim, H. J., Hirayama, H., Kim, S., Han, K. J., Zhang, R., & Choi, J. W. (2017). Review of near-field wireless power and communication for biomedical applications. *IEEE Access, 5*, 21264.
26. Cleveland, R. F., Sylvar, D. M., & Ulcek, J. L. (1997). *Evaluating compliance with FCC guidelines for human exposure to radiofrequency electromagnetic fields* (Vol. 65). Washington, D.C.
27. Heetderks, W. J. (1988). RF powering of millimeter- and submillimeter-sized neural prosthetic implants. *IEEE Transactions on Biomedical Engineering, 35*(5), 323–327.
28. Von Arx, J. A., Najafi, K. (1999, June). A wireless single-chip telemetry-powered neural stimulation system. *1999 IEEE international solid-state circuits conference digest of technical paper ISSCC first Ed* (cat No99CH36278) (pp. 214–5). San Francisco, CA, USA
29. Neagu, C. R., Jansen, H. V., Smith, A., Gardeniers, J. G. E., & Elwenspoek, M. C. (1997). Characterization of a planar microcoil for implantable microsystems. *Sensors and Actuators A: Physical, 62*(1–3), 599–611.
30. Ali H, Ahmad TJ, Khan SA. 2009. Inductive link design for medical implants. In: *IEEE symposium on industrial electronics and applications* (pp. 694–699). Kuala Lumpur.
31. Li, X., Tsui, C. Y., & Ki, W. H. (2015). A 13.56 MHz wireless power transfer system with reconfigurable resonant regulating rectifier and wireless power control for implantable medical devices. *IEEE Journal of Solid-State Circuits, 50*(4), 978–989.
32. Jow, U., & Ghovanloo, M. (2008). Design and optimization of printed spiral coils for efficient transcutaneous inductive power transmission. *Optimization, 1*(3), 193–202.
33. Parramon, J., Doguet, P., Marin, D., Verleyssen, M., Munoz, R., Leija, L., & Valderrama, E. (1997). ASIC-based batteryless implantable telemetry microsystem for recording purposes. In: *Engineering in medicine and biology society, 1997 proceedings of the 19th annual international conference of the IEEE* (pp. 1–2225). Chicago, IL, USA

34. Monti, G., Tarricone, L., & Trane, C. (2012). Experimental characterization of a 434 MHz wireless energy link for medical applications. *Progress in Electromagnetics Research C, 30*(June), 53–64.
35. Lee, S. Y., Su, M. Y., Liang, M. C., Chen, Y. Y., Hsieh, C. H., Yang, C. M., Lai, H. Y., Lin, J. W., & Fang, Q. (2011). A programmable implantable microstimulator soc with wireless telemetry: Application in closed-loop endocardial stimulation for cardiac pacemaker. *IEEE Transactions on Biomedical Circuits and Systems., 5*, 511–522.
36. Lee, H. M., & Ghovanloo, M. (2013). A power-efficient wireless capacitor charging system through an inductive link. *IEEE Transactions Circuits Syst II Express Briefs., 60*(10), 707–711.
37. Abiri, P., Abiri, A., Packard, R. R. S., Ding, Y., Yousefi, A., Ma, J., Bersohn, M., Nguyen, K.-L., Markovic, D., Moloudi, S., & Hsiai, T. K. (2017). Inductively powered wireless pacing via a miniature pacemaker and remote stimulation control system. *Scientific Reports, 7*(1), 6180.
38. Kurs, A., Karalis, A., Moffatt, R., Joannopoulos, J. D., Fisher, P., & Soljacic, M. (2007). Wireless power transfer via strongly coupled magnetic resonances. *Science, 317*(5834), 83–86.
39. Ramrakhyani, A. K., Mirabbasi, S., & Chiao, M. (2011). Design and optimization of resonance-based efficient wireless power delivery systems for biomedical implants. *IEEE Transactions on Biomedical Circuits and Systems, 5*(1), 48–63.
40. Lee, B., Kiani, M., & Ghovanloo, M. (2016). A triple-loop inductive power transmission system for biomedical applications. *IEEE Transactions on Biomedical Circuits and Systems, 10*(1), 138–148.
41. Kiani, M., Jow, U. M., & Ghovanloo, M. (2011). Design and optimization of a 3 coil inductive link for efficient wireless power transmission. *IEEE Transactions on Biomedical Circuits and Systems, 5*(6), 579–591.
42. Zhong, W., Lee, C. K., & Ron Hui, S. Y. (2013). General analysis on the use of tesla's resonators in domino forms for wireless power transfer. *IEEE Transactions on Industrial Electronics, 60*(1), 261–270.
43. Pozar, D. M. (2009). Microwave engineering. *Wiley*. https://www.wiley.com/en-us/Microwave+Engineering%2C+4th+Edition-p-9780470631553.
44. Nikoletseas, S., Yang, Y., & Georgiadis, A. (2016). *Wireless power transfer algorithms, technologies and applications in ad Hoc communication networks*. Wireless Power Transfer Algorithms, Technologies and Applications in Ad Hoc Communication Networks. https://www.springer.com/gp/book/9783319468099
45. Zeng, Y., Clerckx, B., & Zhang, R. (2017). Communications and signals design for wireless power transmission. *IEEE Transactions on Communications, 65*(5), 2264–2290
46. Brown, W. (1969). *Microwave to DC converter*. US Patent 3,434,678.
47. Bakogianni, S., & Koulouridis, S. (2015). Sub-1 GHz far-field powering of implantable medical devices: Design and safety considerations. In: *IEEE Antennas and Propagation Society, AP-S International Symposium (Digest)* (pp. 942–943). Vancouver, BC, Canada
48. Liu, C., Guo, Y. X., Sun, H., & Xiao, S. (2014). Design and safety considerations of an implantable rectenna for far-field wireless power transfer. *IEEE Transactions on Antennas and Propagation, 62*(11), 5798–5806.
49. Sun, Y., Greet, B., Burkland, D., John, M., Razavi, M., & Babakhani, A. (2017). Wirelessly powered implantable pacemaker with on-chip antenna. In: *IEEE MTT-S international microwave symposium digest* (pp. 1242–1244). Honololu, HI, USA
50. Ho, J. S., Yeh, A. J., Neofytou, E., Kim, S., Tanabe, Y., Patlolla, B., Beygui, R. E., & Poon, A. S. Y. (2014). Wireless power transfer to deep-tissue microimplants. *Proceedings of the National Academy of Sciences of the United States of America, 111*(22), 7974–7979.
51. Kim, S., Ho, J. S., Chen, L. Y., & Poon, A. S. Y. (2012). Wireless power transfer to a cardiac implant. *Applied Physics Letters, 101*(7), 1–5.

Electrocardiogram: Acquisition and Analysis for Biological Investigations and Health Monitoring

Tai Le, Isaac Clark, Joseph Fortunato, Manuja Sharma, Xiaolei Xu, Tzung K. Hsiai, and Hung Cao

1 Introduction

1.1 Background

Cardiovascular disease (CVD) is the leading cause of morbidity and mortality in the developed world and accounted for more than nine million deaths in 2016 [1]. Adult mammalian ventricular cardiomyocytes (CMs) have almost zero capacity to divide, and thus the proliferation is insufficient to overcome the significant loss of myocardium from ischemia-induced infarct [2–4]. Cardiac arrhythmic disease alone contributed about 350,000 deaths annually in the USA. Although causative genes for some syndromes such as long QT syndrome, catecholaminergic ventricular tachycardia, and Brugada syndrome have been partially discovered [5], genetic basis for the majority of cardiac arrhythmia remains poorly understood [6, 7]. Most known causative genes were discovered through human genetic studies of patient

T. Le · I. Clark · J. Fortunato
Electrical Engineering and Computer Science, University of California Irvine, Irvine, CA, USA

M. Sharma
Electrical Engineering, University of Washington, Seattle, WA, USA

X. Xu
Cardiology, Mayo Clinic, Rochester, MN, USA

T. K. Hsiai
Cardiology, University of California Los Angeles, Los Angeles, CA, USA

H. Cao (✉)
Electrical Engineering and Computer Science, University of California Irvine, Irvine, CA, USA

Biomedical Engineering, University of California Irvine, Irvine, CA, USA
e-mail: hungcao@uci.edu

H. Cao et al. (eds.), *Interfacing Bioelectronics and Biomedical Sensing*,
https://doi.org/10.1007/978-3-030-34467-2_5

cohorts. The advent of next-generation sequencing technology enabled more efficient approaches such as the genome-wide association studies (GWAS) [8, 9]. However, because of the statistics nature of the GWAS approach, it is difficult to assign unambiguous causality to candidate loci [10].

Further, many CVDs may occur due to an inherited disorder or environmental factors such as infections or drug misuse from parents [11]. Heart defects are among the most common birth defects and the leading cause of birth defect-related mortality [12]. Annually, about 1% of newborns have some form of congenital heart disease (CHD) [12] and that rate has been increasing epidemically. According to a thorough investigation among infants of Gulf War Veterans (GWVs), it was reported that infants conceived postwar to male GWVs had a significantly higher prevalence of tricuspid valve insufficiency and aortic valve stenosis compared to infants conceived postwar to Non-Deployed Veterans (NDV) males [13]. Another case-control study revealed that the rate of congenital anomalies in Vietnam veterans' children was high, attributing to the Agent Orange (or dioxin) exposure of their parent [14]. Since transcriptome and epigenome modifications can be inherited transgenerationally [15], it motivates the question whether substance abuse or toxin exposure of the modern lifestyle are the underlying reasons via transgenerational epigenetic inheritance? The answer may provide insights into the understanding and early detection of CHD caused by epigenetic factors from parental exposure and open a new road map to discover therapeutic potentials.

1.2 The Studied Animal Model: Zebrafish

Unlike adult mammals, some fish and amphibians possess cardiac regeneration capacity throughout their adult life. Zebrafish (*Danio rerio*) hearts fully recover after 20% ventricular amputation [16, 17], thereby providing a pivotal model system for heart regeneration studies [18, 19]. Recently, Porrello and colleagues discovered the transient cardiac regenerating capacity of 1-day-old neonatal mice, but it is lost 7 days after birth [20]. The insight of mammalian hearts as having virtually no regenerative capacity is now revised by recent animal and human studies, in which scientists found new CMs may arise from existing CMs and progenitor or stem cells [21–25]. The discovery of specific genes' roles toward myocardium regeneration in studies with animal models would suggest methods to activate limited regeneration capacity in the human heart, potentially providing novel cardiac therapies [26].

The power of a forward genetic approach in discovering new genes has already been proven in lower model organisms such as yeast, *Drosophila*, and *C. elegans* [27–29]. However, because of the high demand on colony management efforts, this approach is very difficult in vertebrates, especially on finding genes for adult phenotypes such as cardiac arrhythmia. While rodents have been widely used owing to several advantages, such as the ease to transport, breed, provide for, and develop as well as the high degree of genome similarities with humans [30], they are expensive to maintain and to use in large-scale genetic or chemical screens. Further, they are not readily assessable to studies at the embryonic stages. The zebrafish model has been used to elucidate various aspects of gene function that can be directly related to human genetics and diseases due to their highly conserved genome [31]. Additional advantages include small size, low cost for maintenance, short generation time, and optical transparency. Zebrafish have long been used as model system for understanding human cardiac development and disease. The zebrafish model also provides a forward genetic approach to reveal the genetic basis underlying molecular mechanisms of numerous heart diseases [32–34]. In brief, zebrafish have been widely used to understand mechanisms of cardiac disease and regeneration [16, 23], complex neurobehavioral phenotypes [35], to study epigenetics of drug abuse [36] and to investigate the transgenerational inheritance of neurological and cardiac disorders induced by parental exposure [15, 37].

1.3 Electrocardiogram

For humans, the measurement of ECG has been in existence for over a century as the method to diagnose the heart's performance as well as one of the key indicators in vital sign monitoring. ECG signals with P waves, QRS complexes, and T waves contain the detailed performance of one's entire heart. In fact, monitoring of ECG provides essential information for diagnosis, prognosis, and early detection/prevention of heart disease. Nevertheless, it has been manually analyzed subjectively by caregivers; thus only extreme cases can be found.

For zebrafish, the use of ECG is still in the early stage, and our team has been pioneering the efforts [31, 38, 39]. In 2004, Forouhar et al. reported an ECG signals of an embryonic zebrafish, revealing similar atrial and ventricular electrical signals as found in a human ECG but encountering challenges in detecting the T wave [40]. Through the use of oral perfusion and injected muscle paralytics, MacRae et al. obtained ECGs of an adult zebrafish with a distinct P wave and QRS complex [41]. In the past several years, we have demonstrated ECG acquisition in zebrafish using flexible microelectrode arrays (MEA), revealing distinct P waves, QRS complexes, and T waves resembling that of humans (Fig. 1) [39, 42–44]. Besides, the average heart rate of humans is ~60–70 beats per minute (bpm), zebrafish's is ~100–120 bpm, and that of rodents could be several hundred bpm, rendering an evidence that findings in cardio-studies using zebrafish would be more proper to translate for humans.

1.4 The Structure of This Chapter

In the next sections of this chapter, the nature of ECG (II) will be followed by methods for ECG acquisition (III); ECG acquisition, processing, and analysis in zebrafish (IV); ECG monitoring by wearable devices in humans (V); discussion and outlook (VI); and conclusion (VII).

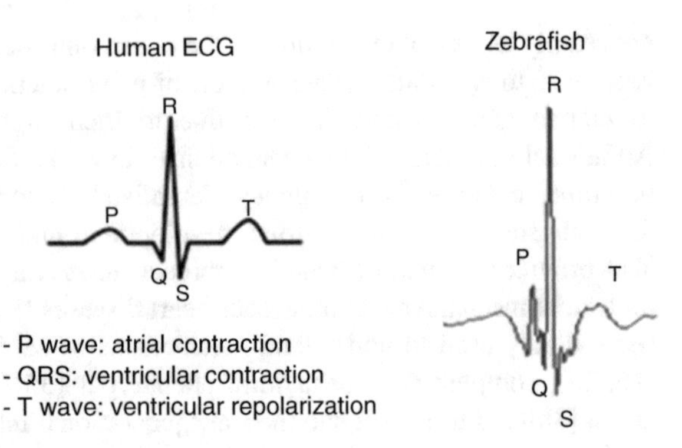

Fig. 1 Comparison between human and zebrafish ECG signals. (Adapted from Ref. [39])

2 The Nature of Electrocardiogram (ECG)

In the heart, there are cardiac muscle cells, namely, cardiomyocytes and cardiac pacemaker cells. Together they synchronously function to make the electrical and mechanical coupling, resulting in heart contraction.

2.1 *Pacemaker Action Potential*

Initiation of each contraction within the heart is driven by the cardiac pacemaker cells. These pacemaker cells are responsible for generating a constant cycle of action potentials (AP) that depolarize and repolarize the cell membrane potentials. Once the signal is generated, it then propagates through gap junctions that connect the cardiac cells and allow for the flow of ions between them [44]. Unlike cardiomyocytes, pacemaker cells do not have a resting membrane potential. Pacemaker cells are in constant flux due to the opening and closing of voltage-gated ion channels. The channels that control the cycle are the potassium voltage-gated channels, calcium voltage-gated channels, and sodium channels. The sodium channels are responsible for the non-resting membrane potential and initiate each AP. Their AP can be split into three phases (4, 0, and 3) seen in Fig. 2 [45, 46].

Phase 4: The cell begins each AP at around −60 mV. The sodium channels slowly cause the membrane potential to rise via an influx of Na^+.

Phase 0: When depolarization of the cell has reached a threshold of −40 mV, the voltage-gated calcium channels open, resulting in an influx of Ca^{++} that more rapidly increases the membrane potential.

Phase 3: As the membrane potential approaches +20 mV, Ca^{++} channels close and K^+ channels open. This results in release of K^+ and causes rapid decline in membrane potential. When the decrease reaches −60 mV the K^+ channels will close and the cycle repeats [47].

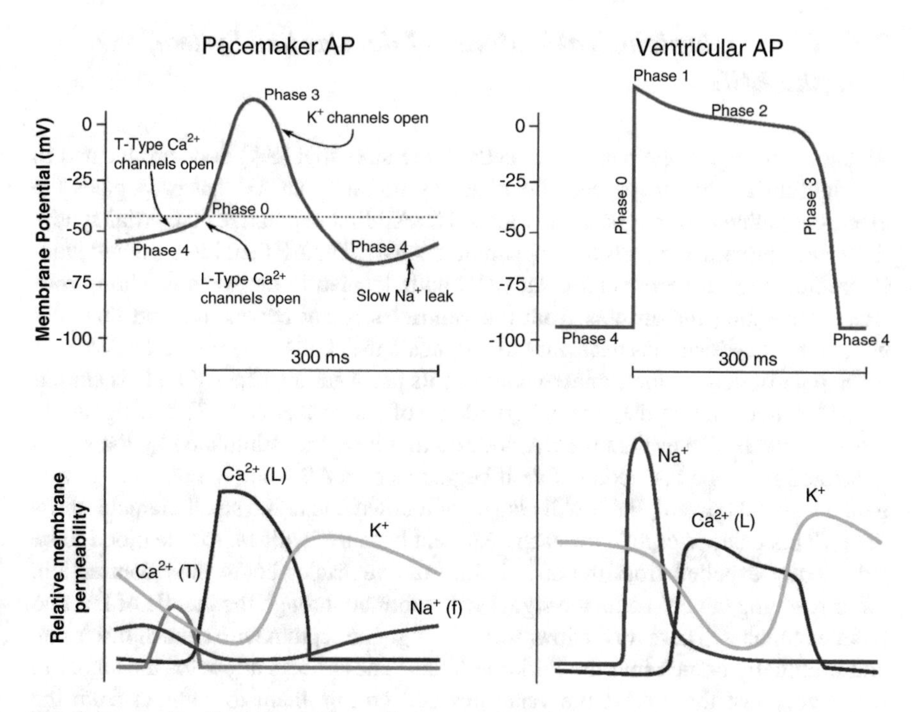

Fig. 2 Pacemaker AP & ventricular cardiomyocyte AP [45]

2.2 *Cardiomyocyte Action Potential*

Unlike pacemaker cells, cardiomyocytes have a resting membrane potential of roughly −90 mV. Their unique AP can be split into five phases (0, 1, 2, 3, and 4) as can be seen in Fig. 2.

Phase 0: A depolarization wave comes into the cell through the gap junctions. When the membrane potential reaches a threshold of roughly −70 mV, the voltage-gated Na+ channels open. Na+ will enter the cell driving the membrane potential up at a rapid rate.

Phase 1: Shortly after the membrane potential reaches +20 mV, the voltage-gated Na + channels will close, preventing the continued flow of Na+, and K+ voltage-gated channels will open. K+ exits the cell and reduces membrane potential.

Phase 2: At around +5 mV, Ca++ voltage-gated channels will open allowing Ca++ to flow into of the cell balancing the effects of K+ entering the cell and causing the membrane potential to plateau at +5 mV.

Phase 3: Shortly after opening, the Ca++ voltage-gated channels will close. With Ca++ no longer flowing into the cell, the membrane potential will continue to fall due to the outflux of K+.

Phase 4: Eventually the cell will reach an electrochemical equilibrium of −90 mV and the resting membrane potential of the cell [47].

2.3 Electrophysiological Pathway of the Cardiac System, the ECG

As shown in Fig. 3, the pacemaker cells in the sinoatrial (SA) node are located in the right atrium and begin each heartbeat by initiating an AP that propagates the electrical pathway of the heart (P wave). The AP first depolarizes the atrial syncytium and causes atriums to contract simultaneously. The AP then travels down junctional fibers toward atrioventricular (AV) node located in the septum. The fibrous tissue separating the atriums from the ventricles is not conductive and does not contain the gap junctions necessary to propagate the signal. Therefore, the AV node is the only option for the signal to continue its propagation. The AV node is similar to the SA node in that they are both made up of pacemaker cells. The firing rate in the AV node is slower than the SA node so that it will be stimulated by the signal generated from the SA node before it begins its own AP. The AV node will propagate the signal onward, but it will also slow it down due to the small diameter of its fibers. This delay is crucial for each heartbeat because it allows for the blood to be sufficiently expelled from the atria before the ventricles begin their contraction. After reaching the AV node, the signal will continue through the bundle of His and bundle branches. These travel downward through the septum with a much faster rate and eventually branch into the Purkinje fibers. These fibers allow for the signal to travel very fast throughout the ventricles and prompt them to contract from the bottom-up. In combination with the whirl-like structure of the cardiac muscle fibers, this results in twisting contraction motion that forces the blood out of the ventricles and into the aorta and pulmonary trunk [44]. This entire phase represents the QRS complex of an ECG. Later stages of the ECG signal represent the repolarization phase of the ventricles (T waves) [48].

In summary, the ECG signal is assessed with three primary phases: the P wave, the QRS complex, and the T wave. Each of these components corresponds to a specific phase in the heart rhythm. The P wave occurs when the SA node initiates the signal, depolarizes the atria, and appears as a small bump raising the voltage difference. The delay between the P wave and the QRS complex is due to the AV node slowing the signal down before it reaches the ventricles. The QRS complex itself corresponds to the depolarization of the ventricles. The QRS complex is much larger than the P wave due to the ventricles having greater mass than the atriums. It appears as a dramatic spike and fall corresponding to the activity in the ventricles. At the same time the QRS complex takes place, the atriums depolarize as well. This is somewhat obscured however, due to the greater electrical activity taking place in the ventricles. The final portion of the signal, the T wave, occurs when the ventricles repolarize [44]. As seen in the bottom left of Fig. 3, considering the heart as a dipole source, acquisition of ECG via direct contact electrodes would be dependent of the locations and direction, and that's the reason for us to have the 12-lead ECG as the gold standard for humans in clinical settings.

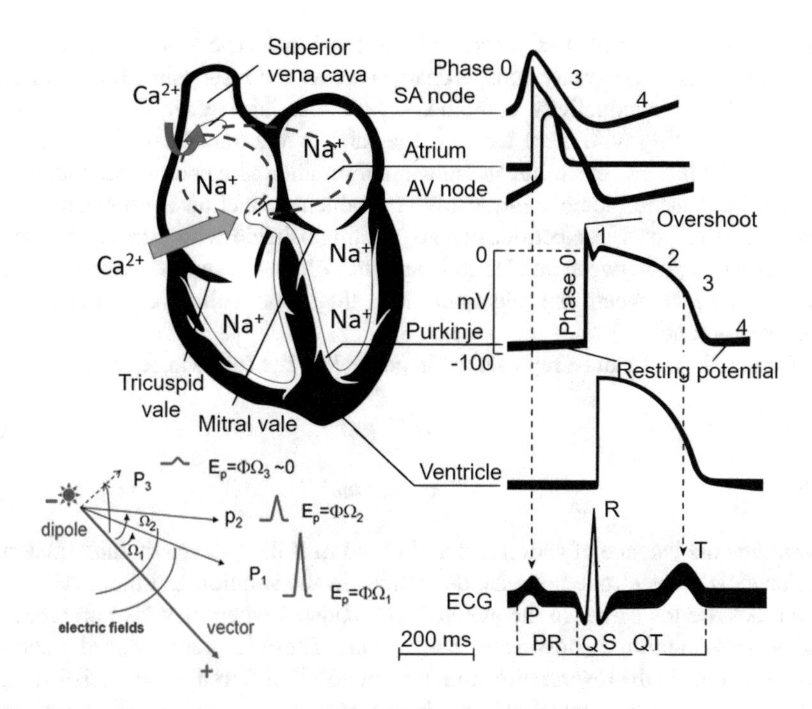

Fig. 3 Electrical pathway and signals of the heart resulting in ECG and vector directions from the heart considered as a dipole source (bottom left). E_p denotes electric potential, Ω is the angle, and Φ is the strength of charge surface. The lower left panel is extracted from [39]

3 ECG Acquisition Methods

It is well-known that the ECG signals can be captured using either a two-electrode or a three-electrode configuration to amplify the differential in potentials between two electrode locations. Such a system includes amplifiers and filters to get the desired outputs [49, 50]. The acquired signals are often further improved using digital filters.

3.1 Contact Electrode

3.1.1 Wet Electrode

Wet electrodes are most widely used to measure human biopotentials, such as ECG, electroencephalogram (EEG), and electromyogram (EMG), to name a few, owing to the great adhesion to skin and thus minimizing the electrode-tissue impedance. Using wet electrode, ECG signals with high signal-to-noise ratio (SNR) can be achieved. Usually, Ag/AgCl electrodes with wet conductive gels are used due to their potential stability. The process of sensing bioelectric signals was carefully described in [51]. Generally, ions as charge carriers are responsible for the mechanism of electric conductivity in the body. Therefore, bioelectric signals measured involve interacting with these ionic charge carriers and transducing ionic currents throughout wires and electronic instrumentation. When an electrical conductor contacts to the aqueous ionic solutions, the transducing function is established. This mechanism makes wet electrodes clearly distinguishable with others. In order to transfer a charge between an electrode and the solution, a reaction would occur at the interface between the electrode and the ionic solution, namely, redox (oxidation-reduction).

These reactions can be represented in general by the following equations:

$$C \rightleftharpoons C^{n+} + ne^{-} \quad (1)$$

$$A^{m-} \rightleftharpoons A + me^{-} \quad (2)$$

where n is the valence of cation material *C* and m is the valence of anion material, *A*. For most of electrode systems, the cations in the solution and the metal of the electrodes are the same, so the atoms *C* are oxidized when they give up electrons and go into solution as positively charged ions. These ions are reduced when the process occurs in the reverse direction. It is found that different characteristic potentials occur for different materials, and the Ag/AgCl electrode is one that has characteristic like a perfectly nonpolarizable electrode, which is practical for use in many biomedical applications.

To analyze the performance of wet electrodes in general and Ag/AgCl electrodes in particular, several electrical models were described [52, 53]. Chi et al. comprehensively described various types of electrodes including wet Ag/AgCl electrodes, seen in Fig. 4 [53]. Between the cardiac activity and the input of the circuit, there are different layers, and each is equivalent to a RC parallel component. Therefore, the transducing function can be calculated by the equivalent skin-electrode model. The gel layer acts as a junction between the electrode and skin, which helps reduce the skin-electrode impedance, consequently enhancing the biopotential signal SNR. Several studies [49, 54–57] were conducted to investigate alternative options such as dry contact and noncontact electrodes, using the wet Ag/AgCl electrodes as a benchmark to compare the performance. Though possessing their advantages, such as simplicity, ease of use, providing favorable SNR, and being disposable [58], wet/gel electrodes also have drawbacks. For instance, the gel could cause allergic reactions or skin irritation [49, 58]. Further, they also have a limited shelf life due to dehydration, which affects impedance, thus generating noise. Finally, the dehydration issues make these electrodes unsuitable for long-term continuous measurement [49].

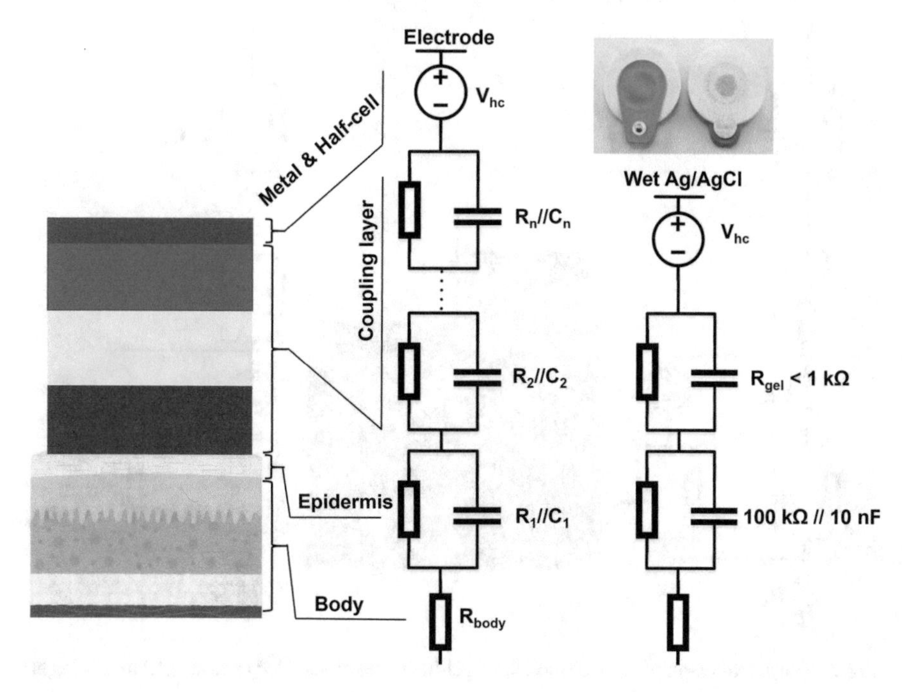

Fig. 4 The equivalent model of the skin-electrode interface of gel-based electrodes

3.1.2 Dry Electrode

The definition of dry electrodes is controversial. Dry electrodes can be classified either as spiky, capacitive, noncontact, or other heterogeneous ones, which refer to all gel-less cases or as direct contact but gel-less [59]. Here, we categorize dry electrodes in the group of contact electrodes but gel-less. Generally, in dry electrode configurations, skin sweat plays the role as the medium to conduct biopotential signals. When the dry electrode touches the skin directly, it may provide both galvanic and capacitive electrical current pathways between the skin surface and electrode. Different from wet electrodes, dry electrodes are usually reusable. Without an electrolyte, the electrode-skin impedance of dry electrodes is comparably higher than that of wet electrodes. Besides, the gap between metal and skin results in some significant capacitive effects. Figure 5 illustrates the equivalent circuit of dry electrodes in comparison with wet electrodes [60].

It is seen that the circuit model of the electrolyte is described as a pure resistor, yet that of the dry electrode-skin interface is modeled as an RC circuit, indicating the skin-electrode impedance of dry electrodes is higher than their counterparts', possibly resulting in lower SNR. However, dry electrodes outperform in terms of long-term monitoring as the interface impedance is more stable over time in many practical scenarios [54].

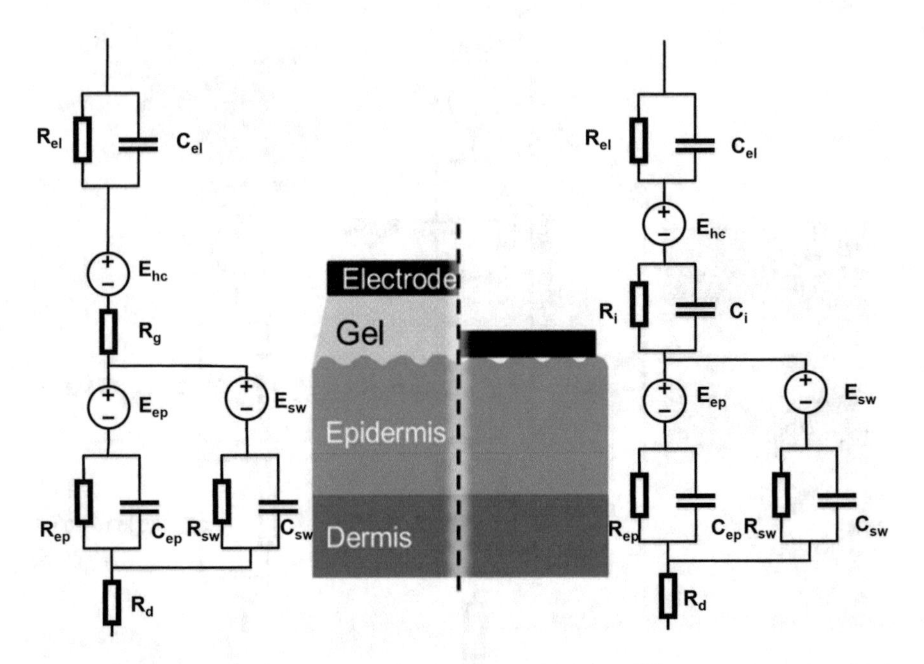

Fig. 5 The electrode-skin interface models of gel-based electrodes (left) and dry electrodes (right)

In order to improve the skin-electrode interface issue, several ways have been applied. First, some wait time (at least 20 min) or reverse iontophoresis [61] can be applied to induce sweat; thus a "self-electrolyte" is initiated. Second, the outermost layer of the skin, namely, stratum corneum, can be removed/scratched [62]. Third, the shape of dry electrodes can be modified to be spiky by micro−/nanofabrication; thus microneedles on the surface of the electrode can penetrate through the skin in a painless way [63]. However, these methods could be inconsistent and may still make users uncomfortable.

3.2 Noncontact Electrodes (NCE)

Noncontact electrodes (NCEs) are promising as it can overcome the drawbacks of contact electrodes in users' comfort and inconsistency. Furthermore, as they are "noncontact", it holds potentials to integrate them inside garment, making them attractive for next-generation wearable devices. NCEs operate based on displacement currents "going" through the capacitor formed by themselves and the body. In order to make the configuration, we can simply put an insulation layer between the electrode and skin. The equivalent circuit of NCEs is shown in Fig. 6 [59].

The insulated NCE and skin are akin to the two parallel plates of a capacitor and the cloth (or insulating polymer) acts as the dielectric. The coupling capacitor in this case is usually very low, in the range of 1–50 pF, resulting in an extremely high

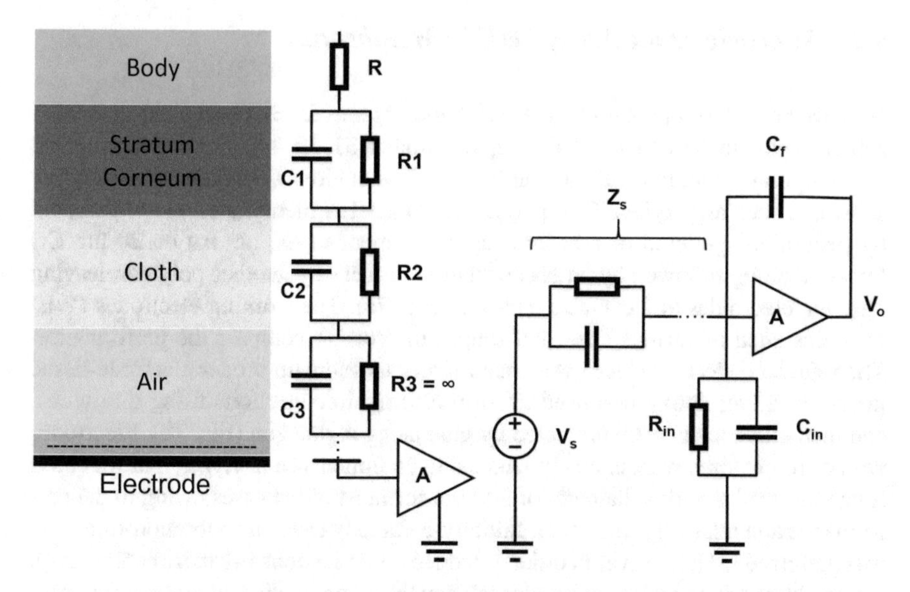

Fig. 6 The equivalent model of the skin-electrode interface and the electronics – analog front end for biopotential acquisition

impedance. Thus, there are several challenges to make the performance of NCEs comparable with wet electrodes. Additionally, due to motional artifacts in practical settings, the gap of the coupling capacitor varies, resulting low SNR in NCE measurement. Therefore, extensive effort in noise reduction for the NCE approach has been carried out and investigated [59].

A significant attempt to enhance the signal quality measured by NCEs was initiated by Chi and colleagues [64, 65]. The team proposed an integrated noncontact sensor front-end amplifier with fully bootstraps internal and external parasitic impedances and their device outperformed INA 116 (ultralow bias current IC from *Texas Instrument*). They also presented an input capacitance cancellation method [64] so that the stability of gain amplifier was preserved when different gaps between the signal source and electrode were employed. In our group, we have been focusing in securing the sensor patch [66] in special cases, and lately we attempted to include an accelerometer to detect the motion, and thus it can be filtered out from the acquired ECG.

4 ECG Acquisition, Processing, and Analysis in Zebrafish

There are a number of challenges for ECG acquisition in zebrafish due to the small size and the aquatic environment. An adult zebrafish is as small as a pinky of a child; thus it would be difficult to apply the NCE approach as the coupling capacitance will be extremely low. For contact electrodes, gel is not needed owing to the nature of the fish skin.

4.1 Microelectrode Array (MEA) Membranes

We have been developing 4-channel MEA membranes for ECG acquisition in adult zebrafish akin to the 12-lead ECG setup in humans [31, 39, 44], yielding high spatial and temporal resolution with favorable SNR. The devices were based on a polymer substrate, such as parylene C or polyimide. The MEA membranes were fabricated by patterning sputtered or e-beam evaporated metals (200 nm Au on 20 nm Cr) by wet etching followed by an encapsulation process by another polymer, leaving only the electrodes and contacts exposed (Fig. 7a). The working electrodes (WE) were designed in various sizes and shapes in order to compare the performance. The reference electrode (RE) was much larger to maintain proper electrode-tissue interface. Silver epoxy was used to form electrical connections using thin wires, and then all contacts were protected by glue using a glue gun (Fig. 7b). Electrodes with different sizes were characterized using an impedance analyzer, and we chose round electrodes with a diameter of 300 μm for most of the cases owing to the performance and relatively small size. Initially, a Faraday cage on a vibration-free table was preferred [31]; however in order to reduce complications in experimental setup, we conducted experiments in regular lab benches then applied rigorous signal processing to enhance SNR using wavelet filtering and the thresholding technique [31, 39, 44]. Using our MEA membranes and systems, we were able to distinguish the cryo-injured and control hearts via ECG analysis (Fig. 8a). Further, site-specific ECGs were revealed similar to the case in humans with 12-lead ECG (Fig. 8b). As can be seen in Fig. 8b, the electrode D was positioned close to the injury site, and ST depression indicating ventricular injury was obvious while that obtained at electrode A (on the other side of the heart) showed normal ECG features [44].

Our MEA membranes have provided a novel tool to study the zebrafish heart; however measurement was done with fish under anesthesia (Fig. 7d), indicating the obtained ECG signals were not intrinsic. Further, the manual one-by-one intermittent measurement is cumbersome and fails to provide (1) high throughput and (2) continuous ECG recording over a long period of time, making it not suitable for a host of biological studies. We have been recently developing wireless ECG jackets which ultimately can provide ECG monitoring of multiple freely swimming fish [67, 68]. In this system, a customized cylinder-shape housing was used; thus a big solenoid transmitter coil (TX) can be position (Fig. 9). A tiny solenoidal coil (RX) can be worn by the fish, and an inductive coupling wireless power transfer (WPT)

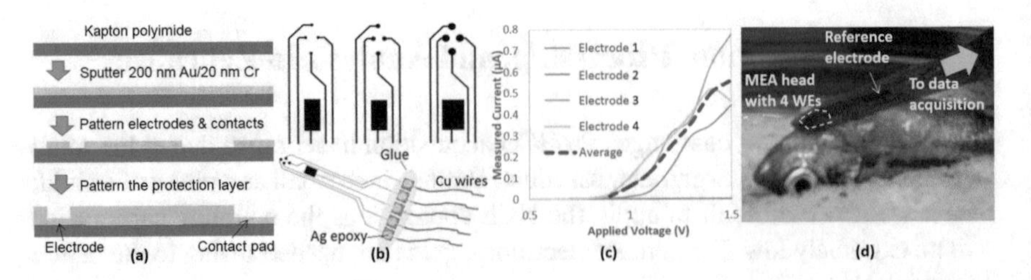

Fig. 7 (**a**) Fabrication processes of the MEA membranes; (**b**) different electrode sizes and the complete device; (**c**) impedance characterization curves of one 300-μm MEA membrane; (**d**) the MEA on fish under anesthesia. (Taken from Ref. [31])

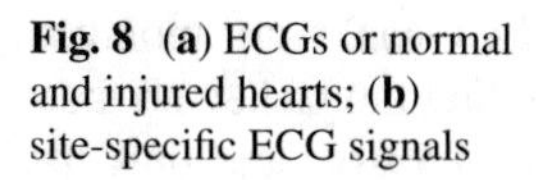

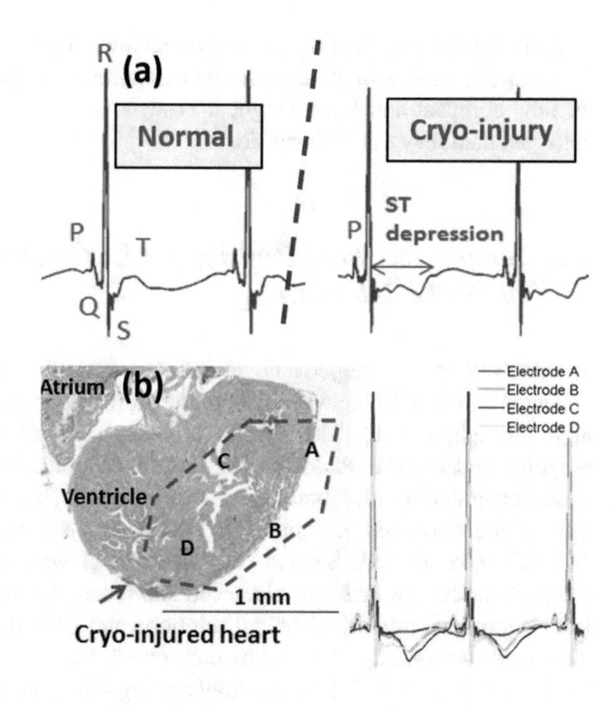

Fig. 8 (**a**) ECGs or normal and injured hearts; (**b**) site-specific ECG signals

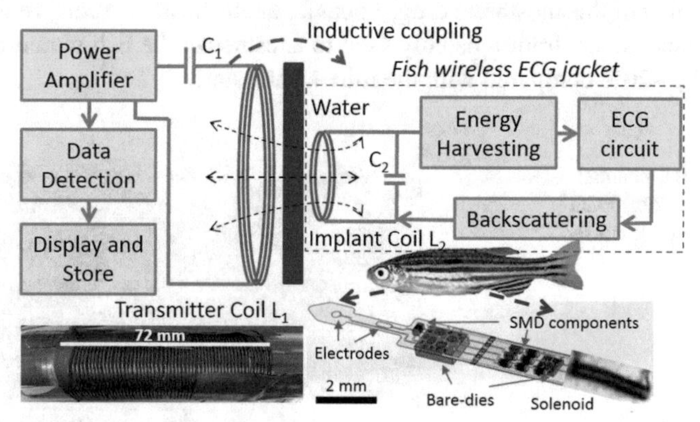

Fig. 9 Block diagram and design WPT and data communication with TX and RX as solenoids [67]

could be achieved. The MEA membrane was modified to house electronic components and the RX. The acquired ECG was sent back via backscattering mechanism [69]. We have been trying with the RX as a regular solenoid (1-mm diameter, 10-mm long) and an L-shape solenoid, and the TX, obtaining >500 μW in most of the cases, in both air and water environment [67, 70]. Based on our calculation, that would guarantee sufficient power to operate at least a one-channel ECG jacket using discrete electronic components. With an integrated chip designed and implemented, multiple-channel ECG jackets can be achieved. In our latest attempt [68], we tested with a simulated ECG and the signal was successfully sent back via the inductive link.

Although we are close to our goal of having a wireless ECG jacket for monitoring of awake adult zebrafish, we have an additional concern that the jacket, no matter how compact and light-weight it is, would still irritate the fish, and thus the collected data may not be intrinsic.

4.2 Simple-Yet-Novel Housing for ECG Measurement of Awake Zebrafish

We have recently developed a novel system, with which data can be simultaneously collected from multiple awake fish, thereby providing intrinsic zebrafish ECG with high throughput [31]. The system includes multiple small housings made of polydimethylsiloxane (PDMS) using 3D-printed molds (Fig. 10). Flexible electrodes were integrated in the bottom of the apparatus, providing comfort and thus minimizing unwanted artifacts. First, the housings were filled with system water and fish were loaded. Second, the water level was securely reduced thus the electrode-body interface was enhanced and recording could be started. The awake ECG comprised artifacts from gill motions and even the beating heart; however, they were successfully removed by our powerful de-noising technique using wavelet filtering (Fig. 10) [44]. Interestingly, using this system, we were able to characterize the effect of the anesthetic drug, Tricaine, on the heart rhythm. We found 50% of the Tricaine concentration usually used to anesthetize the fish would reduce the heart rate by ~20% compared with the fully awake state.

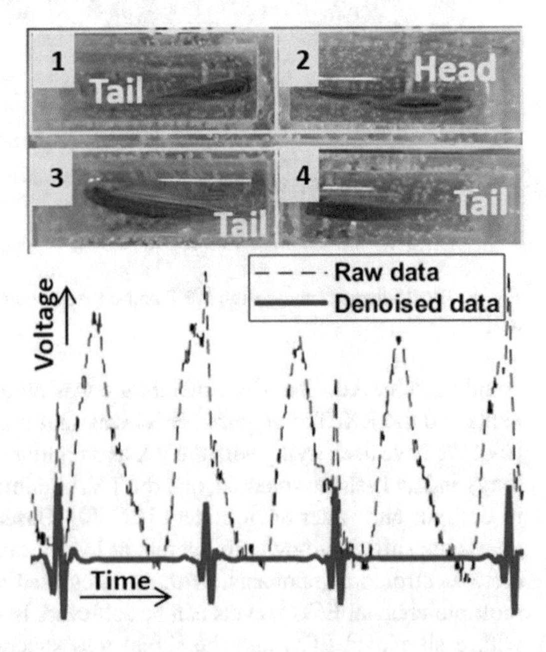

Fig. 10 (Top) 4-channel apparatus. (Bottom) ECG signals, raw and de-noised

4.3 Zebrafish ECG Analysis

The processing and analysis schemes are currently done offline and manually, making it impossible to study large-scale data with hundreds of subjects. As we have been delivering great strides on sensing and acquisition systems, there is a dire need to establish a universal platform to collect data with automated real-time processing and analyses. Recently, we have built a universal graphical user interface (GUI) aiming for those. The program was developed using LabView (*National Instruments*, Austin, Texas) in order to exploit existing powerful features, such as multi-channel data acquisition, filtering, peak detection, and capabilities for integration and expansion. The GUI can be translated to an execution file, which would be easily used by non-tech personnel. The GUI can be used in real-time as well as offline mode to analyze saved data. Further, we also implemented this GUI to be customizable; thus users can change settings for specific projects, and we are aiming to connect this GUI with a secure cloud-server. In the GUI, features of P, Q, R, S, and T waves can be detected. Additionally, the intervals (such as RR, PR, QT, etc.) and amplitudes are also calculated and stored, and some analysis is displayed. In the first generation, our GUI was able to detect several important arrhythmic features. As an example, Fig. 11 shows the *sinus arrest* pattern was detected in the ECG obtained from a fish mutant.

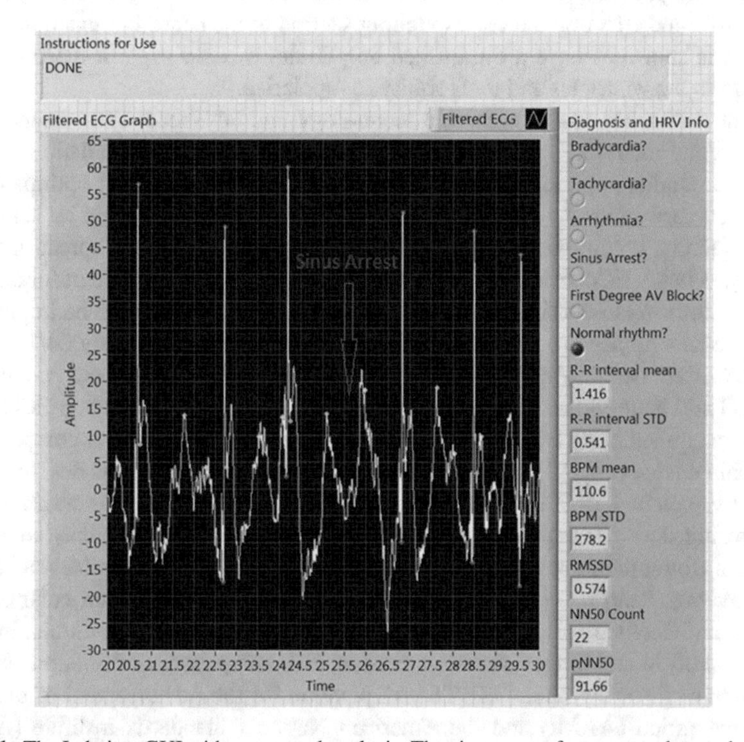

Fig. 11 The Labview GUI with automated analysis. The sinus arrest feature was detected

5 ECG monitoring in humans

For humans, ECG monitoring has been used for decades in clinical settings with 12-lead ECG as the gold standard. It is used for vital sign monitoring as well as for cardiac event diagnoses. Trained medical personnel have experience in identifying diseases and conditions based on ECG signals, but even some untrained personnel can identify when an ECG signal appears abnormal. The standard ECG signal has a relatively constant frequency and amplitude for each of its distinct waves. While each patient will likely have slight variations in their ECG signal, a healthy heart will tend to produce consistent signals that resemble the standard ECG. Major differences from the standard can therefore be detected even by untrained personnel who can then inform the trained personnel of the changes.

It is not exactly practical, however, to have personnel constantly monitoring a patient's ECG signal. As such, computers are often used to detect significant changes in the signal. If significant changes in frequency or amplitude are detected, the computer will notify medical personnel. Computer analysis of ECG signals is often used in research as well. Computer programs can be written to average signals to establish a norm for a particular subject and then compare that norm to the subject's ECG after some procedure has been done. Researchers can then learn what kind of effect the procedure had on the heart based on the differences between the signal before and after the procedure. Changes to specific waves can often suggest that the procedure had some effect on the corresponding section of the heart. For example, if the T wave is shown to have a diminished amplitude, it could suggest there is some damage to the ventricles and their ability to repolarize.

With the recent rise of wearable technology and advanced data science techniques, continuous ECG data of numerous patients and healthy populations can be obtained, studied, and analyzed in out-of-clinic settings, garnering optimism that distanced care and early detection of cardiac disease or events could be achieved. Consequently, ubiquitous healthcare systems have been gradually spread into the society. A bulky ECG machine is used to be an indispensable equipment for diagnosis symptoms related to human's cardiac system. However, it would be impossible for patients who need to be observed their heart activity continuously 24/7 in their home. Certain health conditions and clinical studies require continuous monitoring of ECG and battery-operated devices are worn by the patients to obtain ECG. One of the examples is an ECG Holter system. The Holter monitor is a compact and wearable device (Fig. 12) that keeps track of your heart signals. This device can be worn by patient 1 or 2 days depending the doctor's decision. If standard Holter monitoring does not capture your irregular heartbeat, your doctor may suggest a wireless Holter monitor, which can work for weeks [71]. Holter monitors can determine average heart rate (HR) and HR range, quantify atrial and ventricular ectopy counts, and detailed episodes of supraventricular or ventricular arrhythmia, including the shortest and longest duration, burden, HR, and pattern of initiation and termination of the arrhythmia [72]. The Association for the Advancement of Medical Instrumentation (AAMI) and the American National Standards Institute (ANSI) provide the technical standards of various ECG recording devices.

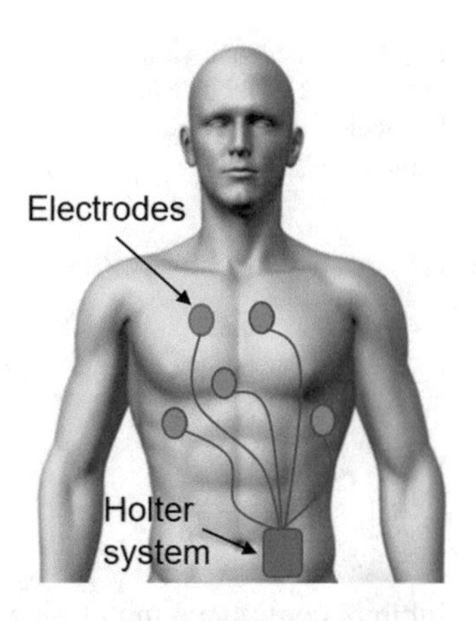

Fig. 12 Holter monitoring

Zeng et al. presented a wearable mobile healthcare system aiming at providing long-term continuous monitoring of vital signs for high-risk cardiovascular patients [73]. The authors aimed to not only monitor ECG signals but also measure respiration and activity. The device comprises a portable patient unit (PPU) and a wearable shirt (WS). Owing to integrating fabric sensors and electrodes endowed with electro-physical properties into the WS, long-term continuous monitoring can be realized without making patients feel uncomfortable and restricting their mobility.

6 Discussion and Outlook

6.1 Current Wearable Technology

The recent advances in smartphone technologies have set a new road map, pushing forward clinical monitoring toward home care and distanced care [74, 75]. Initially, the ECG systems came with bulky data logging hardware, making them obtrusive to use and wear (e.g., *ReadMyHeart*, DailyCare Biomedical Inc.). Smartphones by communicating with the wearables have replaced the storage hardware with cloud-based storage, considerably reducing their sizes and thus making them user-friendly. The "smart" technology revolution has helped develop Bluetooth- or Wi-Fi-enabled wearable devices that interact with smartphone to record heart rate, ECG, respiratory patterns, and other vital health parameters [76]. Apple watch, Fitbit wearables, and others are devices that provide 24-hour monitoring of heart rate and track sleep patterns and other health parameters. Similarly, smartphone-based ECG devices are being designed as well. By connecting to a smartphone, these wearables not only record data but also help in analyzing and interpreting the trend using various algorithms. By moving the power-intensive calculation to the smartphone, low-cost battery-enabled devices are being developed. The influx of Bluetooth-low-energy (BLE) chips in the market has further helped in reducing the power consumption of these wearables. However, the cross-electrode configuration for recording of ECG

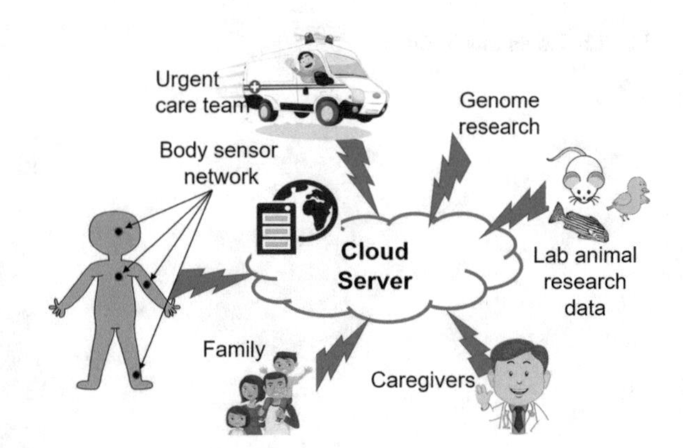

Fig. 13 Conceptual view of the future healthcare and biomedical research system

inhibits continuous monitoring without being unobtrusive. At present, most of the devices prompt user to record ECG by monitoring other vital parameters. For example, *Kardia Band*, a wrist band compatible with Apple watches, continuously monitors heart rate and compares it with user's activity; if there is an anomaly, the user would receive a notification to record ECG [77]. It provides an easy way to record ECG and interprets the ECG as well for normal or abnormal features but doesn't record continuously as required in certain conditions. *AliveCor* is another pocket size by the same company device that provides ECG measurement using smartphone. In our group, we have been investigating same side electrode measurements to provide ECG data, but it requires clinical testing for further validation of noninvasive continuous monitoring of ECG [66]. We have been also extensively exploiting the NCE approach for various applications, such as firefighter ECG monitoring [78] and home-based monitoring of fetal and maternal ECG [50]. We envision once we can overcome the motion-artifact issue in the NCE approach, it will open new doors to bring the technology to a wide range of populations for daily use.

Cloud-based monitoring is crucial in providing m-health services, as caretakers can monitor the database and provide required assistance based on the data reducing patient's physical presence at clinics (Fig. 13) [79]. Automated health monitoring, decision-making, distanced-care smartphones, and embedded-based systems are being used to calculate the various parameters from the ECG signal, for instance, R-R interval and ST interval, to evaluate various heart conditions. This provides automated health monitoring reducing the burden of nurses. ECG interpretation is one of the critical factors in evaluating heart condition, and automation can reduce faulty interpretations and help in better and quicker decision-making.

6.2 Connecting Findings in Animal Research with Diagnosis and Prognosis in Humans

Our team has been trying to translate findings in zebrafish research to human monitoring and investigations of therapeutic potentials. For instance, in the Xu lab at Mayo Clinic, numerous heart disease mutant lines have been produced, and our team has been collecting the abnormal patterns in ECG associated with specific diseases. We are also studying the continuous ECG signals over time in order to find anomalies appeared before cardiac events. This holds promise to achieve early detection of disease and events and eventually may help reduce the mortality rate in humans. As seen in Fig. 13, we envision studies of genomics and pharmacogenomics should be included so that personalized medicine and precision medicine can be achieved for a particular person or a group of population sharing similar genetics. Ultimately, a human would be monitored by multimodal sensing systems continuously 24/7, and all information is uploaded to the central cloud securely with his/her privacy encrypted. The health information would be automatically analyzed in order to diagnose and predict any anomaly. This information could also and only be accessed by professional caregivers and close family members and friends. However, in order to reach that stage, we need the power of advanced data science techniques to transfer, correlate, study, secure, process, and analyze the massive data obtained from multiple sensors, millions of people, and billions of animal subjects.

6.3 The Promise of Artificial Intelligence

Recently, machine learning (ML), as a part of artificial intelligence, has been employed to great effect in a variety of fields [80]. The rise of machine learning to prominence is due in large part to the Internet and evolution of big data which simultaneously increased the need for large-scale data processing techniques while providing the scale necessary for their further refinement and development. This coevolution of ML and big data has led to a further increase in the data creation as companies and researchers look to explore new possibilities. ML is experiencing widespread adoption and shows promise in the field of medicine while posing unique challenges to the clinician and researcher.

IBM estimated that 90% of the world's data have been created in the last 2 years, with 2.5 quintillion bytes of data created every day. That is approximately 2.5 million terabytes or almost ten million 256 gigabyte iPhone X's worth of data. This explosion of computer-available data, aptly termed "Big Data," has helped to spur to widespread adoption of ML. Big Data and ML go hand in hand for two main reasons. First, many ML algorithms require large amounts of data to perform optimally. Second, ML algorithms provide highly accurate and insightful predictive measures when applied to the incredible size and complexity of Big Data, where many standard statistical methods fall short.

The medical field is not immune to the rise of Big Data and ML. The ability of ML algorithms to identify patterns in vast quantities of complex, heterogeneous data has prompted many researchers to begin looking outside of standard data collection channels to improve diagnostic and predictive measures. ML has the potential to be able to mine large numbers of data sources for information, such as insurance claims databases and hospital and clinician databases, to develop better models for prognoses and diagnosis [81]. At the macro-level, the increasing ubiquity of ML and developments in cloud computing, mobile devices, and the

Internet-of-Things (IoT) have helped to spur the creation and coalescence of even more medical-related data. Data-centered consumer products are becoming increasingly common. Wearable devices like the Fitbit and Apple watch allow users to record biometric data throughout the day, and smartphone apps help to track and interpret the collected data, gaining insights about things like sleep, heart rate, and daily activity level. Genetics and genomics have also seen growth in data-centered consumer products. Consumers hoping to gain valuable insights can now pay to have their DNA sequenced by companies like *Helix*, and *23 and Me*, among others. This has allowed these companies to create large databases of DNA as well as the physical attributes of its users. In research and education, large open databases of health-related information have been assembled which allow researchers in a variety of fields to gain access to datasets which they may have otherwise been unable to obtain [82].

ML generally refers to the utilization of certain statistical techniques which identify patterns in large datasets and make predictions about future data. At a high level, there are three main categories of machine learning: supervised learning, unsupervised learning, and semi-supervised learning [83]. Supervised learning requires human beings to provide insights to the machines, such as labeling data; examples of these are many deep neural networks and support vector machines. Unsupervised learning attempts to discover latent features within the data to identify features and make classifications, and example of this would be clustering. Semi-supervised learning uses a combination of both supervised and unsupervised techniques. ML is often best applied when there are relatively large datasets (often more than 500 individual samples) with multiple predictor variables that it is theorized may predict an outcome [84]. ML is different from traditional statistical modeling in a few main ways. There is no need for a specific hypothesis of an association between an independent and dependent variable. Instead, hypotheses are generally made about associations between groups of variables, and the ML algorithms search for patterns in sets of variables to identify the outcome [84].

While there are many different ML approaches being employed, artificial neural networks (ANNs) have been getting an increasing amount of attention. ANNs are modeled after the structure of biological neural networks where interconnected neurons are layered on top of one another [83]. Deep neural networks (DNNs) employ many of these layers, and researchers have found that these approaches can yield impressive results when properly trained. For instance, Google released a neural network capable of classifying 1000 classes with a Top 5 error rate of 3.58% [85]. Neural networks constructed in the field of cardiology have similarly astounding accuracy levels, some yielding an upward of 98% accuracy [86–88]. While ANNs can have extremely high accuracy, they are also some of the most data-hungry algorithms. The neural network used in [88] required more than 80,000 samples to train, and Google's image recognition network used an astounding 1.2 million labeled images [85]. How many samples are required varies from case-to-case, but a general starting point is that the number of data samples should be greater than the number of features in the dataset. This is one of the reasons why it has been argued that the benefits of larger, better datasets greatly outweigh the benefits of better algorithms for learning [89]. ML has been successfully employed in a variety of healthcare-related settings. ML has been applied to genetic sequences to identify protein structure and function, enabling diets to be more individually tailored. It has also allowed for the restoration of motor function to a quadriplegic patient's hand by reading signals from the human motor cortex [86]. In the field of cardiology, researchers have employed machine learning to phenotype noninvasive LVFP left ventricular (LV) filling pressures, predict acute coronary syndrome in emergency departments, as well as classify abnormal ECG patterns [86–88, 90]. In our group, we have attempted to use ANNs to classify abnormal ECG patterns of different arrhythmic zebrafish mutant [31], yielding accuracy of ~95%. This garners optimism that we can collect and classify ECG anomalies from our zebrafish mutant lines using ML to establish an ECG-anomaly library which can be used for numerous studies and health monitoring applications.

While we could use the ECG data from zebrafish to train the NN model, its accuracy of predicting human disease is questionable as the humans' heart rhythm is slower and the ECG waveforms are wider. On the other hand, the labeled human ECG data (especially those with heart disease) are scarce. Currently, we address this problem by taking advantage of transfer learning techniques [91], which are able to adapt the model trained under one domain to another domain using a new dataset. Particularly for this case, we first feed the labeled human ECG data into the zebrafish NN model to adjust the weights of the last layers, so that the output labels are adapted to humans. Then, the same human data would be used to adjust the weights of all layers by training NN again with lower learning rate.

7 Conclusion

Acquisition and analysis of electrocardiogram have been utilized for many decades for healthcare and biological study purposes. Nevertheless, only till recently we could be able to obtain remote precise ECG data in out-of-clinic settings, and we could have the computation power to analyze large-scale data. With the fast growth of wearable technologies, IoT systems, and data science, we would be able to provide health monitoring for both patients and healthy populations in an effective way that we have been dreaming of, via mobile health, distanced care, personalized medicine, and even efficient preventive care. However, monitoring of ECG alone may not be enough. Gathering multimodal sensory data of EEG, EMG, blood, body temperature, and dietary information, to name a few, from a network of body sensors may provide novel insights in monitoring and prediction as the human body should be seen as a whole with multiple interacting biological and physiological components and processes. Further, animal research still plays the utmost important role not only in studying the basic principles and elucidating therapeutic potentials but also in collecting transferrable massive data of disease models.

Acknowledgement This work is financially supported by the NSF CAREER Award #1917105 (H.C.) and NIH R41 #OD024874 (H.C.).

References

1. W. H. Organization. (2018). *The top 10 causes of death*. Available: https://www.who.int/news-room/fact-sheets/detail/the-top-10-causes-of-death
2. Hsieh, P. C., Segers, V. F., Davis, M. E., MacGillivray, C., Gannon, J., Molkentin, J. D., et al. (2007, August). Evidence from a genetic fate-mapping study that stem cells refresh adult mammalian cardiomyocytes after injury. *Nature Medicine, 13*, 970–974.
3. Bergmann, O., Bhardwaj, R. D., Bernard, S., Zdunek, S., Barnabe-Heider, F., Walsh, S., et al. (2009, April 3). Evidence for cardiomyocyte renewal in humans. *Science, 324*, 98–102.
4. Bersell, K., Arab, S., Haring, B., & Kuhn, B. (2009, Jul 23). Neuregulin1/ErbB4 signaling induces cardiomyocyte proliferation and repair of heart injury. *Cell, 138*, 257–270.
5. Giudicessi, J. R., & Ackerman, M. J. (2013, Janunary). Genetic testing in heritable cardiac arrhythmia syndromes: differentiating pathogenic mutations from background genetic noise. *Current Opinion in Cardiology, 28*, 63–71.
6. Haïssaguerre, M., Derval, N., Sacher, F., Jesel, L., Deisenhofer, I., de Roy, L., et al. (2008). Sudden cardiac arrest associated with early repolarization. *New England Journal of Medicine, 358*, 2016–2023.
7. Lubitz, S. A., & Ellinor, P. T. (2015, May). Next-generation sequencing for the diagnosis of cardiac arrhythmia syndromes. *Heart Rhythm : The Official Journal of the Heart Rhythm Society, 12*, 1062–1070.
8. Christophersen, I. E., Magnani, J. W., Yin, X., Barnard, J., Weng, L.-C., Arking, D. E., et al. Fifteen genetic loci associated with the electrocardiographic P wave clinical perspective. *Circulation: Genomic and Precision Medicine, 10*, e001667, 2017.
9. Nielsen, J. B., Fritsche, L. G., Zhou, W., Teslovich, T. M., Holmen, O. L., Gustafsson, S., et al. (2017). Genome-wide study of atrial fibrillation identifies seven risk loci and highlights biological pathways and regulatory elements involved in cardiac development. *The American Journal of Human Genetics, 102*(1), 103–115.
10. Visscher, P. M., Wray, N. R., Zhang, Q., Sklar, P., McCarthy, M. I., Brown, M. A., et al. (2017). 10 years of GWAS discovery: biology, function, and translation. *The American Journal of Human Genetics, 101*, 5–22.
11. Pajkrt, E., Weisz, B., Firth, H. V., & Chitty, L. S. (2004). Fetal cardiac anomalies and genetic syndromes. *Prenatal Diagnosis, 24*, 1104–1115.
12. Miniño, A. M., Heron, M. P., Murphy, S. L., & Kochanek, K. D.. (2007). Deaths: Final data for 2004, ed: Department of Health and Human Services, Centers for Disease Control and Prevention, National Center for Health Statistics.
13. Araneta, G., Rosario, M., Schlangen, K. M., Edmonds, L. D., Destiche, D. A., Merz, R. D., et al. (2003). Prevalence of birth defects among infants of Gulf War veterans in Arkansas, Arizona, California, Georgia, Hawaii, and Iowa, 1989–1993. *Birth Defects Research Part A: Clinical and Molecular Teratology, 67*, 246–260.
14. Donovan, J. W., Maclennan, R., & Adena, M. (1984). Vietnam service and the risk of congenital anomalies. A case-control study. *Obstetrical & Gynecological Survey, 39*, 24–25.
15. Lombó, M., Fernández-Díez, C., González-Rojo, S., Navarro, C., Robles, V., & Herráez, M. P. (2015). Transgenerational inheritance of heart disorders caused by paternal bisphenol A exposure. *Environmental Pollution, 206*, 667–678.
16. Poss, K. D., Wilson, L. G., & Keating, M. T. (Dec 13 2002). Heart regeneration in zebrafish. *Science, 298*, 2188–2190.
17. Raya, A., Consiglio, A., Kawakami, Y., Rodriguez-Esteban, C., & Izpisua-Belmonte, J. C. (2004). The zebrafish as a model of heart regeneration. *Cloning and Stem Cells, 6*, 345–351.
18. Poss, K. D., Wilson, L. G., & Keating, M. T. (2002). Heart regeneration in zebrafish. *Science, 298*, 2188–2190.
19. Raya, Á., Consiglio, A., Kawakami, Y., Rodriguez-Esteban, C., & Izpisúa-Belmonte, J. C. (2004). The zebrafish as a model of heart regeneration. *Cloning and Stem Cells, 6*, 345–351.

20. Porrello, E. R., Mahmoud, A. I., Simpson, E., Hill, J. A., Richardson, J. A., Olson, E. N., et al. (2011). Transient regenerative potential of the neonatal mouse heart. *Science, 331*, 1078–1080.
21. Huang, G. N., Thatcher, J. E., McAnally, J., Kong, Y., Qi, X., Tan, W., et al. (2012). C/EBP transcription factors mediate epicardial activation during heart development and injury. *Science, 338*, 1599–1603.
22. Kikuchi, K., Holdway, J. E., Werdich, A. A., Anderson, R. M., Fang, Y., Egnaczyk, G. F., et al. (2010). Primary contribution to zebrafish heart regeneration by gata4+ cardiomyocytes. *Nature, 464*, 601–605.
23. Lien, C. L., Harrison, M. R., Tuan, T. L., & Starnes, V. A. (2012). Heart repair and regeneration: Recent insights from zebrafish studies. *Wound Repair and Regeneration, 20*, 638–646.
24. Narula, J., Haider, N., Virmani, R., DiSalvo, T. G., Kolodgie, F. D., Hajjar, R. J., et al. (1996). Apoptosis in myocytes in end-stage heart failure. *New England Journal of Medicine, 335*, 1182–1189.
25. Olivetti, G., Abbi, R., Quaini, F., Kajstura, J., Cheng, W., Nitahara, J. A., et al. (1997). Apoptosis in the failing human heart. *New England Journal of Medicine, 336*, 1131–1141.
26. Rosenzweig, A. (2012). Cardiac regeneration. *Science, 338*, 1549–1550.
27. Forsburg, S. L. (Sep 2001). The art and design of genetic screens: Yeast. *Nature Reviews. Genetics, 2*, 659–668.
28. St Johnston, D. (Mar 2002). The art and design of genetic screens: Drosophila melanogaster. *Nature Reviews. Genetics, 3*, 176–188.
29. Jorgensen, E. M., & Mango, S. E. (May 2002). The art and design of genetic screens: Caenorhabditis elegans. *Nature Reviews. Genetics, 3*, 356–369.
30. Angel, P. M., Nusinow, D., Brown, C. B., Violette, K., Barnett, J. V., Zhang, B., et al. (2011, December 22). Networked-based characterization of extracellular matrix proteins from adult mouse pulmonary and aortic valves. *Journal of Proteome Research, 10*, 812–823.
31. Lenning, M., Fortunato, J., Le, T., Clark, I., Sherpa, A., Yi, S., et al. (2018). Real-time monitoring and analysis of Zebrafish electrocardiogram with anomaly detection. *Sensors, 18*, 61.
32. Ding, Y., Liu, W., Deng, Y., Jomok, B., Yang, J., Huang, W., et al. (2013, February 15). Trapping cardiac recessive mutants via expression-based insertional mutagenesis screening. *Circulation Research, 112*, 606–617.
33. Ding, Y., Long, P. A., Bos, J. M., Shih, Y. H., Ma, X., Sundsbak, R. S., et al. (2016). A modifier screen identifies DNAJB6 as a cardiomyopathy susceptibility gene. *JCI Insight, 1*(14), e88797. https://doi.org/10.1172/jci.insight.88797.
34. Clark, K. J., Balciunas, D., Pogoda, H. M., Ding, Y., Westcot, S. E., Bedell, V. M., et al. (2011, Jun). In vivo protein trapping produces a functional expression codex of the vertebrate proteome. *Nature Methods, 8*, 506–515.
35. Mathur, P., & Guo, S. (2010). Use of zebrafish as a model to understand mechanisms of addiction and complex neurobehavioral phenotypes. *Neurobiology of Disease, 40*, 66–72.
36. Ninkovic, J., & Bally-Cuif, L. (2006). The zebrafish as a model system for assessing the reinforcing properties of drugs of abuse. *Methods, 39*, 262–274.
37. Knecht, A. L., Truong, L., Marvel, S. W., Reif, D. M., Garcia, A., Lu, C., et al. (2017). Transgenerational inheritance of neurobehavioral and physiological deficits from developmental exposure to benzo [a] pyrene in zebrafish. *Toxicology and Applied Pharmacology, 329*, 148–157.
38. Cao, H., Yu, F., Zhao, Y., Zhang, X., Tai, J., Lee, J., et al. (2014). Wearable multi-channel microelectrode membranes for elucidating electrophysiological phenotypes of injured myocardium. *Integrative Biology: Quantitative Biosciences from Nano to Macro, 6*, 789–795.
39. Yu, F., Zhao, Y., Gu, J., Quigley, K. L., Chi, N. C., Tai, Y.-C., et al. (2012). Flexible microelectrode arrays to interface epicardial electrical signals with intracardial calcium transients in zebrafish hearts. *Biomedical Microdevices, 14*, 357–366.
40. Forouhar, A., Hove, J., Calvert, C., Flores, J., Jadvar, H., & Gharib, M. (2004). Electrocardiographic characterization of embryonic zebrafish. In *Engineering in Medicine and Biology Society, 2004. IEMBS'04. 26th Annual International Conference of the IEEE*, 2004, pp. 3615–3617.

41. Milan, D. J., & MacRae, C. A. (2005). Animal models for arrhythmias. *Cardiovascular Research, 67*, 426–437.
42. Sun, P., Zhang, Y., Yu, F., Parks, E., Lyman, A., Wu, Q., et al. (2009). Micro-electrocardiograms to study post-ventricular amputation of zebrafish heart. *Annals of Biomedical Engineering, 37*, 890–901.
43. Yu, F., Huang, J., Adlerz, K., Jadvar, H., Hamdan, M. H., Chi, N., et al. (2010). Evolving cardiac conduction phenotypes in developing zebrafish larvae: Implications to drug sensitivity. *Zebrafish, 7*, 325–331.
44. Shier, D. N., Butler, J. L., & Lewis, R. (2011). *Hole's essentials of human anatomy and physiology*. McGraw-Hill Higher Education. Pennsylvania Plaza New York City.
45. Widmaier, E. P., Raff, H., & Strang, K. T. (2006). *Vander's human physiology: The mechanisms of body function* (Vol. 10, pp. 454–455). New York: McGraw-Hill.
46. Natalie casebook. Available: http://www.nataliescasebook.com/tag/cardiac-action-potentials
47. Klabunde, R. E. (2011). *Cardiovascular physiology concepts*: Wolters Kluwer Health/ Lippincott Williams & Wilkins.
48. Sharma, M., Barbosa, K., Ho, V., Griggs, D., Ghirmai, T., Krishnan, S. K., et al. (2017). Cuff-less and continuous blood pressure monitoring: A methodological review. *Technologies, 5*, 21.
49. Le, T., Han, H. D., Hoang, T. H., Nguyen, V. C., & Nguyen, C. K. (2016). A low cost mobile ECG monitoring device using two active dry electrodes. In *2016 IEEE Sixth International Conference on Communications and Electronics (ICCE)*, 2016, pp. 271–276.
50. Sharma, M., Ritchie, P., Ghirmai, T., Cao, H., & Lau, M. P. (2017) Unobtrusive acquisition and extraction of fetal and maternal ECG in the home setting. In *SENSORS, 2017 IEEE*, 2017, pp. 1–3.
51. Neuman, M. R. (1998). Biopotential amplifiers. In J. G. Webster (Ed.), *Medical instrumentation: Application and design* (pp. 233–286). New York: Wiley.
52. Merletti, R., Botter, A., Troiano, A., Merlo, E., & Minetto, M. A. (2009). Technology and instrumentation for detection and conditioning of the surface electromyographic signal: State of the art. *Clinical Biomechanics, 24*, 122–134.
53. Chi, Y. M., Jung, T. P., & Cauwenberghs, G. (2010). Dry-contact and noncontact biopotential electrodes: Methodological review. *IEEE Reviews in Biomedical Engineering, 3*, 106–119.
54. Tseng, K. C., Lin, B. S., Liao, L. D., Wang, Y. T., & Wang, Y. L. (2014). Development of a wearable mobile electrocardiogram monitoring system by using novel dry foam electrodes. *IEEE Systems Journal, 8*, 900–906.
55. Liao, L.-D., Wang, I.-J., Chen, S.-F., Chang, J.-Y., & Lin, C.-T. (2011). Design, fabrication and experimental validation of a novel dry-contact sensor for measuring electroencephalography signals without skin preparation. *Sensors, 11*, 5819.
56. Lin, B. S., Chou, W., Wang, H. Y., Huang, Y. J., & Pan, J. S. (2013). Development of novel non-contact electrodes for mobile electrocardiogram monitoring system. *IEEE Journal of Translational Engineering in Health and Medicine, 1*, 1–8.
57. Ribeiro, D. M. D., Fu, L. S., Carlos, L. A. D., & Cunha, J. P. S. (2011). A novel dry active biosignal electrode based on an hybrid organic-inorganic interface material. *IEEE Sensors Journal, 11*, 2241–2245.
58. Wang, Y., Pei, W., Guo, K., Gui, Q., Li, X., Chen, H., et al. (2011, October 19). Dry electrode for the measurement of biopotential signals. *SCIENCE CHINA Information Sciences, 54*, 2435.
59. Sun, Y., & Yu, X. B. (2016). Capacitive biopotential measurement for electrophysiological signal acquisition: A review. *IEEE Sensors Journal, 16*, 2832–2853.
60. Cömert, A., Honkala, M., & Hyttinen, J. (2013, April 08). Effect of pressure and padding on motion artifact of textile electrodes. *Biomedical Engineering Online, 12*, 26.
61. Bandodkar, A. J., & Wang, J. (2014). Non-invasive wearable electrochemical sensors: A review. *Trends in Biotechnology, 32*, 363–371.
62. Anna, G., Stefan, H., & Jörg, M. (2007). Novel dry electrodes for ECG monitoring. *Physiological Measurement, 28*, 1375.
63. Lopez-Gordo, M. A., Sanchez-Morillo, D., & Valle, F. P. (2014, July 18). Dry EEG electrodes. *Sensors (Basel, Switzerland), 14*, 12847–12870.

64. Chi, Y. M., & Cauwenberghs, G. (2009). Micropower non-contact EEG electrode with active common-mode noise suppression and input capacitance cancellation. In *2009 Annual International Conference of the IEEE Engineering in Medicine and Biology Society*, 2009, pp. 4218–4221.
65. Chi, Y. M., Maier, C., & Cauwenberghs, G. (2011). Ultra-high input impedance, low noise integrated amplifier for noncontact biopotential sensing. *IEEE Journal on Emerging and Selected Topics in Circuits and Systems, 1*, 526–535.
66. Griggs, D., Sharma, M., Naghibi, A., Wallin, C., Ho, V., Barbosa, K., et al. (2016). Design and development of continuous cuff-less blood pressure monitoring devices. In *SENSORS, 2016 IEEE*, 2016, pp. 1–3.
67. Schossow, D., Ritchie, P., Cao, H., Chiao, J.-C., Yang, J., & Xu, X. (2017). A novel design to power the micro-ECG sensor implanted in adult zebrafish. In *Antennas and Propagation & USNC/URSI National Radio Science Meeting, 2017 IEEE International Symposium on, 2017*, pp. 1681–1682.
68. Gruber, S., Le, T., Huerta, M., Wilson, K., Yang, J., Xu, X., et al.. (2018). Characterization of passive wireless electrocardiogram acquisition in Adult Zebrafish. In *2018 IEEE International Microwave Biomedical Conference (IMBioC)*, 2018, pp. 115–117.
69. Cao, H., Landge, V., Tata, U., Seo, Y.-S., Rao, S., Tang, S.-J., et al. (2012). An implantable, batteryless, and wireless capsule with integrated impedance and pH sensors for gastroesophageal reflux monitoring. *IEEE Transactions on Biomedical Engineering, 59*, 3131–3139.
70. Gruber, S., Schossow, D., Lin, C.-y., Ho, C. H., Jeong, C., Lau, T. L., et al. (2017). *Wireless power transfer for ECG monitoring in freely-swimming Zebrafish*, presented at the IEEE Sensors, Glasgow, Scotland, 2017.
71. Brunger, J. M., Zutshi, A., Willard, V. P., Gersbach, C. A., & Guilak, F. (2017). CRISPR/Cas9 editing of murine induced pluripotent stem cells for engineering inflammation-resistant tissues. *Arthritis & Rheumatology, 69*, 1111–1121.
72. Dimarco, J. P., & Philbrick, J. T. (1990). Use of ambulatory electrocardiographic (Holter) monitoring. *Annals of Internal Medicine, 113*, 53–68.
73. Zheng, J. W., Zhang, Z. B., Wu, T. H., & Zhang, Y. (2007). A wearable mobihealth care system supporting real-time diagnosis and alarm. *Medical & Biological Engineering & Computing, 45*, 877–885.
74. Piwek, L., Ellis, D. A., Andrews, S., & Joinson, A. (2016). The rise of consumer health wearables: Promises and barriers. *PLoS Medicine, 13*, e1001953.
75. Milošević, M., Shrove, M. T., & Jovanov, E. (2011). "Applications of smartphones for ubiquitous health monitoring and wellbeing management," *JITA-Journal of Information Technology and Applications 1*, 7–15.
76. Lee, Y.-G., Jeong, W. S., & Yoon, G. (2012). Smartphone-based mobile health monitoring. *Telemedicine and e-Health, 18*, 585–590.
77. Cohrs, K. M., Dancy, J., Besko, D. P., Lohrman, L. L., & Miller, R. M. (2017). *Combined strap and cradle for wearable medical monitor*. ed: Google Patents, 2017.
78. Le, T., Huerta, M., Moravec, A., & Cao, H. (2018). Wireless passive monitoring of electrocardiogram in firefighters. In *2018 IEEE International Microwave Biomedical Conference (IMBioC)*, 2018, pp. 121–123.
79. Benharref, A., & Serhani, M. A. (2014). Novel cloud and SOA-based framework for E-Health monitoring using wireless biosensors. *IEEE Journal of Biomedical and Health Informatics, 18*, 46–55.
80. Deo, R. C. (2015). Machine learning in medicine. *Circulation, 132*, 1920–1930.
81. Obermeyer, Z., & Emanuel, E. J. (2016). Predicting the future — Big data, machine learning, and clinical medicine. *New England Journal of Medicine, 375*, 1216–1219.
82. (2018). *NIH Data Sharing Repositories* [Product, Program, and Project Descriptions]. Available: https://www.ncbi.nlm.nih.gov/pubmed/
83. Géron, A. l. (2017). *Hands-on machine learning with Scikit-Learn and TensorFlow: Concepts, tools, and techniques to build intelligent systems* (1st ed.). Sebastopol: O'Reilly Media.

84. Waljee, A. K., & Higgins, P. D. R. (2010, 06/03/online). Machine learning in medicine: A primer for physicians. *The American Journal Of Gastroenterology, 105*, 1224.
85. Szegedy, V. V. C., Ioffe, S., Shlens, J., & Wojna, Z. (2015). Rethinking the inception architecture for computer vision, 2015.
86. Pandit, D., Zhang, L., Aslam, N., Liu, C., Hossain, A., & Chattopadhyay, S. (2014) An efficient abnormal beat detection scheme from ECG signals using neural network and ensemble classifiers. In *The 8th International Conference on Software, Knowledge, Information Management and Applications (SKIMA 2014)*, 2014, pp. 1–6.
87. Gupta, A., Thomas, B., Kumar, P., Kumar, S., & Kumar, Y. (2014). Neural network based indicative ECG classification, In *2014 5th International Conference – Confluence The Next Generation Information Technology Summit (Confluence)*, 2014, pp. 277–279.
88. Jun, H. J. P. T. J., Minh, N. H., Kim, D., & Kim, Y. H. (2016). Premature ventricular contraction beat detection with deep neural networks. In *15th IEEE International Conference on Machine Learning and Applications (ICMLA),* pp. 859–864.
89. Halevy, A., Norvig, P., & Pereira, F. (2009). "The unreasonable effectiveness of data," *IEEE Intelligent Systems, 24*, 8-12.
90. Park, J. Y., Noh, Y.-K., Choi, B. G., Rha, S.-W., & Kim, K. E. (2015). TCTAP A-010 a machine learning-based approach to prediction of acute coronary syndrome. *Journal of the American College of Cardiology, 65*, S6.
91. Pan, S. J., & Yang, Q. (2010). A survey on transfer learning. *IEEE Transactions on Knowledge and Data Engineering, 22*, 15.

Flexible Intravascular EIS Sensors for Detecting Metabolically Active Plaque

Yuan Luo, Rene Packard, Parinaz Abiri, Y. C. Tai, and Tzung K. Hsiai

1 Introduction

1.1 Atherosclerosis

Atherosclerosis refers to the local built-up of fats, cholesterols, and other substance on the artery walls, commonly known as plaques. It is a chronic inflammatory disease with both innate and adaptive immune pathways involving a complicated interplay between lipid metabolism, inflammation, macrophage activation, and collagen breakdown [1, 2]. Oxidized low-density lipoprotein (oxLDL) and macrophage infiltrates contribute to pro-inflammatory states relevant to the initiation of atherosclerotic lesions [3]. These lesions can be classified as stable or unstable plaque. The latter is often non-obstructive by conventional X-ray angiography or intravascular ultrasound (IVUS) and is prone to rupture, leading to acute coronary syndromes or stroke [4–7].

Unstable plaque or otherwise known as thin-cap fibroatheroma represents a plaque status that is characterized by a thin fibrous cap (<65 μm), a large necrotic core, an excessive amount of macrophages, and limited luminal narrowing [3]. The rupture of atherosclerotic plaques remains a leading cause of mortality in developed countries [8]. Detection of atherosclerotic lesions prone to rupture is of utmost importance for patients with acute coronary syndromes or stroke [9].

Y. Luo · Y. C. Tai
Medical Engineering, California Institute of Technology, Pasadena, CA, USA

R. Packard · P. Abiri
Department of Bioengineering, University of California, Los Angeles, Los Angeles, CA, USA

T. K. Hsiai (✉)
Medical Engineering, California Institute of Technology, Pasadena, CA, USA

Department of Bioengineering, University of California, Los Angeles, Los Angeles, CA, USA
e-mail: thsiai@mednet.ucla.edu

H. Cao et al. (eds.), *Interfacing Bioelectronics and Biomedical Sensing*,
https://doi.org/10.1007/978-3-030-34467-2_6

However, detection and diagnosing the non-obstructive, albeit pro-inflammatory, lesions during catheterization remains a clinical challenge. Up to date, little is known about the numerous variables affecting assessment of atherosclerotic lesions. Investigations have been centered around atherosclerotic lesion's lipid pool content, calcification, fibrous cap, and intima/media thickness [7]. The vascular oxidative hypothesis of unstable plaque is that atherosclerotic lesion is deemed a dynamic process [10]. Reactive oxygen species via NADPH oxidase enzyme system and the release of matrix metalloproteinase (MMP) by the inflammatory cells contribute to the vulnerability of a rupture-prone plaque [11]. OxLDL induces transformation of macrophages to lipid-laden foam cells [12]. These macrophage-derived foam cells are trapped by interaction with oxLDL and can be mobilized by dynamic exposure in response to key antioxidants such as apocynin and N-acetylcysteine (NAC) [13]. Growing evidence suggests that oxLDL and thin-cap fibroatheromas (TCFA) rich in macrophage/foam cells are prone to mechanical stress and destabilization [14, 15]. Severe stenosis reduces arterial blood flow and causes high tensile and compressive stress in the stenotic plaque [16]. Oxidative stress and pro-inflammatory states modulate mechanical vulnerability of plaque [17]. Oxidative stress is involved in the oxidation/modification of low-density lipoprotein [18, 19], which occurs at high levels in the atherogenic prone regions of aorta and plays a critical role in pro-inflammatory states [20]. Mounting evidence supports that oxLDL and macrophage/foam cell-rich shoulder areas are more prone to disruption, leading to thrombus formation and embolic events [14].

Serum biomarkers have been proposed for predicting pro-inflammatory states, including C-reactive protein, matrix metalloproteinases, CD40 ligand, Lp-PLA2 (lipoprotein-associated phospholipase A2), and myeloperoxidase. Specifically, circulating Lp-PLA2 is bound predominantly to low-density lipoprotein (LDL) (~80%) which transmigrated to the subendothelial space of atherosclerosis-prone regions [21]. On the other hand, detection techniques often adopted currently, including fractional flow reserve (FFR), IVUS, or optical coherence tomography (OCT), are limited to anatomic and hemodynamic characteristics of atherosclerotic plaques [22]. Despite the advent of computerized tomographic (CT) angiography, high-resolution magnetic resonance imaging (MRI) [23], near-infrared fluorescence (NIRF) [24], and time-resolved laser-induced fluorescence spectroscopy [25], real-time interrogation of metabolically active, lipid-rich plaques remains an unmet clinical need.

There are numerous excellent literature reviews on various aspects of atherosclerosis. This chapter does not intend to provide detailed treatment from pathological and diagnostic perspectives, and the authors refer the reader to reports by other great scholars in this field from all the cited reference in the end.

1.2 Electrochemical Impedance Spectroscopy and Its Relevance to Atherosclerosis

Biological samples (e.g., tissues and organs) are composed of conductive (e.g., ionic solution) and capacitive elements (e.g., cellular membranes). In general, these samples can be considered as a network of resistors and capacitors, which can be conveniently represented as the electrochemical impedance Z of the samples. Electrochemical impedance spectroscopy (EIS) is a technique that measures electrochemical impedance in response to applied alternating current (AC) excitation with varying frequency. Equivalently, EIS can also be understood as the macroscopic representation of the electric field and current density distribution within the samples of interest. Such distribution can be described with the harmonic Maxwell's equations (when a sinusoidal signal is applied), which, under the quasi-electrostatic limit, reduces to [26]:

$$\nabla \cdot \left(\sigma^{*} \nabla V\right) = 0 \tag{1}$$

where $\sigma^* = \sigma_s + i\omega\varepsilon_s$ is referred to as the complex conductivity of the sample and σ_s and ε_s represent the normal conductivity and permittivity of the sample, respectively. ω is the angular frequency of the applied excitation signal, $i = \sqrt{-1}$, and V denotes the voltage distribution on the sample. Subsequently, the impedance of the sample, Z, can expressed as:

$$Z = \frac{\Delta V}{\int_s \vec{J} \cdot dS} \tag{2}$$

where $\vec{J} = \sigma^* \cdot \nabla V$ is the total current density, S denotes the electrode-tissue interface area, and ΔV represents the voltage difference across the two measuring electrodes.

Due to distinct material properties including ion concentrations in intra/extracellular fluids, membrane characteristics, cell structure, and density, among other factors, biological samples exhibit different EIS response, which has gained increasing attention as the foundation of biosensing approach in recent years [27]. Previously proposed applications of EIS include assessment of cellular viability of human cancer cells [28] and amyloid β-sheet misfolding [29]. The dramatic difference in cell types, molecular composition, and tissue properties between atherosclerosis lesion and the normal artery has been recognized as a critical basis for differentiation. Pre-atherosclerotic lesions harbor pro-inflammatory substrates, namely, oxLDL and macrophage-derived foam cells, which have distinct endoluminal electrochemical impedance and can be quantified by electrochemical impedance spectroscopy (EIS) [30]. Lipid-free tissue (in the intravascular diagnostics context, there are blood, artery tissue, etc.) is known to be a viable electrical conductor for its high water (approximately 73%) and electrolytes (ions and proteins) content. On the other hand, lipid tissue is anhydrous and, thus, a poor conductor. For this reason, EIS can be used to measures the intrinsic electrochemical properties of the tissue; namely, water content, electrolyte concentration, vascular calcification, and cholesterol/lipid content influence the frequency-dependent changes in impedance. These findings spurs the recent interest in leveraging EIS on atherosclerosis detection, and the focus of the remaining sections will be to present a detailed discussion on the recent development in this front.

1.3 Equivalent Circuit Model for EIS

Equivalent circuit model, which treats individual elements in a complex biological sample as simple circuit components, is a useful tool in analyzing impedance data from EIS measurement. This section provides a general example of how to model the complicated electrical behaviors of endoluminal environment. As shown in Fig. 1a, a cross-sectional perspective of the circuit configuration provides the operational principle underlying the endoluminal EIS interrogation. Imaging the electrodes that is attached on an inflatable medical balloon are deployed within the artery; four main types of tissue contribute to the aggregated impedance values, namely, blood, aortic wall, atherosclerotic plaque, and perivascular fat circumscribing the vessel. In this context, a simple circuit block is first applied to generalize the impedimetric behavior of each individual tissues, consisting of a parallel circuit with two paths: (1) a resistive element (R_1) in series with a capacitive element (C) to model the cells in that tissue and (2) a pure resistive element (R_2) to model the extracellular materials (red frame in Fig. 1b) [31]. Another key electrical component in this setting is the electrode-tissue interface, which is modeled using the constant phase element (CPE) to take into account the nonlinear double-layer capacitance behavior. The impedance of the interface can be expressed as:

$$Z_{\mathrm{CPE}} = \frac{1}{Y(i\omega)^a} \tag{3}$$

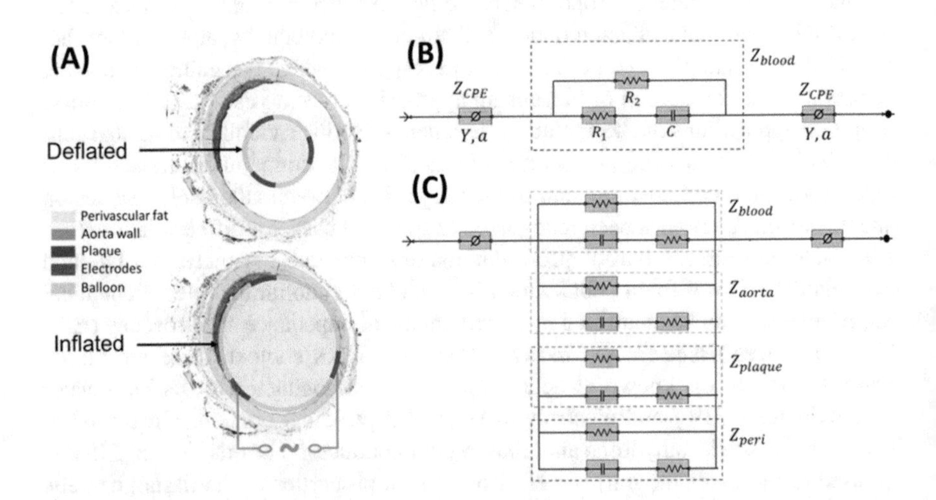

Fig. 1 EIS equivalent circuit model: (**a**) a cross-sectional perspective of the imaginary scenario where EIS electrodes are deployed inside the blood vessel. (**b**) The equivalent circuit includes the blood as primary component upon balloon deflation. (**c**) The equivalent circuit further includes the aorta, plaque, blood, and perivascular fat all as the circuit components upon balloon inflation. (Adapted from [32])

where Y denotes the empirical admittance value and a is a constant between 0 and 1, representing the non-ideal interface effects. When the balloon is deflated (Fig. 1a), blood is considered as the primary component (Z_{blood}) in the circuit model as other tissues are shielded by the presence of blood in contact with the electrodes (Fig. 1b). When the balloon is inflated, the endoluminal surface is in contact with the electrodes. As a result, all of the tissue types contributed to the path of the current flow, accounting for the parallel circuit configuration for the blood, plaque, vessel wall (aorta), and perivascular fat ($Z_{blood}//Z_{plaque}//Z_{aorta}//Z_{peri}$) (Fig. 1c).

2 Electrochemical Impedance Spectroscopy Implementation

2.1 *Four-Point EIS*

One of the earliest attempts of implementation in using EIS to detection atherosclerosis is established in the standard 4-point configuration. Figure 2 presents the comparison between a 4-point configuration and the more intuitive 2-point configuration impedance measurement. In the 2-point setting (Fig. 2a, c), the excitation signal (e.g., AC current) is injected through the two available electrodes, and the measurement of the resulting voltage response will also be acquired through the same pair of electrodes. Such an experimental condition inevitably includes the impact of the contact impedance between the electrodes and the measured sample

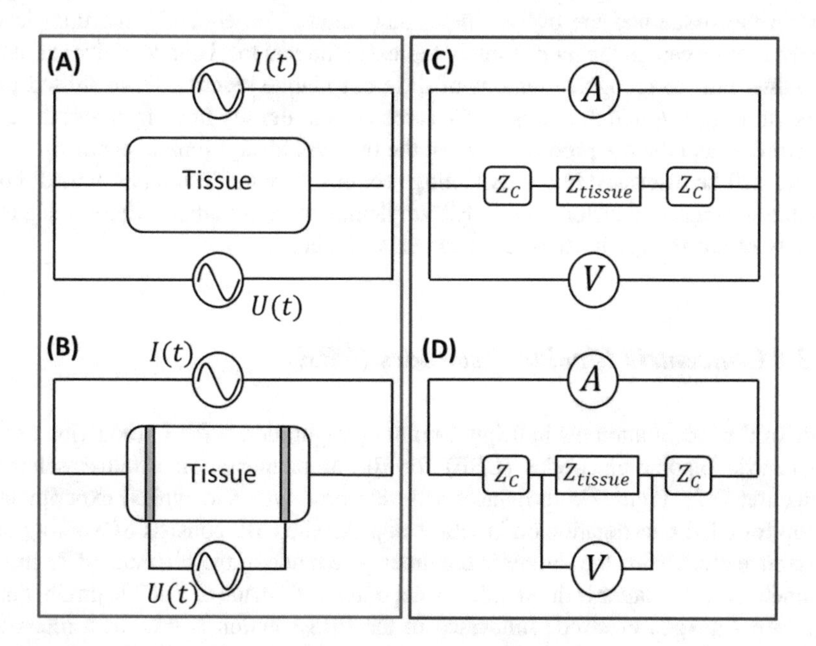

Fig. 2 Schematic comparison between 2-point and 4-point EIS

when deriving the overall impedance value. On the other hand, the main advantage for using 4 individual electrodes is that we can choose one pair of electrode as the current injecting electrodes and other two as the voltage measuring electrodes (Fig. 2b, d). As long as the input impedance of the voltage measuring equipment is significantly higher than the sample, there will be effectively no current flowing into that pair of electrodes, hence minimizing the influence of the contact impedance. A group of German researchers first fabricate miniature 4-point configuration electrodes with platinum deposited on polyimide flexible substrate and attached them on a medical balloon [33, 34]. They demonstrated the different impedance response from normal sample compared to sample with atherosclerosis lesion from in vitro settings. The same device design was also tested to assess EIS in live New Zealand White rabbit models.

However, one of the key issues for the 4-point EIS measurement is the existence of the negative sensitivity field [35]. The formulation of the sensitivity field (ζ) of a 4-point configuration has been detailed in previous works [36] and can be written as:

$$\zeta = \frac{\overrightarrow{J_1} \cdot \overrightarrow{J_2}}{I^2} \tag{4}$$

where J_1 and J_2 are the current density at the current injection electrode pair and the voltage measuring electrode pair, respectively, and I denotes the actual current being injected.

As seen in Eq. (4), $\overrightarrow{J_1} \cdot \overrightarrow{J_2}$ can possibly become a negative value in certain regions of the tissue, indicating a reverse correlation between the local conductivity change within the tissue and the overall impedance value. Consequently, the impedance measurement can no longer genuinely represent the real local conductivity variation and thus fails to recognize regions of different plaque severity. Once shifted to a 2-point design, J_1 and J_2 represent the same current density flowing across the two electrodes, and the dot product between the two will always remain positive.

As will be discussed in the remaining sections, 2-point design has gained more attention in recent development of EIS implementation for atherosclerosis due to a variety of advantages it can offer as can be seen below.

2.2 *Concentric Bipolar Electrodes (CBE)*

One of the recent attempts in using 2-point configuration is the introduction of the concentric bipolar electrodes (CBE) for EIS implementation in atherosclerosis detection [37]. Figure 3a, b depicts a device prototype and typical experimental setup for CBE interrogation on in vitro samples. The CBE consists of working and reference electrodes; the former is the inner pole made of the platinum at 75 μm in diameter, and the latter is the stainless steel outer shell made of a at 300 μm in diameter. An Ag/AgCl electrode immersed in the PBS solution is used as a reference electrode. Distinct from the linear 4-point electrode arrays, the unique feature of the concentric electrodes includes the constant and symmetric displacement between

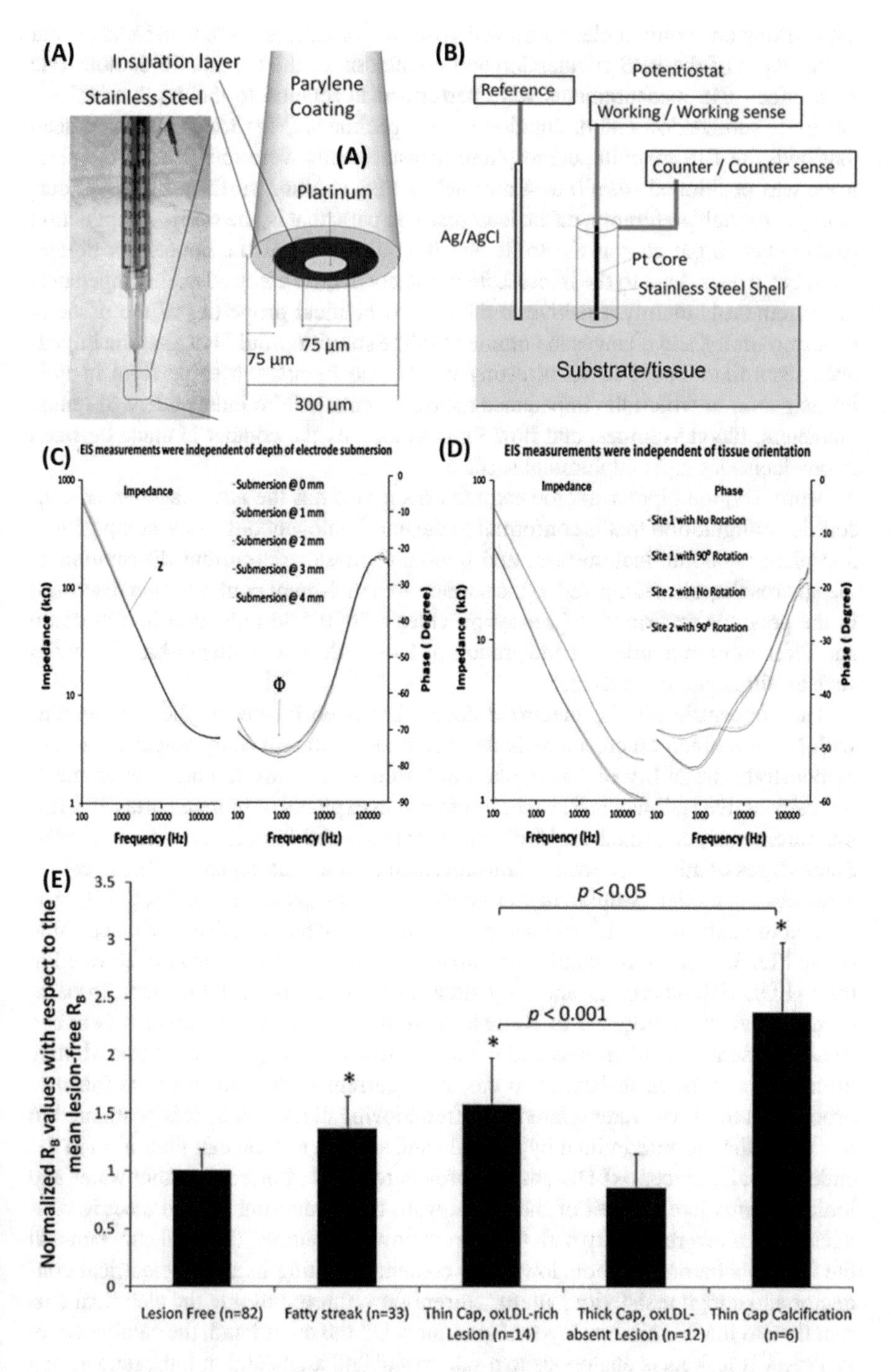

Fig. 3 (**a**) Schematic illustration of the geometry and dimension of the concentric bipolar electrode and (**b**) of experimental setup. EIS (Magnitude & Phase) measurements in relation of (**c**) the depth of microelectrode submersion and (**d**) orientation of the specimens. (**e**) Sensitivity of EIS measurement in differentiating samples in different Stary stage. Adapted from [37, 38]

the working and counter electrodes which allows for EIS measurement independent of the depth of the PBS submersion and orientation of the tissues. To demonstrate such effect, EIS measurements were performed in relation to the depth of microelectrode submersion and orientation of the specimens (Fig. 3c, d). It can be seen that both the EIS magnitude and phase measurements were identical as the electrode was positioned from 0 to 4 mm below PBS solution surface. Electrical currents go through preferably via the least resistive path, that is, the shortest conducting paths between the central electrode and the outer shell of the concentric bipolar microelectrodes. Due to the microscale of the concentric electrodes, the impedance measurement is mainly sensitive to the electrochemical properties of the tissue at close proximity, and changes in volume of saline solution would not alter the impedance recording. These in vitro testing can be the foundation to perform in vivo investigation in which the impedance measurements will be independent of lumen diameters, blood volumes, and flow rates as long as the contact is made between microelectrodes and endoluminal surface.

Moreover, the bipolar microelectrode sensor also has the advantages in its concentric configuration that is conformal to the non-homogeneous tissue composition, non-planar endoluminal surface, and non-uniform electric current distribution of the atherosclerosis. Compared to the surface area of 4-point configuration discussed in the previous section, CBE sensor provides a 2000-fold reduction in dimension and offer more potential for integration of EIS with other catheter-based devices such as ultrasonic transducers.

The concentric bipolar electrode device has been further applied to samples with better defined atherosclerosis features under different Stary stages to further demonstrate the utility of this method in a following study to characterize metabolically active lesions via EIS measurements in explants of human aorta [38]. EIS measurements performed on 15 coronary, carotid, and femoral arteries at various Stary stages of atherosclerotic lesions revealed distinct EIS signals [39]. An equivalent circuit model (similar to the general concept presented in Sect. 1.3) was applied to analyze the EIS data and provide a convenient effective resistance value (R_B in Fig. 3e) for cross-sample comparison. R_B value was significantly elevated in the oxLDL-rich atheromas and fatty streaks compared to oxLDL-absent fibroatheromas, and the difference in R_B values were statistically insignificant between oxLDL-absent fibroatheromas and the lesion-free regions (Fig. 3e). As previously discussed, the tissue resistance to electrical current is dependent on its intrinsic property in terms of water content and free-moving electrolytes. It is postulated in this study that despite intimal hyperplasia and smooth muscle cell migration to the endoluminal surface, oxLDL-absent fibroatheromas harbor comparable water and ionic contents like the rest of the vessel wall, thus rendering a good electric conductor and a low resistivity path for current flow. In contrast, the lipid core beneath the fibrous atheroma harbors low water content, resulting in a poor electrical conductor and a high resistivity path for current flow, thus confining the electrical current flow to the thin fibrous layer of atheroma. On the other hand, the calcified core in Type VII lesions is analogous to a salt crystal and an insulator, thus, rendering a high resistivity path. In this context, endoluminal EIS signals were elevated in the presence of both active lipids and calcification, while the normalized R_B was significantly higher in the calcific atheromas (Fig. 3e).

These set of studies provides new electrochemical insights into the applications of EIS measurements to detect active metabolic conditions in the mechanically unstable atherosclerotic lesions. Electrochemical characterization of fibrous atheromas in terms of impedance spectroscopy and bioactive lipids offers a novel entry point to identify the high-risk and rupture-prone plaques. For in vitro biosensing applications using CBE, a large surface area provided by the inert platinum or carbon electrode is commonly used as the counter electrode, providing high double-layer capacitance. Hence, the overall contact impedance contributed by counter electrode can be ignored. For intravascular EIS applications requiring high spatial resolution, the confined space in the catheters warrants close packaging of both the counter and working electrodes. Consequently, EIS measurements must account for the electrochemical interference at both the counter and working electrode interfaces.

2.3 CBE for In Vivo Animal Study

The advent of flexible and stretchable electronics provides advanced sensing of biomechanical and biophysical signals from tissues or organ systems otherwise difficult with conventional technologies [40]. Combining the concentric bipolar electrode device with flexible and stretchable concept to create in vivo interrogation device is the next logical advancement toward real clinical application. One of the difficulties in advancing the EIS concept toward in vivo animal testing is integrating EIS sensors with catheter-based devices. A custom-designed fabrication and assembly process of integration has been adopted [41]. It was mainly composed of four steps: (1) fabricating the highly stretchable silicone balloon (Fig. 4b-a), (2) assembling the sealed tubing connecting the balloon to a syringe (Fig. 4b-a), (3) microfabricating the impedance sensor (Fig. 4b-b), and (4) securing the impedance sensor to the balloon (Fig. 4b-c). The final device prototype is shown in Fig. 4c with closeup SEM images illustrating the concentric electrode configuration (Fig. 4d). The justification of adopting a dilatable balloon catheter approach is rather straightforward. Percutaneous transluminal coronary angioplasty (PTCA), or balloon dilation of the coronary arteries, has been routinely conducted in patient for 40 years. This is originally used as a standalone procedure [42] and subsequently combined with coronary stent deployment [43]. The safety of balloon dilation of arteries to treat atherosclerotic lesions is well documented in humans [44] and experimental models [45], with success determined by a post-procedure angiogram [46].

The novelty of this device lies in the intravascular deployment of flexible and stretchable EIS sensor to assess oxLDL-rich atherosclerotic lesions. The concentric bipolar microelectrodes provide a sensitive and reproducible strategy to detect electrochemical impedance at a close proximity to the lesions. The features of flexible and stretchable electronics have made real-time endoluminal assessment of lipid-rich lesions possible. Before in vivo deployment, the catheter-based EIS sensor was

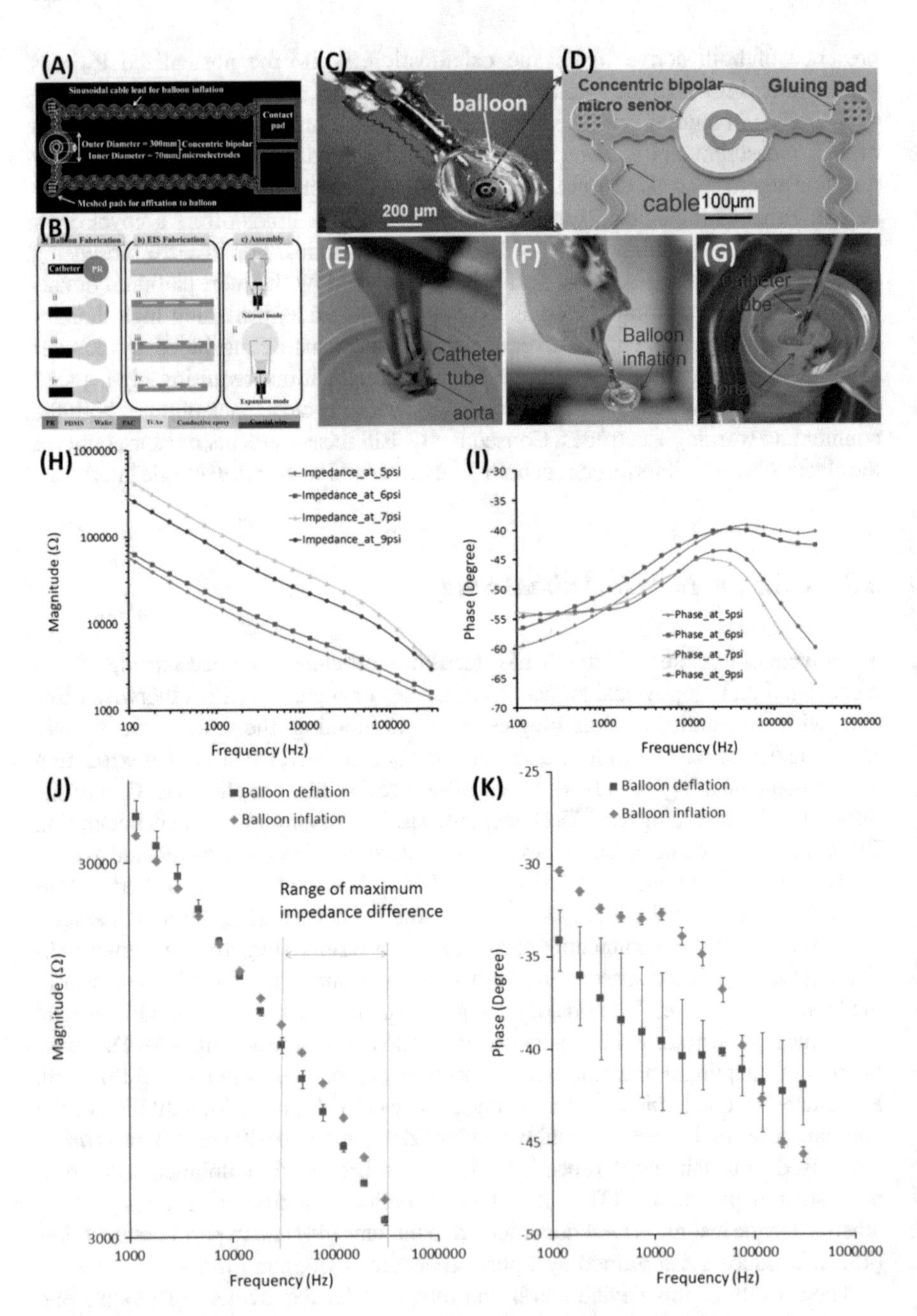

Fig. 4 (**a**) Schematics of concentric bipolar electrode design. (**b**) The complete fabrication and assembly process for catheter-based EIS device. (**c**) Image of the EIS sensor mounted on a balloon. (**d**) SEM of the finished EIS sensor. (**e**) The balloon impedance sensor was inserted into the rabbit aorta. (**f**) Demonstration of balloon inflation prior to impedance assessment. (**g**) Demonstration of intravascular balloon inflation. EIS ((**h**) magnitude and (**i**) phase) characterization in response to different balloon inflation pressure. In vivo EIS ((**j**) magnitude and (**k**) phase) in the rabbit carotid arteries. (Adapted from [41])

deployed into the phosphate saline buffer (PBS)-filled explants of aorta in which EIS measurements were performed to determine the optimal inflation pressures (Fig. 4e–g). At a balloon inflation pressure below 6 psi, the balloon was not fully inflated, and sensor was not in contact with the endoluminal surface (Fig. 4h, i). At an inflation pressure of 7 psi, the inflated balloon enabled the microelectrodes to make contact with the vessel wall, resulting in a significant change in the impedance spectrum compared to below 6 psi. At 9 psi, a slight decrease in magnitude and phase was observed, implicating a higher pressure applied by the balloon to the vessel wall causing local vessel wall deformation. Hence, our observations suggest that balloon inflation pressure at 7 psi allowed for both a complete surface contact and a minimum pressure to the arterial wall. While the difference in viscosity between PBS and blood could affect the extent of balloon inflation, this difference is experimentally negligible. This catheter-based balloon EIS sensor has then been deployed in vivo to interrogate carotid arteries of NZW rabbits (Fig. 4j, k). The impedance magnitude of blood is significantly lower than that of the vascular tissue due to its electrolyte-rich fluid property. For this reason, we can detect a significant increase in impedance magnitude when the inflated balloon makes contact between sensor and vascular wall. The impedance (magnitude and phase) of before and after the balloon inflation is compared. Most notably, balloon inflation induced an increase in impedance magnitude and a decrease in phase toward the higher frequency range. This observation is consistent with previous findings in in vitro testing that the EIS was most sensitive in detecting biological tissue compositions in the higher frequency range from 10 to 300 kHz.

2.4 Two-Point Symmetric Configuration

More recently, the 2-point symmetric configuration for EIS sensing has been proposed and tested in live NZW model [47]. A device prototype with 2-point symmetric design is presented in Fig. 5a–c. Several microelectrodes pairs in an array have been micro-fabricated with varying distances between different electrode pairs. The flexible substrate adopted here is parylene C (PAC) for its supreme biocompatibility and chemical inertness [48]. A 3-D rendering of the deployment of the EIS device with the balloon in gray, plaque in green, and vessel wall segment in blue is also presented. This is followed by a 2-D side view of the device showing where the microelectrodes are placed and a 2-D cross-sectional view of the device in contact with an atherosclerotic plaque segment with electric fields generated between the microelectrodes. When comparing the 2-point symmetric sensor with the concentric bipolar electrodes, the major advance is the fact that it allows for deep tissue penetration for intraplaque interrogation. As can be seen in the previous sections, the typical spacing between the two concentric electrodes was 100 μm, which confines the current traveling to a very small thickness, roughly around the fibrous cap area, whereas the spacing between the 2-point round electrodes is at least 400 μm, allowing for deep current penetration in the 10–100 kHz frequency regime. Also, the contact impedance between the two concentric bipolar electrodes

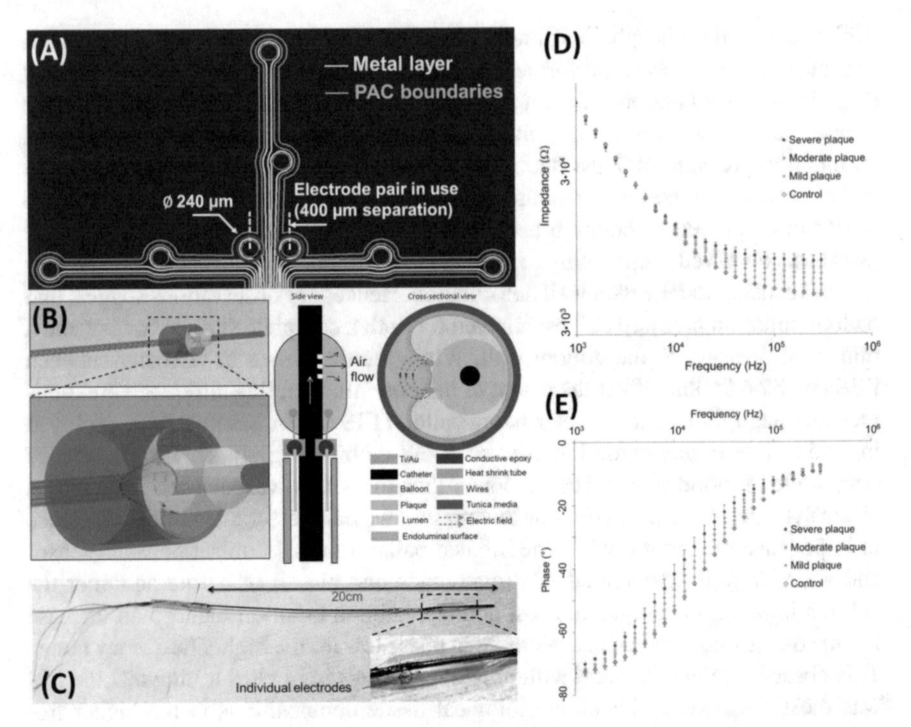

Fig. 5 (**a**) Design schematic of impedance sensing electrode array. (**b**) 3-D rendering of the deployed EIS sensor in contact with a plaque, and 2-D side-view of the device illustrating micro-electrode placement. (**c**) An image of the actual EIS sensor with closeup view of flexible electrodes attached on the inflated balloon. (**d**) Impedance magnitude and (**e**) phase measurement in control aortas and aortic segments with mild, moderate, and severe plaque. (Adapted from [47])

varies due to the different electrode area, giving rise to different voltage drops at each double layer and introducing common mode noise into the detected signal. This is apparently not an issue in the symmetric design. When trying to use an equivalent circuit model approach to further analyze the EIS data and possibly eliminate the impact of contact impedance, the symmetric design has fewer unknown variables than the concentric design, which reduces the difficulty in the fitting process while using the equivalent circuit.

For in vivo rabbit testing, the flexible 2-point electrodes were mounted on an inflatable balloon for endoluminal EIS quantification of lipid burden. Sites of mild, moderate, and severe atherosclerotic plaque burden in previously described specific segments of the aorta in the rabbit [49] were verified by fluoroscopy and IVUS for gross anatomy, by histology for lipid presence, and by immunohistochemistry for foam cells or macrophages. The EIS impedance profiles obtained were significantly different over the 10–300 kHz frequency sweep, with high impedance in the severe plaque area of the descending thoracic aorta, intermediate impedance in the moderate plaque area of the proximal abdominal aorta, and low impedance in the mild plaque area of the distal abdominal aorta (Fig. 5d, e). These results support the application of EIS to quantify intraplaque lipid burden in real time.

Flexible and stretchable electronics is an emergent intravascular approach for preclinical models. Flexible hybrid electronics embedded in polydimethylsiloxane (PDMS) and attached to the skin have been tested for electrocardiogram and temperature monitoring [50]. Implantable flexible electrode arrays have been embedded in PAC for its biocompatibility [48] and investigated for retina implants to restore vision [51]. In terms of safety issues during device deployment, we can consider the following: Young's modulus of nylon for the balloon is 2.7~4.8 GPa [52], matching with that of PAC (3.2 GPa) [53]. There will be no strain mismatch with the flexible PAC electrodes affixed onto the balloon when the balloon is inflated for optimal endoluminal contact. Furthermore, Young's modulus of the blood vessel ranges from 0.2 to 0.6 MPa [54] resulting in $<35\%$ deformation in the normal direction of blood flow. Hence, the flexible and stretchable 2-point EIS balloon catheter provides safe endoluminal contact.

2.5 3-D EIS Interrogation in NZW Rabbit Model

All of the electrode configurations discussed above share the same feature of being a localized interrogation device, namely, the EIS sensing can only be focused on one specific sites on a typical artery wall due to the concentrated electrode dimensions and distances (in the order of a few hundred μm). However, the formation of atherosclerotic plaques is rather unpredictable as which portion of the circular endoluminal surface will be located. Therefore, intravascular deployment of a spot-focused device could potentially miss specific lesion. With the successful demonstration of 2-point symmetric configuration in lesion measurement, a newly designed 6-point EIS sensor featured six individual electrodes that were circumferentially mounted on an inflatable balloon (Fig. 6a–c) has been further implemented to address this particular issue [32]. In one device prototype for in vivo study, the individual electrodes are identical in dimensions (600 μm × 300 μm). Specifically, there are three electrodes embedded in each row, and the distance between the two rows is 2.4 mm (Fig. 6b). Within each layer, the 3 electrodes are equidistantly placed around the circumference of the balloon at 120° separation from each other (Fig. 1c). This 6-point configuration optimizes the contact with the endoluminal surface for EIS measurements. Furthermore, the 6 electrodes allow for 15 different combinations of 2-point electrodes for three-dimensional endoluminal interrogation.

The 6-point electrode configuration enabled the demonstration of 15 EIS permutations (3 + 6 + 6) consisting of 3 2-point electrodes that are vertically linked between the two rows (Fig. 6d-i), 6 2-point electrodes that are linked circumferentially within rows (Fig. 6d-ii), and 6 2-point electrodes that are cross-linked diagonally between the 2 rows (Fig. 6d-iii). This novel combination of 15 permutations pave the way for flexible 3-D interrogation and impedimetric mapping of the arterial segment over 3 sub-segments, as illustrated (Fig. 6d–f). For the 3-D mapping, each color represents impedance values using a distinct electrode permutation with lighter colors indicating lower impedances and darker colors higher impedances, as

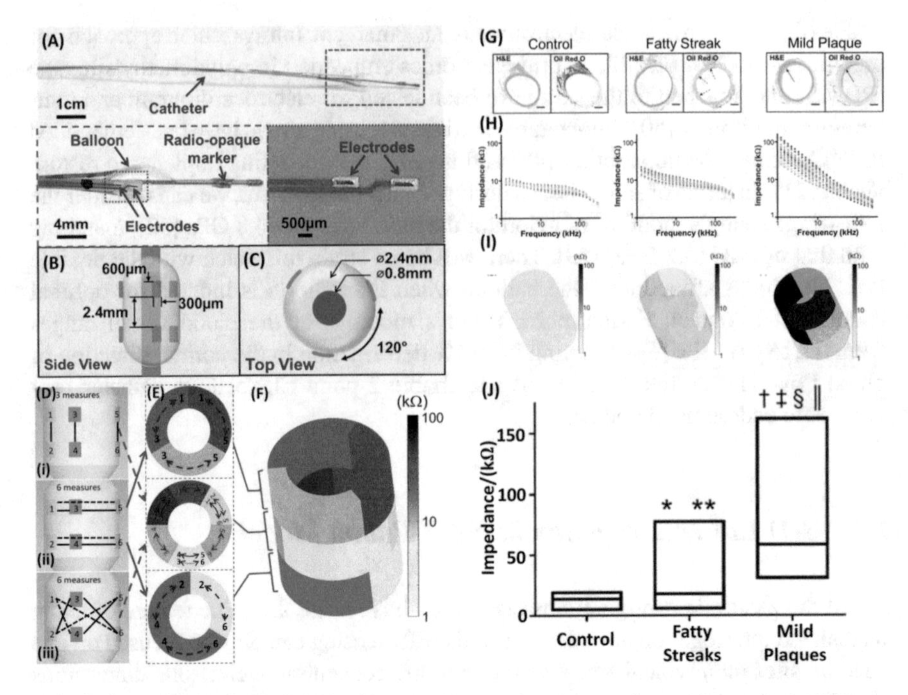

Fig. 6 (**a**) Photographs showing the 3-D EIS device prototype with zoomed-in views of balloon-inflation, providing details of the individual electrodes, balloon, and radio-opaque markers. (**b**) Top and (**c**) side views showing the dimensions of the 6-point sensor upon balloon inflation. (**d–f**) Detailed illustration of 3-D EIS mapping scheme using the 15 permutations of the 6-point device. Experimental results in control aortas, fatty streaks, and mild atherosclerosis plaques: (**g**) Histology; (**h**) EIS magnitude; (**i**) 3-D mapping; (**j**) summary of median and range. (Adapted from [32])

illustrated using a logarithmic scale. This user-friendly readout permits rapid clinical interpretation of lesion types, detection of clinically silent atherosclerosis, and physician adoption based on usability.

Representative real-time EIS measurements of impedance and phase are compared among the three segments of the aorta and correlated with histological evidences (Fig. 6g–i). Three kinds of tissues can be found: control, fatty streaks, and mild plaques. The key finding in this study can be summarized by displaying the 15 permutations obtained at 1 kHz as the medians and the 2 extreme values (minimum–maximum) of the impedance range (Fig. 6j). The control aortas have a median impedance of 13.79 kΩ and a narrow range from a minimum of 5.22 kΩ to a maximum of 19.13 kΩ. Fatty streaks identified in the abdominal aorta at the level of renal artery bifurcation area demonstrate a median (17.75 kΩ) and range of values (minimum 5.79–maximum 77.05 kΩ). In mild plaque segments from the thoracic aorta, the median is 58.32 kΩ and minimum 30.86 kΩ, maximum 161.74 kΩ). There is a significant difference in impedance medians ($P = 0.016$) and value distributions ($P = 0.024$) between control and fatty streak samples. This difference is even larger when comparing control to mild atherosclerotic plaque ($P < 0.001$ for differences in both medians and value distributions).

From these results, we can see that this novel 6-point configuration advances disease detection to flexible 3-D interrogation of early atherosclerotic lesions that harbor distinct electrochemical properties otherwise invisible by current imaging modalities such as invasive angiography. Atherosclerosis develops over decades [8] with evidence of early lesions, or fatty streaks, present in autopsy series of young adults who died in their early 20s [55]. Invasive coronary angiography is considered the gold standard of coronary artery disease determination. Whereas this technique permits visualization of established atherosclerotic plaques, it does not have the necessary spatial resolution to detect early stages of the disease [56]. Thus, 3-D EIS mapping offers the possibility to detect metabolically active atherosclerotic lesions, albeit angiographically invisible, for possible early medical intervention and prevention of acute coronary syndromes or stroke.

Let us return to the early 4-point EIS configuration, its drawback becomes obvious when extended to accommodate multiple measuring sites, as in the case of the 6-point configuration presented in this section. There would be 12 electrodes needed for 4-electrode systems, thereby greatly complicating the possible electrode placement as well as the electrical connections to the measurement instruments. Regarding the electrode-tissue interface impedance, as shown in Fig. 6a, there is a clear shift of the impedance value throughout the frequency spectrum between balloon deflation and inflation. These impedance values are composed of the interface impedance as well as the tissues under interrogation. It suggests that the interface impedance is dominated by the tissue impedance. Hence, the impedance measurement reflects varying responses from the underlying tissues (atherosclerotic plaques, aorta, etc.) and can be utilized in evaluating different tissue compositions. It is worth noting that if the electrodes were to be further miniaturized, which will increase the electrode-tissue interface impedance, the tissue impedance might not be the dominating component in the 2-electrode configuration. Further treatment, such as electroplating platinum black onto the electrode to reduce the interface impedance, would be necessary to achieve high measurement specificity.

EIS sensing devices in previous sections only focus on the localized detection that merely detects a small region of the entire endovascular segment where the atherosclerotic lesions are often eccentric and multiple. The 6-point configuration further advances EIS sensing capability to optimize 3-D detection of small, angiographically invisible, atherosclerotic lesions. The addition in number of active electrodes (from 2 to 6) engendered 15 different permutations to extend the EIS measurements from a localized site to an entire circumferential ring of the aorta. The local EIS measurements are then reconstructed into 3-D impedimetric mapping to significantly enhance the visualization quality and translational applicability of the impedance data. This finding signifies a major characteristic shift from healthy arteries to ones with subclinical atherosclerosis and therefore can serve as a detection criterion. The wider range of impedance value arises from the fact that the existence of eccentric and multiple atherosclerotic lesions around the endoluminal surface increases the overall impedance variation compared to a homogeneous healthy artery.

3 Conclusion and Future Outlook

This chapter presents the fundamentals of EIS and its application in atherosclerosis focusing on a variety of electrode configuration in recent development for achieving better detection performance. The evolution from the early 4-point configuration, to the concentric bipolar electrodes, to 2-point symmetric and eventually 6-point configuration that enables the 3-D EIS interrogation is discussed in great details. With all the above discussed exciting results, the next question to answer is the potential of EIS toward clinical application. The intravascular assessment of coronary arteries in human with varying levels of intraplaque lipid burden has reached clinical use [22]. There are studies in humans demonstrating that plaques with larger lipid and oxLDL area are more prone to vulnerability and rupture with ensuing clinical events such as myocardial infarction [14]. These comparisons with pathology studies in humans provide the basis for future applicability of EIS to distinguish stable from vulnerable atherosclerotic plaques. Future clinical adaptation in humans of EIS sensor would also require intravascular advancement of the catheter and verification of balloon inflation under angiographic guidance, with close monitoring of impedance characteristics to differentiate the distinct patterns obtained when the sensor is only in contact with blood as opposed to the endoluminal vessel wall.

From technological advancement perspective, the next development for the EIS sensor will be to further increase the number of individual electrodes, which will significantly increase the number of EIS measurement permutation. This will lead to finer angular coverage and better spatial resolution for the 3-D EIS mapping. On the other hand, the potential to integrate the EIS sensing with other established techniques will also bring significant advancement in the overall diagnostic capability. For instance, flow fraction reserve (FFR) has been routinely applied for determining the indication for intervention [57]. Under the current guideline, there could be potential failure in recognizing rupture-prone plaque that does not cause significant hemodynamic occlusion. The combination of EIS and FFR can offer a solution in that EIS helps to identify lipid-rich core even it is still not significant enough to be detected by FFR. On the other hand, fibrous cap oxLDL-rich atheroma generates echolucency while the calcified lesions engenders high echogenicity [58]. Encouraging findings from the high-frequency dual ultrasound (US) and optical coherence tomographic (OCT) probe demonstrate the feasibility to detect both high resolution thin-cap fibroatheroma (TCFA) and intimal hyperplasia [59]. Integrating the US–OCT system with EIS modality to detect active lipids and macrophages in the vessel wall will further enhance the capability to characterize intimal thickening, calcification, thin fibrous cap, calcification, and metabolic states.

References

1. Bentzon, J. F., Otsuka, F., Virmani, R., & Falk, E. (2014). Mechanisms of plaque formation and rupture. *Circulation Research, 114*, 1852–1866.
2. Yahagi, K., Kolodgie, F. D., Otsuka, F., Finn, A. V., Davis, H. R., Joner, M., & Virmani, R. (2016). Pathophysiology of native coronary, vein graft, and in-stent atherosclerosis. *Nature Reviews Cardiology, 13*, 79.
3. Vengrenyuk, Y., Carlier, S., Xanthos, S., Cardoso, L., Ganatos, P., Virmani, R., Einav, S., Gilchrist, L., & Weinbaum, S. (2006). A hypothesis for vulnerable plaque rupture due to stress-induced debonding around cellular microcalcifications in thin fibrous caps. *Proceedings of the National Academy of Sciences, 103*, 14678–14683.
4. Libby, P., Ridker, P. M., & Hansson, G. K. (2011). Progress and challenges in translating the biology of atherosclerosis. *Nature, 473*, 317.
5. Little, W. C., Constantinescu, M., Applegate, R. J., Kutcher, M. A., Burrows, M. T., Kahl, F. R., & Santamore, W. P. (1988). Can coronary angiography predict the site of a subsequent myocardial infarction in patients with mild-to-moderate coronary artery disease? *Circulation, 78*, 1157–1166.
6. Ambrose, J. A., Tannenbaum, M. A., Alexopoulos, D., Hjemdahl-Monsen, C. E., Leavy, J., Weiss, M., Borrico, S., Gorlin, R., & Fuster, V. (1988). Angiographic progression of coronary artery disease and the development of myocardial infarction. *Journal of the American College of Cardiology, 12*, 56–62.
7. Fuster, V., Badimon, L., Badimon, J., & Chesebro, J. (1992). The pathogenesis of coronary artery disease and the acute coronary syndromes (1). *The New England Journal of Medicine, 326*(4), 242–250. doi: 10.1056, NEJM199201233260406.[Abstract][Cross Ref].
8. Libby, P. (2013). Mechanisms of acute coronary syndromes and their implications for therapy. *New England Journal of Medicine, 368*, 2004–2013.
9. Anderson, J. L., Adams, C. D., Antman, E. M., Bridges, C. R., Califf, R. M., Casey, D. E., Chavey, W. E., Fesmire, F. M., Hochman, J. S., & Levin, T. N. (2007). ACC/AHA 2007 guidelines for the management of patients with unstable angina/non–ST-elevation myocardial infarction—executive summary: A report of the American College of Cardiology/American Heart Association Task Force on Practice Guidelines (Writing Committee to Revise the 2002 Guidelines for the Management of Patients With Unstable Angina/Non–ST-Elevation Myocardial Infarction) Developed in Collaboration with the American College of Emergency Physicians, the Society for Cardiovascular Angiography and Interventions, and the Society of Thoracic Surgeons Endorsed by the American Association of Cardiovascular and Pulmonary Rehabilitation and the Society for Academic Emergency Medicine. *Journal of the American College of Cardiology, 50*, 652–726.
10. Heistad, D. D. (2003). Unstable coronary-artery plaques. *New England Journal of Medicine, 349*, 2285–2287.
11. Galis, Z. S. (2004). Vulnerable plaque: The devil is in the details. *Circulation, 110*(3), 244–246.
12. Brown, M. S., & Goldstein, J. L. (1983). Lipoprotein metabolism in the macrophage: Implications for cholesterol deposition in atherosclerosis. *Annual Review of Biochemistry, 52*, 223–261.
13. Park, Y. M., Febbraio, M., & Silverstein, R. L. (2009). CD36 modulates migration of mouse and human macrophages in response to oxidized LDL and may contribute to macrophage trapping in the arterial intima. *The Journal of Clinical Investigation, 119*, 136–145.
14. Ehara, S., Ueda, M., Naruko, T., Haze, K., Itoh, A., Otsuka, M., Komatsu, R., Matsuo, T., Itabe, H., & Takano, T. (2001). Elevated levels of oxidized low density lipoprotein show a positive relationship with the severity of acute coronary syndromes. *Circulation, 103*, 1955–1960.
15. Chinetti-Gbaguidi, G., Baron, M., Bouhlel, M. A., Vanhoutte, J., Copin, C., Sebti, Y., Derudas, B., Mayi, T., Bories, G., & Tailleux, A. (2011). Human atherosclerotic plaque alternative macrophages display low cholesterol handling but high phagocytosis because of distinct activities of the PPARγ and LXRα pathways. *Circulation Research*. CIRCRESAHA. 110.233775.

16. Tang, D., Yang, C., Kobayashi, S., Zheng, J., & Vito, R. P. (2003). Effect of stenosis asymmetry on blood flow and artery compression: A three-dimensional fluid-structure interaction model. *Annals of Biomedical Engineering, 31*, 1182–1193.
17. Schwartz, S. M., Galis, Z. S., Rosenfeld, M. E., & Falk, E. (2007). Plaque rupture in humans and mice. *Arteriosclerosis, Thrombosis, and Vascular Biology, 27*, 705–713.
18. Berliner, J. A., & Heinecke, J. W. (1996). The role of oxidized lipoproteins in atherogenesis. *Free Radical Biology and Medicine, 20*, 707–727.
19. Asatryan, L., Hamilton, R. T., Isas, J. M., Hwang, J., Kayed, R., & Sevanian, A. (2005). LDL phospholipid hydrolysis produces modified electronegative particles with an unfolded apoB-100 protein. *Journal of Lipid Research, 46*, 115–122.
20. Navab, M., Berliner, J. A., Watson, A. D., Hama, S. Y., Territo, M. C., Lusis, A. J., Shih, D. M., Van Lenten, B. J., Frank, J. S., & Demer, L. L. (1996). The yin and yang of oxidation in the development of the fatty streak: A review based on the 1994 George Lyman Duff Memorial Lecture. *Arteriosclerosis, Thrombosis, and Vascular Biology, 16*, 831–842.
21. Ross, R. (1999). Pathogenesis of atherosclerosis-atherosclerosis is an inflammatory disease. *American Heart Journal, 138*, S419.
22. Bourantas, C. V., Garcia-Garcia, H. M., Naka, K. K., Sakellarios, A., Athanasiou, L., Fotiadis, D. I., Michalis, L. K., & Serruys, P. W. (2013). Hybrid intravascular imaging: Current applications and prospective potential in the study of coronary atherosclerosis. *Journal of the American College of Cardiology, 61*, 1369–1378.
23. Worthley, S. G., Helft, G., Fuster, V., Fayad, Z. A., Shinnar, M., Minkoff, L. A., Schechter, C., Fallon, J. T., & Badimon, J. J. (2003). A novel nonobstructive intravascular MRI coil: In vivo imaging of experimental atherosclerosis. *Arteriosclerosis, Thrombosis, and Vascular Biology, 23*, 346–350.
24. Jaffer, F. A., Vinegoni, C., John, M. C., Aikawa, E., Gold, H. K., Finn, A. V., Ntziachristos, V., Libby, P., & Weissleder, R. (2008). Real-time catheter molecular sensing of inflammation in proteolytically active atherosclerosis. *Circulation, 118*, 1802–1809.
25. Marcu, L., Fishbein, M. C., Maarek, J.-M. I., & Grundfest, W. S. (2001). Discrimination of human coronary artery atherosclerotic lipid-rich lesions by time-resolved laser-induced fluorescence spectroscopy. *Arteriosclerosis, Thrombosis, and Vascular Biology, 21*, 1244–1250.
26. Castellanos, A., Ramos, A., Gonzalez, A., Green, N. G., & Morgan, H. (2003). Electrohydrodynamics and dielectrophoresis in microsystems: Scaling laws. *Journal of Physics D: Applied Physics, 36*, 2584.
27. Lisdat, F., & Schäfer, D. (2008). The use of electrochemical impedance spectroscopy for biosensing. *Analytical and Bioanalytical Chemistry, 391*, 1555.
28. Hondroulis, E., Zhang, R., Zhang, C., Chen, C., Ino, K., Matsue, T., & Li, C.-Z. (2014). Immuno nanoparticles integrated electrical control of targeted cancer cell development using whole cell bioelectronic device. *Theranostics, 4*, 919.
29. Li, H., Huang, Y., Zhang, B., Yang, D., Zhu, X., & Li, G. (2014). A new method to assay protease based on amyloid misfolding: Application to prostate cancer diagnosis using a panel of proteases biomarkers. *Theranostics, 4*, 701.
30. Konings, M., Mali, W. T. M., & Viergever, M. A. (1997). Development of an intravascular impedance catheter for detection of fatty lesions in arteries. *IEEE Transactions on Medical Imaging, 16*, 439–446.
31. Meissner, R., Eker, B., Kasi, H., Bertsch, A., & Renaud, P. (2011). Distinguishing drug-induced minor morphological changes from major cellular damage via label-free impedimetric toxicity screening. *Lab on a Chip, 11*, 2352–2361.
32. Packard, R. R. S., Luo, Y., Abiri, P., Jen, N., Aksoy, O., Suh, W. M., Tai, Y.-C., & Hsiai, T. K. (2017). 3-D electrochemical impedance spectroscopy mapping of arteries to detect metabolically active but angiographically invisible atherosclerotic lesions. *Theranostics, 7*, 2431.
33. Streitner, I., Goldhofer, M., Cho, S., Thielecke, H., Kinscherf, R., Streitner, F., Metz, J., Haase, K. K., Borggrefe, M., & Suselbeck, T. (2009). Electric impedance spectroscopy of human atherosclerotic lesions. *Atherosclerosis, 206*, 464–468.

34. Süselbeck, T., Thielecke, H., Köchlin, J., Cho, S., Weinschenk, I., Metz, J., Borggrefe, M., & Haase, K. K. (2005). Intravascular electric impedance spectroscopy of atherosclerotic lesions using a new impedance catheter system. *Basic Research in Cardiology, 100*, 446–452.
35. Grimnes, S., & Martinsen, Ø. G. (2006). Sources of error in tetrapolar impedance measurements on biomaterials and other ionic conductors. *Journal of Physics D: Applied Physics, 40*, 9.
36. Geselowitz, D. B. (1971). An application of electrocardiographic lead theory to impedance plethysmography. *IEEE Transactions on Biomedical Engineering, BME-18*, 38–41.
37. Yu, F., Li, R., Ai, L., Edington, C., Yu, H., Barr, M., Kim, E., & Hsiai, T. K. (2011). Electrochemical impedance spectroscopy to assess vascular oxidative stress. *Annals of Biomedical Engineering, 39*, 287–296.
38. Yu, F., Dai, X., Beebe, T., & Hsiai, T. (2011). Electrochemical impedance spectroscopy to characterize inflammatory atherosclerotic plaques. *Biosensors and Bioelectronics, 30*, 165–173.
39. Stary, H. C. (2000). Natural history and histological classification of atherosclerotic lesions: An update. *Arteriosclerosis, Thrombosis, and Vascular Biology, 20*, 1177–1178.
40. Kim, D.-H., Lu, N., Ma, R., Kim, Y.-S., Kim, R.-H., Wang, S., Wu, J., Won, S. M., Tao, H., & Islam, A. (2011). Epidermal electronics. *Science, 333*, 838–843.
41. Cao, H., Yu, F., Zhao, Y., Scianmarello, N., Lee, J., Dai, W., Jen, N., Beebe, T., Li, R., & Ebrahimi, R. (2014). Stretchable electrochemical impedance sensors for intravascular detection of lipid-rich lesions in New Zealand White rabbits. *Biosensors and Bioelectronics, 54*, 610–616.
42. Grüntzig, A. (1978). Transluminal dilatation of coronary-artery stenosis. *The Lancet, 311*, 263.
43. Sigwart, U., Puel, J., Mirkovitch, V., Joffre, F., & Kappenberger, L. (1987). Intravascular stents to prevent occlusion and re-stenosis after transluminal angioplasty. *New England Journal of Medicine, 316*, 701–706.
44. Indolfi, C., De Rosa, S., & Colombo, A. (2016). Bioresorbable vascular scaffolds—Basic concepts and clinical outcome. *Nature Reviews Cardiology, 13*, 719.
45. Iqbal, J., Chamberlain, J., Francis, S. E., & Gunn, J. (2016). Role of animal models in coronary stenting. *Annals of Biomedical Engineering, 44*, 453–465.
46. Levine, G. N., Bates, E. R., Blankenship, J. C., Bailey, S. R., Bittl, J. A., Cercek, B., Chambers, C. E., Ellis, S. G., Guyton, R. A., & Hollenberg, S. M. (2011). 2011 ACCF/AHA/SCAI guideline for percutaneous coronary intervention: A report of the American College of Cardiology Foundation/American Heart Association Task Force on Practice Guidelines and the Society for Cardiovascular Angiography and Interventions. *Journal of the American College of Cardiology, 58*, e44–e122.
47. Packard, R. R. S., Zhang, X., Luo, Y., Ma, T., Jen, N., Ma, J., Demer, L. L., Zhou, Q., Sayre, J. W., & Li, R. (2016). Two-point stretchable electrode array for endoluminal electrochemical impedance spectroscopy measurements of lipid-laden atherosclerotic plaques. *Annals of Biomedical Engineering, 44*, 2695–2706.
48. Chang, J. H.-C., Huang, R., & Tai, Y.-C. (2011). High density 256-channel chip integration with flexible parylene pocket. In *Solid-state sensors, actuators and microsystems conference (TRANSDUCERS), 2011 16th international* (pp. 378–381).
49. Koike, T., Liang, J., Wang, X., Ichikawa, T., Shiomi, M., Sun, H., Watanabe, T., Liu, G., & Fan, J. (2005). Enhanced aortic atherosclerosis in transgenic Watanabe heritable hyperlipidemic rabbits expressing lipoprotein lipase. *Cardiovascular Research, 65*, 524–534.
50. Yeo, W. H., Kim, Y. S., Lee, J., Ameen, A., Shi, L., Li, M., Wang, S., Ma, R., Jin, S. H., & Kang, Z. (2013). Multifunctional epidermal electronics printed directly onto the skin. *Advanced Materials, 25*, 2773–2778.
51. Chang, J. H.-C., Liu, Y., Kang, D., & Tai, Y.-C. (2013). Reliable packaging for parylene-based flexible retinal implant. In *Solid-state sensors, actuators and microsystems (Transducers & Eurosensors XXVII), 2013 Transducers & Eurosensors XXVII: The 17th international conference on* (pp. 2612–2615).

52. Fakirov, S., Evstatiev, M., & Petrovich, S. (1993). Microfibrillar reinforced composites from binary and ternary blends of polyesters and nylon 6. *Macromolecules, 26*, 5219–5226.
53. Lin, J. C.-H., Lam, G., & Tai, Y.-C. (2012). Viscoplasticity of parylene-C film at body temperature. In *Micro electro mechanical systems (MEMS), 2012 IEEE 25th international conference on* (pp. 476–479).
54. Ebrahimi, A. P. (2009). Mechanical properties of normal and diseased cerebrovascular system. *Journal of Vascular and Interventional Neurology, 2*, 155.
55. Enos, W. F., Holmes, R. H., & Beyer, J. (1953). Coronary disease among United States soldiers killed in action in Korea: Preliminary report. *Journal of the American Medical Association, 152*, 1090–1093.
56. Dweck, M. R., Doris, M. K., Motwani, M., Adamson, P. D., Slomka, P., Dey, D., Fayad, Z. A., Newby, D. E., & Berman, D. (2016). Imaging of coronary atherosclerosis—Evolution towards new treatment strategies. *Nature Reviews Cardiology, 13*, 533.
57. De Bruyne, B., Pijls, N. H., Kalesan, B., Barbato, E., Tonino, P. A., Piroth, Z., Jagic, N., Möbius-Winkler, S., Rioufol, G., & Witt, N. (2012). Fractional flow reserve–guided PCI versus medical therapy in stable coronary disease. *New England Journal of Medicine, 367*, 991–1001.
58. Yamada, R., Okura, H., Miyamoto, Y., Kawamoto, T., Neishi, Y., Hayashida, A., Tsuchiya, T., Nezuo, S., & Yoshida, K. (2011). A newly developed radio frequency signal-based intravascular ultrasound tissue characterization: A comparison between imap and integrated backscatter intravascular ultrasound. *Journal of the American College of Cardiology, 57*, E1882.
59. Li, X., Yin, J., Hu, C., Zhou, Q., Shung, K. K., & Chen, Z. (2010). High-resolution coregistered intravascular imaging with integrated ultrasound and optical coherence tomography probe. *Applied Physics Letters, 97*, 133702.

Epidermal EIT Electrode Arrays for Cardiopulmonary Application and Fatty Liver Infiltration

Yuan Luo, Parinaz Abiri, Chih-Chiang Chang, Y. C. Tai, and Tzung K. Hsiai

1 Introduction

Electrical impedance tomography (EIT) is a noninvasive medical imaging technique by which an image of the tissue electrical properties is derived from surface electrode measurements [1]. The first clinical images obtained using EIT were produced by the Sheffield group for pulmonary function [2]. Compared with other more commonly adopted imaging modalities such as X-ray computed tomography (CT), magnetic resonance imaging (MRI), ultrasound imaging, etc., EIT has the advantage of little to none safety concern for its extremely low current requirement (~5–10 mA), significantly lower material cost and very compact and portal hardware for its main implementation depends only on standard electronics, and long-term continuous monitoring capability for its low power consumption and hardware stability. Although first applied in geographical exploration for minerals, over the past three decades, EIT has seen ever-increasing medical imaging applications including cardiopulmonary, breast tissue, and brain function [3]. Respiratory monitoring has been established by transthoracic impedance pneumography [4, 5]. This chapter aims at providing a general discussion for the fundamental principles of EIT as well as the imaging implementation. Moreover, two particular applications, one well-established area cardiopulmonary monitoring using EIT and one recent developing area fatty liver infiltration monitoring, will be discussed.

Y. Luo · Y. C. Tai
Medical Engineering, California Institute of Technology, Pasadena, CA, USA

P. Abiri · C.-C. Chang
Department of Bioengineering, University of California, Los Angeles, CA, USA

T. K. Hsiai (✉)
Medical Engineering, California Institute of Technology, Pasadena, CA, USA

Department of Bioengineering, University of California, Los Angeles, CA, USA
e-mail: thsiai@mednet.ucla.edu

H. Cao et al. (eds.), *Interfacing Bioelectronics and Biomedical Sensing*,
https://doi.org/10.1007/978-3-030-34467-2_7

1.1 Fundamental Principle of EIT

Numerous experimental evidence have demonstrated that different biological samples exhibit different electrical conductivity (σ) and permittivity (ε) under different physiological and pathological circumstances [6]. From classical electromagnetism perspective, the mathematical description of the response (electrical current/voltage distribution) from a biological sample under electrical excitation has been well formulated. In a typical EIT measurement setting, sinusoidal signal with frequency ranging from 1 k to 1 M Hz is applied to the sample. Therefore, time-harmonic Maxwell equation is used to describe the response from the sample. We confine our mathematical description within the sample of interest and define this specific domain as Ω. Next, we can write electric $\vec{E}$ and magnetic field $\vec{H}$ distribution as:

$$\vec{E}(\vec{r},t) = \mathrm{Re}\left\{\left\|\vec{E}(\vec{r})\right\| e^{i\omega t}\right\} \tag{1}$$

$$\vec{H}(\vec{r},t) = \mathrm{Re}\left\{\left\|\vec{H}(\vec{r})\right\| e^{i\omega t}\right\} \tag{2}$$

where $\vec{r}$ represents the vector connecting any point of interest within the domain to the origin, t denotes time, ω is the angular frequency of the signal, and $i = \sqrt{-1}$. Next, we have the relation among electric field and magnetic field as:

$$\nabla \times \vec{H}(\vec{r}) = -i\omega\varepsilon \vec{E}(\vec{r}) + J_f(\vec{r}) \tag{3}$$

$$\nabla \times \vec{E}(\vec{r}) = -i\omega\mu \vec{H}(\vec{r}) \tag{4}$$

where ε and μ is the permittivity and permeability of the sample, J_f is the independent current not arising from magnetization-bound current or polarization-bound charge movement [7]. Under typical frequency range and dimension scale in EIT, $\omega\mu\vec{H}(\vec{r})$ is negligible [1]; therefore, $\nabla \times \vec{E}(\vec{r}) = 0$, which then leads to the following relation:

$$\vec{E}(\vec{r}) = -\nabla V(\vec{r}) \tag{5}$$

where $V(\vec{r})$ is the voltage distribution within the sample. We have one more physical relation to be used, the Ohm's law:

$$J_f(\vec{r}) = \sigma \vec{E}(\vec{r}) \tag{6}$$

Combining Eqs. (1), (2), (3), (4), (5), and (6), we can obtain:

$$\nabla\left((\sigma + i\omega\varepsilon)\nabla V(\vec{r})\right) = 0 \tag{7}$$

Conventionally, we define $\sigma^* = \sigma + i\omega\varepsilon$ as the complex conductivity of the sample. Equation (7) is sometimes referred to as the time-harmonic Laplace equation. Next, we consider the boundary condition (parameters in $\partial\Omega$). We define the so-called Dirichlet-to-Neumann (DtN) mapping:

$$\Lambda_{\sigma^*} : V\big|_{\partial\Omega} \to J \tag{8}$$

where $J = \sigma^* \frac{\partial V}{\partial \vec{n}}\Big|_{\partial\Omega}$ is called the Neumann boundary condition (current density on the boundary) and $V|_{\partial\Omega}$ is referred to as the Dirichlet boundary condition (voltage on the boundary). The standard inverse problem of EIT is to find the mapping from Λ_{σ^*} to σ^*, which is first formulated by Calderon [8]. For such a physical-mathematical problem, we define whether it is well-posed according to Jacques Hadamard's three criteria [9]: (1) a solution exists, (2) the solution is unique, and (3) the solution depends continuously on the data. The existence of a solution is really not a concern since we are dealing with real physical problem and there must be a solution. The uniqueness of the solution has been proven under certain sets of reasonable assumption under typical EIT measurement setting [10]. It is the continuity criterion (as well as the nonlinear nature of the Laplace equation) that renders the EIT problem notoriously difficult to solve. One last note about the boundary condition is the consideration of the contact impedance between the electrode and the biological sample stemming from the ionic nature of biological tissue in contact with metal electrodes. It is recommended to use the complete electrode model (CEM) while handling the EIT problem. The readers can find great detail about the CEM in Holder's great book on EIT [1].

1.2 *Nonlinear Inverse Problem for EIT Imaging*

Despite great success in the understanding of the mathematical nature about EIT problem, it is impossible to obtain analytical solution. Nowadays, numerical methods are the predominant approach applied to reconstruct EIT mapping from experimental data. In a typical EIT measurement setup, it is impossible to obtain either the Dirichlet or Neumann boundary condition in a continuous sense. The normal measuring protocol is that an array of discrete electrodes is placed around the target object. A typical electrode placement scheme is illustrated in Fig. 1a. To perform the measurement, certain pair of electrodes (I_{14}) will be chosen as the current injection pair, while voltage values are measured for all the other pairs of electrodes ($V_{14}^1, V_{25}^1, V_{36}^1 \supset, V_{16,3}^1$, Fig. 1b). The current injection pair shifts one position (I_{25}, Fig. 1c), and voltage measurement is repeated to generate another ($V_{14}^2, V_{25}^2, V_{36}^2 \supset, V_{16,3}^2$). Eventually, there will be N^2 voltage data in one set of measurement (N is the number of electrode placed on the surface of the object).

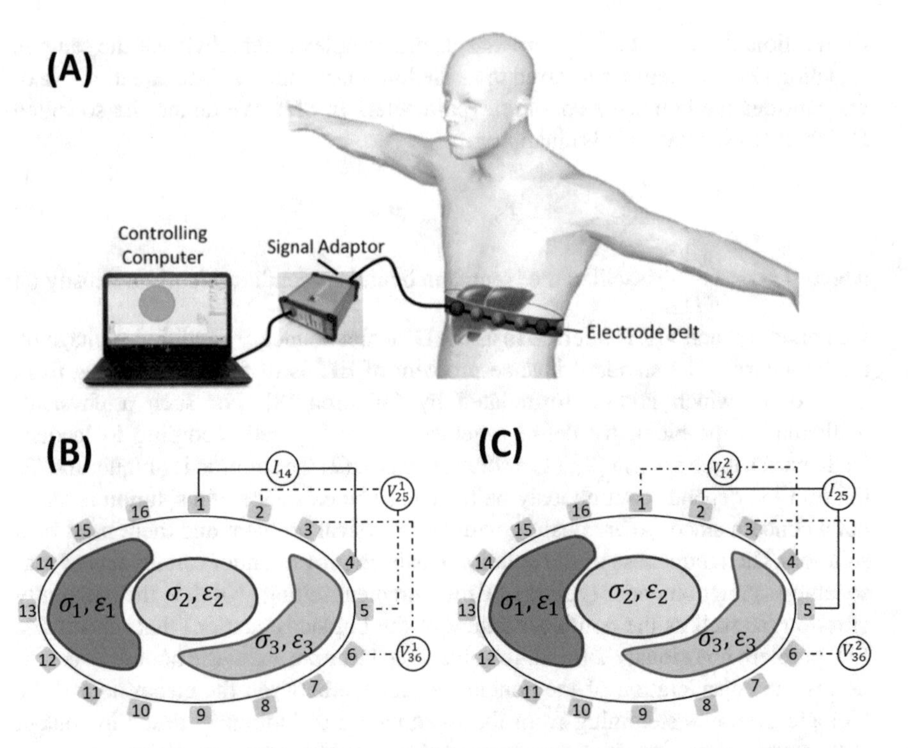

Fig. 1 (**a**) Schematics of typical EIT experimental setup with electrode array attached on tested subject. (Adapted from Luo et al. [11] (B&C) EIT current injection/voltage measurement protocol)

Now let us reconsider Eq. (7). For the ensuing discussion, σ will be used to replace σ^* as conclusions using σ (the static form of σ^* with $\omega = 0$) can be extended to the complex case [12]. EIT is the inverse problem of Eq. (7) in determining σ distribution in Ω, and the continuous boundary condition in Eq. (8) becomes the discrete voltage data measured from the object domain boundaries $\partial\Omega$ and can be expressed as $\vec{V}_m = \left(V_{15}^1, V_{26}^1, V_{37}^1 \ldots, V_{15}^2, V_{26}^2, V_{37}^2 \ldots, V_{15}^N, V_{26}^N, V_{37}^N \ldots\right)$.

Most of nowadays EIT algorithm adopts the Gauss-Newton (GN)-type numerical schemes. Consider an error function of the following form:

$$e = \vec{V}_m - \vec{f}(\sigma) \tag{9}$$

where $\vec{f}(\sigma)$ is implicitly defined as the forward model to obey Eq. (7). To implement the GN solver, we solve σ to minimize the L-2 norm of the error e:

$$\Phi = ||e||^2 = ||\vec{V}_m - \vec{f}(\sigma)||^2 \tag{10}$$

By taking the first order of Taylor series expansion of the forward problem function, we approximated as follows:

$$\vec{f}(\sigma) \cong \vec{f}(\sigma_0) + J(\sigma - \sigma_0) \tag{11}$$

where σ_0 is a reference conductivity value and J is called Jacobian matrix (or sometimes sensitivity matrix) of our inverse problem. The objective function is defined as:

$$\Phi = ||\vec{V}_s - \vec{f}(\sigma) - J(\sigma - \sigma_0)||^2 \tag{12}$$

To minimize Φ, we set $\partial\Phi/\partial\sigma = 0$ to solve for σ from Eq. (12) and obtain:

$$\sigma = \sigma_0 + \left(J^T J\right)^{-1} J^T \left(\vec{V}_m - \vec{f}(\sigma_0)\right) \tag{13}$$

Equation (13) is an unconstrained GN form of the inverse problem with the well-recognized instability issues. We thus introduced a constraint term to sway the solution toward the preferred solution:

$$\Phi = ||e||^2 + \lambda ||\Gamma\sigma||^2 \tag{14}$$

The coefficient, λ, is the regularization parameter to "punish" the large conductivity spikes in the solution space and to balance the trade-off between fitting the error and constraining the solution from the undesired properties. The term, Γ, is introduced to select more properties of the sigma. For a smooth solution, Γ is considered to be a Laplacian operator to punish the non-smoothness of the solution. For an a priori conductivity with similar area, Γ is considered to be a "weighted" Laplacian operator to punish the non-smooth regions. This strategy is applicable in medical imaging where EIT is applied in concert with other imaging modalities to acquire the conductivity information of the organ systems, and the anatomic localization of the organs is available from the a priori information obtained from another imaging modality, such as CT or MRI. Following application of the regulation term to Eq. (14), the solution is defined as follows:

$$\sigma = \sigma_0 + \left(J^T J + \lambda \Gamma^T \Gamma\right)^{-1} J^T \left(\vec{V}_m - \vec{f}(\sigma_0)\right) \tag{15}$$

The scheme of absolute imaging is performed by an iterated approach, assuming an arbitrary conductivity, σ_0, for calculating J, Γ, and $\vec{f}(\sigma_0)$. Using Equation (15), we obtain a new conductivity value, σ_1. The iteration continued until the difference between σ_n and σ_{n+1} reached a minimally desired value.

In some applications where the absolute conductivity value is not needed, but rather the variation of conductivity from a previous distribution is more valuable, we can adopt the differential form of the EIT imaging solution:

$$\delta\sigma = \left(J^T J + \lambda \Gamma^T \Gamma\right)^{-1} J^T \delta\vec{V}_m \tag{16}$$

where $\delta\vec{V}_m$ is the difference between the voltage measurement of two different time points and $\delta\sigma$ is the conductivity variation between the two. Such a differential approach can significantly minimize experimental errors from the measurement system and the object geometry.

1.3 EIDORS Open-Source Tools

An online open-source software suite known as EIDORS [13], used for image reconstruction in electrical impedance tomography and diffuse optical tomography, has been established by a group of researchers as a development tool kit for the EIT community. EIDORS has been under active maintenance to date as a collective effort over the years. All the abovementioned numerical algorithm for EIT reconstruction is available in EIDORS, as well as physical modeling and a variety of data analysis options. As it is open source, EIT researchers around the world can also make their own version of the algorithm available. People with less mathematical background but intending to use EIT in interesting imaging applications can have access to a tremendously helpful tool. These all together make great contribution in promoting the advancement of EIT to a new level.

For EIT simulation, EIDORS allowed us to establish a forward finite element model (with help of Netgen, a mesh generator [14]) with arbitrary geometric boundaries and conductivity values that are assigned within the model in specific constrained volumes. Moreover, the electrode arrays on the surface of the model with specified coordinates, shape, and area can be precisely defined. Voltage values at each electrode are simulated with the current injection boundary condition and with the pre-assigned conductivity value based on Eq. (7). These voltage values served as the input for the subsequent EIT imaging reconstruction. EIDORS also allows the addition of noise mathematically to the simulated data. Next, to perform the EIT imaging, one needs to create a new inverse finite element model to avoid the so-called inverse crime [15], and no specific conductivity value is assigned. A variety of algorithm is available, in particular, the GN-type method with different regularization schemes. For real-imaging application, only the inverse finite element model is required, and the voltage data is directly obtained from the experimental measurements. Distinct geometry of the target object can be established accordingly in EIDORS for the inverse finite element model.

In the next two sections, we will review the exciting results in applying EIT for real medical imaging applications in two areas. We will first discuss EIT for cardiopulmonary monitoring, which is widely considered the most successful application for EIT. Next, we discuss a new application in using EIT for liver fat infiltration detection.

2 EIT for Cardiopulmonary Application

2.1 Motivation

For better diagnosis and therapy of cardiac, circulatory, and ventilatory disorders, it is critical to obtain anatomical and functional information within the thorax. Recently, magnetic resonance imaging (MRI), ultrasound, positron emission tomography (PET), and X-ray computed tomography (CT) are widely applied to acquire anatomical structure or functional imaging or even both simultaneously. We refer the readers further for a detailed review of advantages and drawbacks for each method in the recent publication [16]. EIT has been suggested as an alternative method for thoracic monitoring largely due to all the previous mentioned advantages [1, 16–18].

For cardiopulmonary application, the measurement of EIT is usually conducted by an electrode belt of multiple electrodes placed around the chest wall. The electrode belt is placed in one transverse plane, usually between the fourth and fifth intercostal space [16, 17]. The variation of electrical impedance within the thorax is strongly related to cardiac and ventilatory events. By adopting the differential mode reconstruction as previously discussed, this technique is possible to measure the change in electrical impedance resulting from blood volume change in ventricles and atria during a cardiac cycle with the simple principle that blood is highly conductive substance as compared to the rest of the tissues. For the pulmonary EIT application, the sources of contrast are mainly from the perfusion of the blood or the inhalation of gas, either during spontaneous breathing or during mechanical ventilation (MV), causing an expansion of the alveoli. This results in lengthening as well as the diameter constricting of current paths, which ultimately leads to the variation of electrical impedance in the thorax. Thus, the air-filled lung has a lower conductivity which is linearly related to the degree of inflation, enabling the measurement of pulmonary ventilation and tracking the development of chronic lung diseases. Due to the clinical need for monitoring of lung ventilation and assessment of regional lung function at the bedside, there are many scientists and clinicians dedicating to study EIT in the thorax after Brown et al. [19] published the first paper of EIT application of the human thorax. Recent investigations and advances in different disease categories in adults are reported including monitoring mechanical ventilation of patients, acute respiratory distress syndrome (ARDS), chronic obstructive pulmonary disease (COPD), and cystic fibrosis (CF). Moreover, several commercial EIT devices are currently available and being used on a day-to-day basis for pulmonary application in different ICUs around the world: Goe-MF ® II (CareFusion - Becton, Dickinson and Company, San Diego, United States), Mark 1 ® and Mark 3.5 ® (Maltron International Ltd., Rayleigh, United Kingdom), PulmoVista ® 500 (Dräger Medical GmbH, Lündbeck, Germany), BB ® (Swisstom AG, Landquart, Switzerland), and Enlight ® (Timpel S/A, Sao Paulo, Brazil) [20]. Following is a detailed discussion on multiple examples in using EIT for thoracic monitoring.

2.2 EIT in Mechanically Ventilated Patients During Surgery or ICU

Mechanical ventilation is a well-established technique during surgery and ICU for replacing or assisting patients' breathing. However, it is still necessary to require "lung protective ventilation" (LPV) to reduce cyclic alveolar collapse and overdistension of the lung [21]. The clinical strategy is now mainly based on the application of low tidal volumes and maneuvers designed to increase the functional residual capacity, trying to reduce ventilator-associated/induced lung injury. EIT is capable to capture the change in lung volume and regional distribution of ventilation in real time. Thus, it may assist in defining mechanical ventilation settings, assess the distribution of tidal volume and end-expiratory lung volume (EELV), and contribute to titrate positive end-expiratory pressure (PEEP)/tidal volume combinations [21–23]. In the past decades, there are researches conducted to demonstrate that EIT-guided ventilation results in improved respiratory mechanics, improved gas exchange, and reduced histologic evidence of ventilator-induced lung injury in patients and animals [24–30]. Eronia et al. [22] recently studied on feasibility setting personalized PEEP with electrical impedance tomography in order to prevent lung de-recruitment following a recruitment maneuver. Sixteen patients were enrolled to undergo mechanical ventilation with Pa_{O2}/Fi_{O2} (arterial oxygen partial pressure to fractional inspired oxygen ratio) <300 mmHg. Pa_{O2}/Fi_{O2} improved during PEEP EIT and the driving pressure decreased. Recruited volume correlated with the decrease in driving pressure but not with oxygenation improvement. In conclusion, regional alveolar hyperdistention and collapse was reduced in dependent lung layers and increased in non-dependent lung layers.

2.3 EIT for Pulmonary Perfusion

The capacity of EIT to detect systolic blood volume changes in the lungs offers the possibility of studying the pulmonary perfusion. Eyuboglu et al. [31] showed that ECG-gated dynamic EIT images of the thorax could be performed. Unfortunately, compared to the change in EIT signals due to ventilation, perfusion impedance signal change is significantly smaller in amplitude. Therefore, EIT for ventilation monitoring is on the verge of clinical trials, whereas pulmonary perfusion imaging with EIT still remains a challenge, especially in spontaneously breathing subjects. A recent study [32] proposed by Borges et al. offered a novel method to quantitatively measure regional lung perfusion based on first-pass kinetics of a bolus of hypertonic saline contrast (20%). Regional blood flow measured by EIT using this novel method, in both healthy and injured lung cases, is highly correlated to that measured by single-photon emission computerized tomography (SPECT).

2.4 EIT for Acute Respiratory Distress Syndrome (ARDS)

ARDS is the sudden failure of the respiratory system and is associated with a high mortality rate (27–45% from mild to severe) [33]. It is characterized by diffuse alveolar damage, associated with an increase in alveolar and capillary permeability due to mechanisms of tension, stretching and shear between alveolar units, leading to an accumulation of interstitial and alveolar edema. Moreover, thoracic CT scans have shown a strong heterogeneous distribution of lung aeration and ventilation in diseased lungs [34]. EIT is one of the few clinical tools that could measure the regional ventilation distributions and assess the degree of heterogeneities with real-time capability. The study conducted by Franchineau et al. [35] applied EIT as the bedside tool to optimize PEEP in ARDS patients for personalized titration of ventilation settings. EIT may have the potential to serve as a noninvasive bedside tool to provide real-time monitoring of the PEEP impact in ARDS patients.

2.5 EIT in Chronic Obstructive Pulmonary Diseases (COPD)

COPD is the most common disease involving the pulmonary vascular bed, especially the lung emphysema type. It is the third leading cause of death worldwide [36]. It is mainly characterized by permanent restricted and inflamed bronchi and/or a pathological over-inflation of the lungs. As previous discussed, the change of the blood volume affects the electrical impedance. Since the small pulmonary vascular bed is mainly responsible for blood volume causing impedance change, EIT has the potential to become a promising tool for diagnosis and monitoring COPD patients. The first clinical study investigating the possibilities of EIT to detect the pathological changes of the pulmonary vascular bed of patients was performed by Vonk Noordegraaf et al. [37]. They found that in emphysematous patients, cardiac-gated lung impedance changes are significantly smaller in comparison with healthy subjects. In addition, Vogt et al. [38] employed EIT to visualize the spatial and temporal ventilation distribution in 35 patients with COPD at baseline and 5, 10, and 20 min after bronchodilator inhalation. A positive response is defined by an increase in forced expiratory volume in 1 s (FEV_1). Patients with obstructive lung diseases commonly undergo bronchodilator reversibility testing during the examination of their pulmonary function by spirometry. Based on spirometric FEV_1, significant bronchodilator response was found in 17 patients. There was a significant improvement for the spatial distribution of pixel FEV_1 and tidal volume and temporal distribution in responders. EIT might be able to offer diagnostic and prognostic information derived from reversibility testing by offering the spatial and temporal regional data.

2.6 EIT in Cystic Fibrosis (CF)

Cystic fibrosis is an autosomal recessive disease. The disease first affects the small airways and starts to cause the obstruction and inflammation syndromes in the lung during a later stage. The current diagnostic relies on imaging techniques like CT, which is not appropriate for long-term monitoring of disease progression because of radiation exposure. Thus, EIT has the potential to become a promising tool to track the impedance change due to the tissue fibrosis. Zhao et al. [39] assessed the regional obstruction in CF patients by EIT and compare with the high-resolution computed tomography (HRCT). Five CF patients were routinely scheduled for HRCT examination. EIT measurements were performed on these patients ±2 months during a standard pulmonary function test. Ratios of maximum expiratory flows at 25% and 75% of vital capacity (MEF_{25} /MEF_{75}) with respect to relative impedance change were calculated for regional areas in EIT images. The regional airway obstruction identified in the MEF_{25} /MEF_{75} maps was similar to that found in CT (Fig. 2).

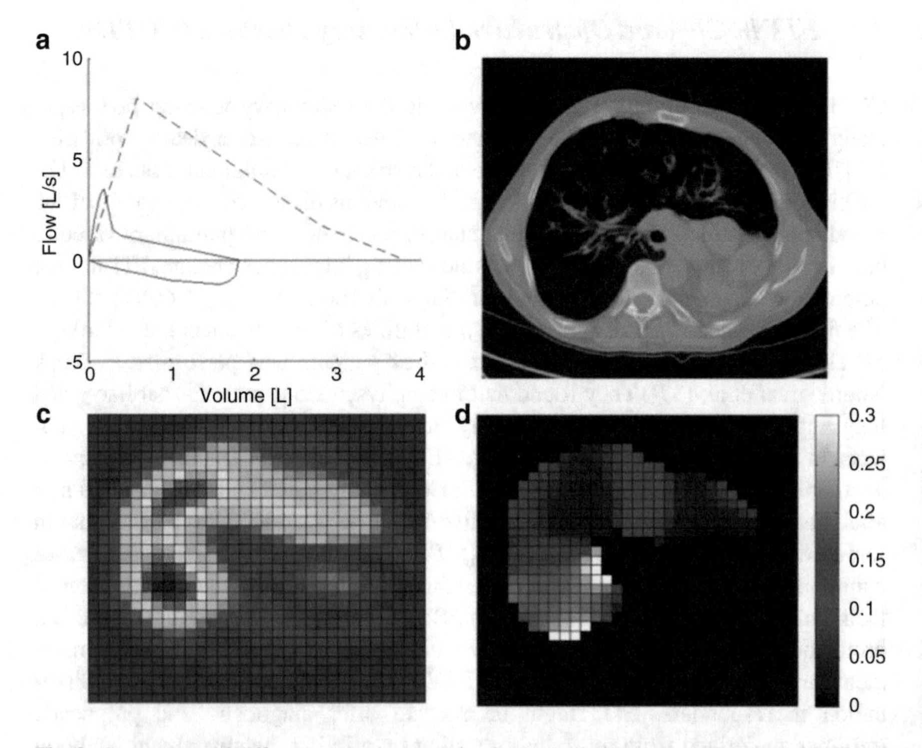

Fig. 2 Comparison of different methods measuring lung ventilation in one CF patient. (**a**) Global flow–volume curve measured by spirometer. (**b**) CT scan at the measurement plane of EIT. (**c**) Functional EIT showing ventilation distribution. High ventilated regions are marked in red. (**d**) Regional obstruction map derived from EIT measurement. The color bar indicates the MEF_{25} / MEF_{75} values in each pixel. Low MEF_{25} /MEF_{75} values are marked in dark blue. (Adapted from Zhao et al. [39])

2.7 *Discussion and Outlook*

Thorax EIT or functional lung imaging is probably the most prominent application of EIT offering real-time information on the local distribution of lung aeration. However, some limitation of EIT must be mentioned. EIT measurement may pose a risk when placed in patients with spinal injury, and the EIT measurement belt could not be placed over areas of damaged or inflamed skin. Moreover, poor-quality EIT images could be obtained from morbid obese patients with body mass index above 50. Measurements of EIT are sensitive to body movements, and hence their use in patients with uncontrolled body movements may not be reliable. Despite the possible limitations of the existing EIT systems, EIT's real-time information is highly relevant for decision-making in acute disease, whereas EIT's long-term monitoring capability is also valuable in chronic diseases. EIT can be a promising alternative for X-ray or CT regarding the applicability, time, and radiation exposure.

3 EIT for Liver Fat Infiltration

3.1 *Motivation*

Over one-third of US adults aged 20 years or older are categorized as obese with a body mass index (BMI) > 30 kg/m^2 [40]. Obesity is associated with an increased risk of nonalcoholic fatty liver disease (NAFLD) [41]. The severity of NAFLD ranges from an increase in intrahepatic triglyceride content (i.e., simple steatosis) to inflammation and fibrosis (i.e., severe steatohepatitis or NASH) [42], with predisposition to an increased risk for developing cirrhosis, hepatocellular carcinoma, type 2 diabetes, and cardiovascular disease [43]. Current NAFLD assessment techniques include liver biopsy, magnetic resonance imaging (MRI), computed tomography (CT), and abdominal ultrasound (US) [44, 45]. While liver biopsy remains a gold standard for the diagnosis of NAFLD, its risk of bleeding and sampling errors limit its clinical practice [44]. Although MRI is widely considered a noninvasive gold standard, patient comfort and high cost limit its application to a highly selected population. CT exposes patients to radiation, whereas abdominal US requires high expertise and is limited in its ability to distinguish stages of NAFLD [44]. The unmet clinical challenges lie in cost lowering and minimizing the operating skills necessary for routine assessment of fatty liver disease for determining patient's need for dietary interventions and monitoring the outcome of clinical interventions.

The fundamental reason that EIT can potentially fulfill this unmet need is the fact that fatty tissue exhibits significantly different electrical properties as compared to the rest of the normal liver tissue. Therefore EIT can theoretically provide a fat content mapping of the liver for the patient based on the reconstructed conductivity mapping. Furthermore, the advantages of EIT imaging (noninvasiveness, safety, low cost, portable, and continuous monitoring) are all desirable features that are hard to match by all current diagnostic methods.

3.2 Simulation Study

Here we discuss an example of using the abovementioned EIDORS library to perform simulation study in the context of fatty liver detection. The purpose of this section is to present a general idea of how EIT simulation study can improve the performance of practical EIT implementation. A two-dimensional (2-D) finite element model (FEM) is established based on the thoracic spine level T11–T12, at which different average tissue conductivities are assigned in accordance with the reported values (Table 1) [46].

The geometry of the EIT output is matched to the perimeter of the cross section in T11–T12 to minimize its impact on the overall conductivity mapping results (Fig. 3a). There are four conditions total in this simulation study: a healthy liver and three progressively increased fatty contents in the liver (Fig. 3ai–iv). These conditions are simulated by varying the conductivity in response to the increase in fat content, as calibrated by the ratios to a reported healthy liver's conductivity of 0.120 S/m [46]. The conductivity of fat is defined as 0.04 S/m [46]. A 15% fat weight ratio in the liver is considered to be the upper percentage of fat content in mild or moderate steatosis [47]. Hence, an 85%/15% weight ratio results in a 10% reduction in conductivity to 0.108 S/m as compared to the normal liver conductivity. For this reason, the initial three simulation scenarios are set to 100%, 95%, and 90% of normal liver conductivity, and the final scenario is set to 70% of normal liver conductivity for severe progression of NAFLD (Fig. 3ai–iv). To simulate the influence of noise during actual measurements, additive white Gaussian noise with signal-to-noise ratio (SNR) of 20 dB to the voltage data has been included. The reconstructed EIT imaging from the simulation model is shown in Fig. 3av–viii. To analyze the conductivity mapping results, the region of interest from the reconstructed images can be determined in two steps: (1) the initial identification of the minimum conductivity value around an approximated liver region within an abdominal cross section and (2) identification of the area that is within 400% of the minimum value to capture the full approximated liver region. All included elements from the area are averaged, and the results are presented as the effective conductivity in the liver region.

No significant difference is observed between the assigned conductivity (Fig. 3ai–vi) and EIT output values (Fig. 3av–viii), indicating reliable EIT computation. In response to noise (SNR = 20 dB) introduced into the reconstructed EIT image, the liver conductivity is only reduced by 3.3%. Furthermore, the impact of inaccuracies in geometric boundaries and the influence of the initial size of the liver on the imaging results are also compared.

Table 1 Average in intro tissue conductivities

Organ	Conductivity (S/m)
Liver	0.120
Spleen	0.132
Spinal cord	0.028
Stomach	0.523

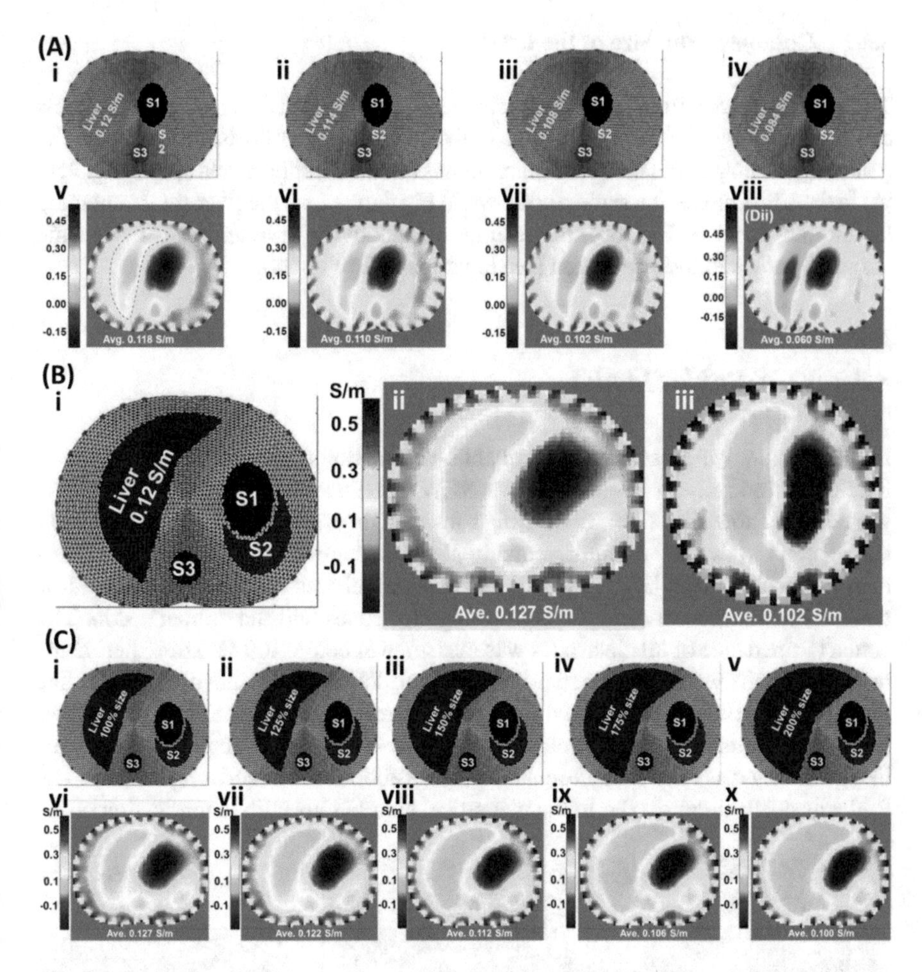

Fig. 3 EIT imaging simulation at thoracic spine level T11–T12 with (**a**) decreasing conductivity for the liver to represent fat content variation (i–iv is the FEM model, v–viii is the reconstruction results); (**b**) different output geometry (ii is thoracic cross section, iii is perfect circle); (**c**) increasing liver size (i–v is the FEM model, vi–x is the reconstruction results). S1, stomach; S2, spleen; S3, spine. (Adapted from Luo et al. [11])

3.2.1 Change of Geometric Boundaries

EIT image reconstructions obtained from two distinct geometric boundaries are further compared: (1) the T11–T12 cross-sectional geometry (Fig. 3bii) and (2) an ideal circular geometry (Fig. 3biii). The forward modeling remains the same for these two conditions. All of the simulation methods, including finite element modeling and data generation, are kept the same for the two geometric boundaries. As illustrated in Fig. 3b, the circular geometry yields a more distorted liver geometry as compared to the T11–T12 cross section in the finite element model. The identified ROI deviates by a higher percentage from the initial assigned value when using the less accurate circular geometry (15% in the circular vs. 5.8% in the oval geometry).

3.2.2 Change in the Size of the Liver

The effects of inaccuracy in liver size can be significant for EIT reconstruction. This can be studied by simulating a change in liver size of up to 200% while keeping the general geometry of the body and other organs (stomach, spleen, and spine (Fig. 3ci–v)) in the FEM model the same. As shown in Fig. 3cvi–x, as the liver size is enlarged (see upper panel in Fig. 3c), the average conductivity deviates further from the assigned values in the simulation (see lower panel in Fig. 3c).

3.3 NZW Rabbit Model

In this section, we present an experimental verification of using EIT to study fatty liver condition in New Zealand White (NZW) rabbits. Age-matched NZW rabbits were fed a chow diet ($n = 3$) or a high-fat diet ($n = 3$) [48]. Thirty-two equally spaced electrodes were circumferentially positioned around a cleanly shaven abdomen to perform liver EIT (Fig. 4a). Conductive gel was applied to ensure a low contact impedance between the connecting electrodes and the animal's skin. The contact impedance of all electrodes was measured at below 200 Ω. The experiments on these rabbits were performed post-mortem following the removal of the cardiac organs (due to other experimental requirement), resulting in partial shift of the liver toward the center of the abdominal cavity (Fig. 4c–f). For EIT computation, an oval-like geometry was used to represent the rabbit's abdominal boundary geometry. The ROI was determined similar to the theoretical simulations with identification of the

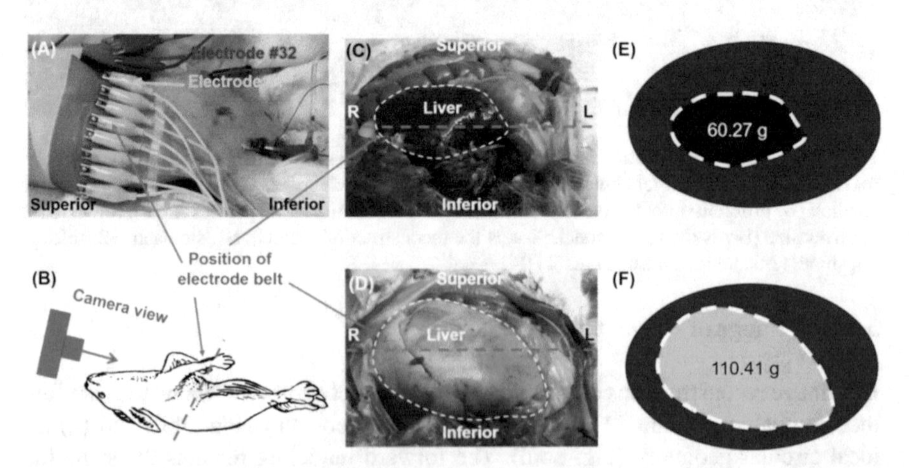

Fig. 4 Rabbit study. (**a**) A belt embedded with 32 electrodes placed around the rabbit's abdomen. (**b**) Schematic diagram depicting the camera perspective for the images captured in (**c**, **d**) to visualize organ arrangement and relative position of the electrode belt along the liver border. The liver from a normal diet-fed NZW rabbit (**c**) was compared with a high-fat fed rabbit (**d**), highlighting the increase in size and change in color from fat accumulation in the fatty liver (**e–f**). (Adapted from Luo et al. [11])

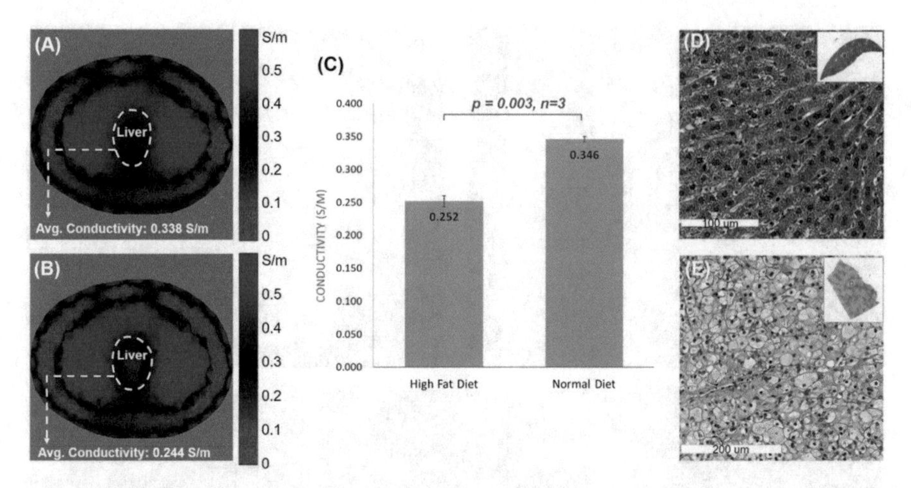

Fig. 5 (**a**, **b**) Comparison of EIT images between normal and high-fat-fed NZW rabbits. (**c**) Statistical analysis indicating significant differences in the average conductivities between the normal and high-fat diet rabbits ($p < 0.03$ vs. normal diet, $n = 3$). (**d**, **e**) Gross histology was assessed using H&E staining of the cross section of the liver at which the electrode belt was positioned to acquire the EIT measurement. (Adapted from Luo et al. [11])

area within 200% of the minimum value. All included elements from the ROI were averaged, and the results were presented as the effective conductivity in the liver region. After euthanasia, the rabbit livers were harvested and stored in 4% paraformaldehyde. Prior to histological staining, the tissues were washed and stored in 70% ethanol. The tissues were stained with hematoxylin and eosin (H&E).

Conductivity values were calculated following the EIT measurements (Fig. 5a, b). There is a significant difference between the normal and fat-fed groups ($p = 0.003$ vs. normal, $n = 3$) (Fig. 5c). The altered anatomical location is reflected in the shifting liver position in the obtained EIT images. Histopathological analysis validated histological difference between the healthy and high-fat livers (Fig. 5d, e). Unlike the normal-appearing hepatocytes-polygonal cells with a central vesicular nucleus and abundant pink cytoplasm (Fig. 5d), a significant accumulation of fat in the fat-fed livers was evidenced by intrahepatic lipid droplets (Fig. 5e).

3.4 *EIT for Clinical Translation*

In this section, the results of EIT technology further tested in a clinical setting are presented. The experimental procedure is as follows: an array of electrodes was circumferentially attached to the outer skin of the abdomen from voluntary human objects, each electrode being equidistance to one another (Fig. 6a, b). The proper vertical level of the electrodes was determined by palpating the position of the liver. The EIT imaging results have been compared to and validated by magnetic

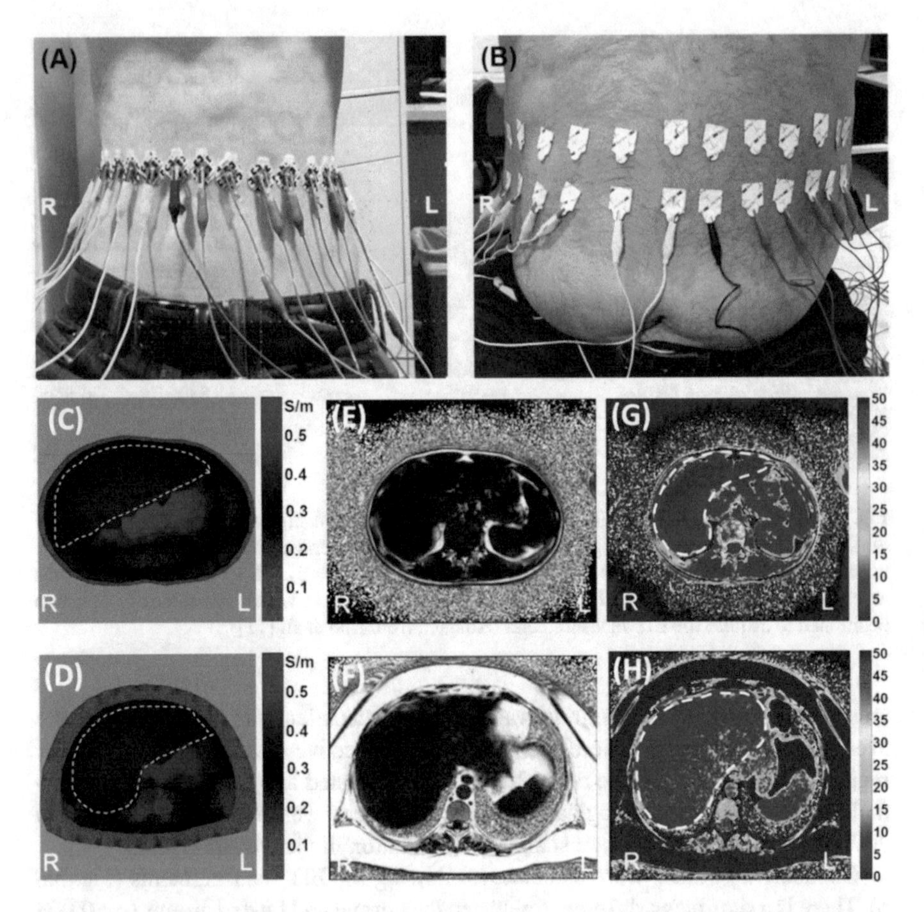

Fig. 6 (**a**, **b**) EIT measurements performed by circumferentially attaching 32 along the abdomen on 2 subjects: low vs. high BMI. (**c**, **d**) The EIT images reconstructed from the measurement in (**a**, **b**); (**e**, **f**) the representative MRI images from the two subjects. (**g**, **h**) The computed FVF values were also compared with the EIT measures. Upper row, low BMI; lower row, high BMI. (Adapted from Luo et al. [11])

resonance imaging (MRI) procedures for fatty liver using the Dixon method [49]. To perform MRI imaging, capsules containing MRI contrast reagent were first attached onto the abdominal skin at the determined vertical level. The electrodes were placed at the exact vertical level of the capsules, the existence of which in the MRI scan helped select the image of interest from a total scan through the entire abdomen of the subjects. The ROI was determined based on the MRI cross-sectional image of the liver. Subjects were asked to remain in a supine position and hold their breath throughout both MRI and EIT measurements to minimize any possible shift or deformation of organs during experiments. The EIT measurements were performed on one subject with low body mass index (BMI < 25 kg/m^2) and one with high BMI (>25 kg/m^2).

Liver MR images were acquired using a 3.0 T MRI scanner. Dixon 3D Volumetric Interpolated Breath-Hold Examination (VIBE) MR imaging was performed using a dual-echo in-phase (In) and opposed-phase (Op) with single breath-hold technique (TR = 3.97 ms, TE (Op/In) = 1.29/2.52 ms, FA = 5°, bandwidth = 1040 Hz/pixel, matrix size = 320 × 260, FOV = 380 × 308.75 mm, slice thickness = 3 mm, and 72 axial slices). Two separate scans were acquired for optimal data quality to avoid possible motion artifacts from a single scan. Fat volume fraction (FVF) maps were generated by using the MRI scans. Using the 2-point Dixon 3D VIBE method, in-phase, opposite-phase, fat, and water images were generated [50]. The FVF maps were generated by combining the fat and water images of the individual using the following equation:

$$\text{FVF} = \frac{F}{W+F} * 100 \tag{17}$$

where F and W denote fat and water content, respectively. A global liver FVF value was quantified by manually segmenting the entire livers in each subject using the opposite-phase images. The segmented liver masks were used to remove non-liver tissue from the FVF maps, and global FVF values were calculated from the individual subjects using MATLAB-based custom software.

The real-time EIT images (Fig. 6c, d) were reconstructed with a priori MRI-acquired images (Fig. 6e, f) to provide boundary conditions including (a) the overall geometric boundary of the abdominal cross sections, (b) distribution of skin and subcutaneous fat in the periphery of the abdomen, and (c) relative location of the organs within the abdomen. The conductivity value of the skin/fat region was set to be 0.1 S/m for the reconstruction (red region in Fig. 6c, d), and the conductivity value within the liver (dotted area in Fig. 6c, d) was set to be of a uniform value. The average conductivity of liver was computed to be 0.331 S/m for the subject with low BMI in Fig. 6a and 0.286 S/m for high BMI in Fig. 6b. Quantification of fatty infiltrate by FVF was consistent with the EIT measurement (Fig. 6g, h). The FVF for the subject with low BMI was computed to be 3.6%, and the FVF for the subject with high BMI was 8.2%. Individuals with 5–15% liver fat content are generally considered to have mild to moderate steatosis or fatty liver [47].

3.5 Discussion and Outlook for EIT in Liver Fat Infiltration

The multi-scale studies presented in this section demonstrate the capability of EIT technology for the detection of hepatic fat content with clinical relevance to monitoring obese individuals at risk for fatty liver disease. The NZW rabbit studies pave the way for distinguishing liver average conductivity in response to normal vs. high-fat diet. The preliminary human testing further differentiates liver conductivity between low vs. high BMI subjects. EIT in the fat-fed NZW rabbit studies did not have a priori information regarding the boundary geometry of the abdomen as well as the liver location, size, and shape. Nevertheless, EIT is still able to quantify a significant decrease in liver conductivity in the high-fat-fed NZW rabbits. The preliminary human BMI correlation provided the basis to apply EIT for quantifying liver lipid content in healthy vs. obese subjects. Simultaneous MRI and EIT measurements demonstrated (a) the improvement in EIT spatial resolution via a priori information acquired by MRI and (b) the correlation between EIT and MRI quantification for FVF values. Thus, this preliminary human study provides the translational implication for further establish EIT as a noninvasive, portable system for assessing and continuous monitoring of liver lipid content. Optimizing EIT would strengthen the translational implications for detecting fatty liver disease.

Solving the absolute conductivity values with the nonlinear inverse problem of EIT remains a great difficulty to enhancing the imaging resolution [51]. It will be highly desirable for a forward model to capture the detailed information about the target objects (such as the liver) except for the internal conductivity distribution for developing a specific and reliable reconstruction algorithm in using EIT for fatty liver. This forward model also requires knowledge of the boundary geometry and electrode positions in the presence of other sources of systematic artifacts in the measured data. In light of the high sensitivity of the inverse problem to modeling errors (e.g., boundary geometry and electrode position errors) [51], the reconstructed images would be compromised as compared to the genuine target objects. In the case of the NZW rabbit model and human studies, EIT is made more difficult by the lack of precise geometry and position of the liver in the presence of heterogeneous conductivities from other organ systems. In addition, the epidermal and subcutaneous adipose layer introduced electrode-tissue contact impedance to potentially reduce the penetration of electrical current into the internal organs. Nevertheless, the preliminary human correlation with MRI provides the basis to overcome these challenges by demonstrating improved resolution compared to the NZW rabbit studies.

While the EIT images in our studies correctly detected relative liver lipid content, the sensitivity and specificity of EIT measurements require further characterization. In the case of animal studies, EIT would be calibrated to the dose-dependent response to high-fat diet in terms of gross histology, liver weight, and fat content. In the case of future human studies, EIT would be correlated to the MRI-based FVF values in the context of subjects' clinical information. One important note, the findings discussed in this section is limited by the small numbers of animal models and human subjects. It will be critical to power up the numbers of animal models and human subjects for confirmation of statistical significance and to include inter- and intrapersonal variability. Such powering up studies would be essential to demonstrate whether the noninvasive EIT approach would enable non-expert operators to perform EIT measurements in an outpatient setting for screening and monitoring. Overall, EIT holds promises as a low-cost screening tool in the outpatient setting and as a periodic monitoring tool for individuals at high risk for nonalcoholic fatty liver disease (NAFLD). The integration of a baseline MRI scan with EIT system further optimizes a priori information for solving the nonlinear inverse problem for the reconstruction algorithm. We envision patients will only need at most on time MRI imaging to establish the baseline information and can continue to rely on EIT for long-term monitoring.

4 Conclusion and Future Direction

In this chapter, we provide a general discussion on EIT technology including the fundamental physics model, numerical methods for its solution, practical simulation, and hardware implementation. We further present two application examples in medical imaging where EIT can exert tremendous impact.

The improvement of imaging quality is still an active research topic in the EIT community. A plethora of methods have been proposed to mitigate imaging errors from the EIT inverse problem. Improvements in the co-registration of EIT with other imaging modalities, such as MRI, would enhance the EIT imaging outcomes [52–54]. Note that this is a different idea compared to using MRI as a priori information in the previous section. Investigators have adapted the use of magnetic resonance electrical impedance tomography (MREIT), in which externally injected current induces internal magnetic flux density information to provide geometric information [55]. To enhance spatial resolution for the ill-posed inverse problem, a shift from the conventional Gauss-Newton methods to optimize the reconstruction algorithm via for solving EIT has been proposed. For instance, a reconstruction method based on the particle swarm optimization (PSO) has been utilized for fast convergence and high spatial resolution [56]. Moreover, solution candidates for the PSO algorithm obtained through the non-blind search have been demonstrated to further strengthen the imaging quality [57]. Furthermore, artificial neural network for post-imaging processing has allowed for auto-segmentation of the target organs to effectively address the inverse problem as compared to the conventional algorithms [58].

On the other hand, identifying potential novel applications of EIT is another route of innovation for this fascinating technology. Although EIT is relatively lacking the absolute imaging accuracy as compared to established modalities like MRI or CT, there is tremendous potential value if the appropriate application is presented. We have already seen the great success of pulmonary monitoring based on EIT due to the tolerance in conductivity mapping using only the differential EIT. We have also seen the potential value of detecting fat content infiltration in liver in a low-maintenance, low-cost, easy-to-use context. While the mathematical solution of EIT remains to be further developed, it can still be a great tool in a variety of medical imaging applications. Conventionally, EIT is implemented by using large electrodes on objects with relatively large sizes, most notably, human bodies. New frontiers become possible with the advent of flexible electronics and microfabrication technology and electrode array can be created with miniaturized size and high conformality on internal tissue/organ, which will potentially facilitate the utilization of EIT for inside the body and localized detection, such as intravascular or gastrointestinal imaging.

References

1. Holder, D. S. (2004). *Electrical impedance tomography: Methods, history and applications*. Bristol: Institute of Physics Publishing.
2. Brown, B. H., & Seagar, A. D. (1987). The Sheffield data collection system. *Clinical Physics and Physiological Measurement.*, *8*, 91.

3. Bayford, R. H. (2006). Bioimpedance tomography (electrical impedance tomography). *Annual Review of Biomedical Engineering, 8*, 63–91.
4. Brown, B. H. (1997). Impedance pneumography, WO. 1997020499 A1.
5. Wilkinson, J., & Thanawala, V. (2009). Thoracic impedance monitoring of respiratory rate during sedation–is it safe? *Anaesthesia, 64*, 455–456.
6. Gabriel, S., Lau, R., & Gabriel, C. (1996). The dielectric properties of biological tissues: II. Measurements in the frequency range 10 Hz to 20 GHz. *Physics in Medicine & Biology, 41*, 2251.
7. Griffiths, D. J. (1999). *Introduction to electrodynamics.* 3rd ed, Upper Saddle River: Prentice Hall.
8. Calderón, A. P. (2006). On an inverse boundary value problem. *Computational & Applied Mathematics., 25*, 133–138.
9. Kirsch, A. (1996). *An introduction to the mathematical theory of inverse problems.* New York: Springer.
10. Nachman, A. I. (1988). Reconstructions from boundary measurements. *Annals of Mathematics., 128*, 531–576.
11. Luo, Y., Abiri, P., Zhang, S., Chang, C.-C., Kaboodrangi, A. H., Li, R., Sahib, A. K., Bui, A., Kumar, R., & Woo, M. (2018). Non-invasive electrical impedance tomography for multi-scale detection of liver fat content. *Theranostics., 8*, 1636.
12. Borcea, L. (2002). Electrical impedance tomography. *Inverse Problems, 18*, R99.
13. Adler, A., & Lionheart, W. R. (2006). Uses and abuses of EIDORS: An extensible software base for EIT. *Physiological Measurement., 27*, S25.
14. Schöberl, J. (1997). NETGEN an advancing front 2D/3D-mesh generator based on abstract rules. *Computing and Visualization in Science, 1*, 41–52.
15. Lionheart, W. R. (2004). EIT reconstruction algorithms: Pitfalls, challenges and recent developments. *Physiological Measurement., 25*, 125.
16. Gong, B., Krueger-Ziolek, S., Moeller, K., Schullcke, B., & Zhao, Z. (2015). Electrical impedance tomography: Functional lung imaging on its way to clinical practice? *Expert Review of Respiratory Medicine., 9*, 721–737.
17. Lobo, B., Hermosa, C., Abella, A., & Gordo, F. (2018). Electrical impedance tomography. *Annals of Translational Medicine., 6*, 26.
18. Kotre, C. (1997). Electrical impedance tomography. *British Journal of Radiology., 70*, S200–S205.
19. Brown, B., Leathard, A., Lu, L., Wang, W., & Hampshire, A. (1995). Measured and expected Cole parameters from electrical impedance tomographic spectroscopy images of the human thorax. *Physiological Measurement., 16*, A57.
20. Frerichs, I., Amato, M. B., Van Kaam, A. H., Tingay, D. G., Zhao, Z., Grychtol, B., Bodenstein, M., Gagnon, H., Böhm, S. H., & Teschner, E. (2017). Chest electrical impedance tomography examination, data analysis, terminology, clinical use and recommendations: Consensus statement of the TRanslational EIT developmeNt stuDy group. *Thorax, 72*, 83–93.
21. Network, A. R. D. S. (2000). Ventilation with lower tidal volumes as compared with traditional tidal volumes for acute lung injury and the acute respiratory distress syndrome. *New England Journal of Medicine, 342*, 1301–1308.
22. Eronia, N., Mauri, T., Maffezzini, E., Gatti, S., Bronco, A., Alban, L., Binda, F., Sasso, T., Marenghi, C., & Grasselli, G. (2017). Bedside selection of positive end-expiratory pressure by electrical impedance tomography in hypoxemic patients: A feasibility study. *Annals of Intensive Care, 7*, 76.
23. Karsten, J., Grusnick, C., Paarmann, H., Heringlake, M., & Heinze, H. (2015). Positive end-expiratory pressure titration at bedside using electrical impedance tomography in post-operative cardiac surgery patients. *Acta Anaesthesiologica Scandinavica, 59*, 723–732.
24. Hinz, J., Moerer, O., Neumann, P., Dudykevych, T., Frerichs, I., Hellige, G., & Quintel, M. (2006). Regional pulmonary pressure volume curves in mechanically ventilated patients with acute respiratory failure measured by electrical impedance tomography. *Acta Anaesthesiologica Scandinavica, 50*, 331–339.

25. Costa, E. L., Borges, J. B., Melo, A., Suarez-Sipmann, F., Toufen, C., Bohm, S. H., & Amato, M. B. (2009). Bedside estimation of recruitable alveolar collapse and hyperdistension by electrical impedance tomography. *Intensive Care Medicine, 35*, 1132–1137.
26. Wolf, G. K., Gómez-Laberge, C., Rettig, J. S., Vargas, S. O., Smallwood, C. D., Prabhu, S. P., Vitali, S. H., Zurakowski, D., & Arnold, J. H. (2013). Mechanical ventilation guided by electrical impedance tomography in experimental acute lung injury. *Critical Care Medicine, 41*, 1296–1304.
27. Luepschen, H., Meier, T., Grossherr, M., Leibecke, T., Karsten, J., & Leonhardt, S. (2007). Protective ventilation using electrical impedance tomography. *Physiological Measurement, 28*, S247.
28. Putensen, C., Wrigge, H., & Zinserling, J. (2007). Electrical impedance tomography guided ventilation therapy. *Current Opinion in Critical Care, 13*, 344–350.
29. Lowhagen, K., Lindgren, S., Odenstedt, H., Stenqvist, O., & Lundin, S. (2011). A new non-radiological method to assess potential lung recruitability: A pilot study in ALI patients. *Acta Anaesthesiologica Scandinavica, 55*, 165–174.
30. Odenstedt, H., Lindgren, S., Olegård, C., Erlandsson, K., Lethvall, S., Åneman, A., Stenqvist, O., & Lundin, S. (2005). Slow moderate pressure recruitment maneuver minimizes negative circulatory and lung mechanic side effects: Evaluation of recruitment maneuvers using electric impedance tomography. *Intensive Care Medicine, 31*, 1706–1714.
31. Eyüboğlu, B. M., Brown, B. H., & Barber, D. C. (1989). In vivo imaging of cardiac related impedance changes. *IEEE EMB Magazine, 8*, 39–45.
32. Borges, J. B., Suarez-Sipmann, F., Bohm, S. H., Tusman, G., Melo, A., Maripuu, E., Sandström, M., Park, M., Costa, E. L., & Hedenstierna, G. (2011). Regional lung perfusion estimated by electrical impedance tomography in a piglet model of lung collapse. *Journal of Applied Physiology, 112*, 225–236.
33. Force, A. D. T., Ranieri, V., & Rubenfeld, G. (2012). Acute respiratory distress syndrome. *Journal of the American Medical Association, 307*, 2526–2533.
34. Gattinoni, L., Pesenti, A., Avalli, L., Rossi, F., & Bombino, M. (1987). Pressure-volume curve of total respiratory system in acute respiratory failure: Computed tomographic scan study. *American Review of Respiratory Disease, 136*, 730–736.
35. Franchineau, G., Bréchot, N., Lebreton, G., Hekimian, G., Nieszkowska, A., Trouillet, J.-L., Leprince, P., Chastre, J., Luyt, C.-E., & Combes, A. (2017). Bedside contribution of electrical impedance tomography to setting positive end-expiratory pressure for extracorporeal membrane oxygenation–treated patients with severe acute respiratory distress syndrome. *American Journal of Respiratory and Critical Care Medicine, 196*, 447–457.
36. Gershon, A., Hwee, J., Victor, J. C., Wilton, A., Wu, R., Day, A., & T. To. (2015). Mortality trends in women and men with COPD in Ontario, Canada, 1996–2012. *Thorax, 70*, 121–126.
37. Noordegraaf, A. V., Kunst, P. W., Janse, A., Marcus, J. T., Postmus, P. E., Faes, T. J., & de Vries, P. M. (1998). Pulmonary perfusion measured by means of electrical impedance tomography. *Physiological Measurement, 19*, 263.
38. Vogt, B., Zhao, Z., Zabel, P., Weiler, N., & Frerichs, I. (2016). Regional lung response to bronchodilator reversibility testing determined by electrical impedance tomography in chronic obstructive pulmonary disease. *American Journal of Physiology-Lung Cellular and Molecular Physiology, 311*, L8–L19.
39. Zhao, Z., Müller-Lisse, U., Frerichs, I., Fischer, R., & Möller, K. (2013). Regional airway obstruction in cystic fibrosis determined by electrical impedance tomography in comparison with high resolution CT. *Physiological Measurement, 34*, N107.
40. Ogden, C. L., Carroll, M. D., Fryar, C. D., & Flegal, K. M. (2015). *Prevalence of obesity among adults and youth: United States, 2011–2014*: US Department of Health and Human Services, centers for disease control and ....
41. Fabbrini, E., Sullivan, S., & Klein, S. (2010). Obesity and nonalcoholic fatty liver disease: Biochemical, metabolic, and clinical implications. *Hepatology, 51*, 679–689.

42. Marchesini, G., Bugianesi, E., Forlani, G., Cerrelli, F., Lenzi, M., Manini, R., Natale, S., Vanni, E., Villanova, N., & Melchionda, N. (2003). Nonalcoholic fatty liver, steatohepatitis, and the metabolic syndrome. *Hepatology, 37*, 917–923.
43. Pagadala, M. R., & McCullough, A. J. (2012). Non-alcoholic fatty liver disease and obesity: Not all about body mass index. *The American Journal of Gastroenterology, 107*(12), 1859–1861. ed: Nature Publishing Group.
44. Mishra, P., & Younossi, Z. M. (2007). Abdominal ultrasound for diagnosis of nonalcoholic fatty liver disease (NAFLD). *The American Journal of Gastroenterology, 102*, 2716.
45. Chen, J., Talwalkar, J. A., Yin, M., Glaser, K. J., Sanderson, S. O., & Ehman, R. L. (2011). Early detection of nonalcoholic steatohepatitis in patients with nonalcoholic fatty liver disease by using MR elastography. *Radiology, 259*, 749–756.
46. Hasgall, P., Di Gennaro, F., Baumgartner, C., Neufeld, E., Gosselin, M., Payne, D., Klingenböck, A., & Kuster, N. (2015). IT'IS database for thermal and electromagnetic parameters of biological tissues, Version 3.0, September 1st; 2015, ed.
47. Leporq, B., Ratiney, H., Pilleul, F., & Beuf, O. (2013). Liver fat volume fraction quantification with fat and water T1 and T2∗ estimation and accounting for NMR multiple components in patients with chronic liver disease at 1.5 and 3.0 T. *European Radiology, 23*, 2175–2186.
48. Packard, R. R. S., Luo, Y., Abiri, P., Jen, N., Aksoy, O., Suh, W. M., Tai, Y.-C., & Hsiai, T. K. (2017). 3-D electrochemical impedance spectroscopy mapping of arteries to detect metabolically active but angiographically invisible atherosclerotic lesions. *Theranostics, 7*, 2431.
49. Szumowski, J., Coshow, W., Li, F., Coombs, B., & Quinn, S. F. (1995). Double-echo three-point-dixon method for fat suppression MRI. *Magnetic Resonance in Medicine, 34*, 120–124.
50. Ding, Y., Rao, S.-X., Chen, C.-Z., Li, R.-C., & Zeng, M.-S. (2015). Usefulness of two-point Dixon fat-water separation technique in gadoxetic acid-enhanced liver magnetic resonance imaging. *World Journal of Gastroenterology: WJG, 21*, 5017.
51. Seo, J. K., & Woo, E. J. (2012). *Nonlinear inverse problems in imaging*. New York: Wiley.
52. Lee, J., Fei, P., Packard, R. R. S., Kang, H., Xu, H., Baek, K. I., Jen, N., Chen, J., Yen, H., & Kuo, C.-C. J. (2016). 4-dimensional light-sheet microscopy to elucidate shear stress modulation of cardiac trabeculation. *The Journal of Clinical Investigation, 126*, 1679–1690.
53. Fei, P., Lee, J., Packard, R. R. S., Sereti, K.-I., Xu, H., Ma, J., Ding, Y., Kang, H., Chen, H., & Sung, K. (2016). Cardiac light-sheet fluorescent microscopy for multi-scale and rapid imaging of architecture and function. *Scientific Reports, 6*, 22489.
54. Crabb, M., Davidson, J., Little, R., Wright, P., Morgan, A., Miller, C., Naish, J., Parker, G., Kikinis, R., & McCann, H. (2014). Mutual information as a measure of image quality for 3D dynamic lung imaging with EIT. *Physiological Measurement, 35*, 863.
55. Ider, Y. Z., & Birgül, Ö. (2000). Use of the magnetic field generated by the internal distribution of injected currents for electrical impedance tomography (MR-EIT). *Turkish Journal of Electrical Engineering & Computer Sciences, 6*, 215–226.
56. Chen, M.-Y., Hu, G., He, W., Yang, Y.-L., & Zhai, J.-Q. (2010). A reconstruction method for electrical impedance tomography using particle swarm optimization. In *Life system modeling and intelligent computing* (pp. 342–350). Berlin, Heidelberg: Springer.
57. Feitosa, A. R., Ribeiro, R. R., Barbosa, V. A., de Souza, R. E., & dos Santos, W. P. (2014). Reconstruction of electrical impedance tomography images using particle swarm optimization, genetic algorithms and non-blind search. In *Biosignals and Biorobotics Conference (2014): Biosignals and Robotics for Better and Safer Living (BRC), 5th ISSNIP-IEEE*, 2014, pp. 1–6.
58. Martin, S. & Choi, C. T. (2015). Electrical impedance yomography: A reconstruction method based on neural networks and particle swarm optimization. In *1st Global Conference on Biomedical Engineering & 9th Asian-Pacific Conference on Medical and Biological Engineering*, 2015, pp. 177–179.

High-Frequency Ultrasonic Transducers to Uncover Cardiac Dynamics

Bong Jin Kang, Qifa Zhou, and K. Kirk Shung

1 High-Frequency Ultrasonic Transducers

Ultrasound is one of the most widely used diagnostic tools which offers real-time grayscale images of anatomical details and blood flow information with Doppler technique [1]. Conventional ultrasound imaging systems typically use frequency ranges from 2 to 15 MHz with the spatial resolution on the order of a few millimeters. In response to the need for improved image resolution, ultrasound imaging is being pushed to higher and higher frequencies. High-frequency (higher than 30 MHz) ultrasound provides a noninvasive imaging method for many clinical and preclinical applications requiring improved spatial resolution. There are a number of clinical problems that may benefit from high-frequency ultrasound imaging [2].

Ultrasound imaging systems require a device called ultrasonic transducer to convert electrical energy into acoustic energy and vice versa. Ultrasonic transducers come in a variety of forms and sizes ranging from single-element transducers for mechanical scanning to array transducers for electronic scanning [1]. Figures 1a and 1b shows a high-frequency single-element transducer and an array transducer, respectively.

Significant progress in the development of high-frequency single-element and array transducers has been achieved in the past few years [3–11]. Single-element ultrasound transducers have been exclusively used in high-frequency ultrasound imaging for a number of years. They have been able to provide an adequate solution in a number of clinical and preclinical applications including small animal imaging. These single-element transducers, however, are less than ideal due to their single geometrical focus and must be mechanically scanned to form an image. Dynamic focusing is a distinct advantage that array transducers possess over single-element

B. J. Kang · Q. Zhou (✉) · K. K. Shung
Department of Biomedical Engineering, University of Southern California,
Los Angeles, CA, USA
e-mail: qifazhou@usc.edu

H. Cao et al. (eds.), *Interfacing Bioelectronics and Biomedical Sensing*,
https://doi.org/10.1007/978-3-030-34467-2_8

Fig. 1 Photographs of high-frequency ultrasound transducers: (**a**) a single-element and (**b**) an array transducer

transducers. Array systems use electronic scanning to form an image slice and therefore can achieve higher frame rates. Also, the ultrasound beam can be steered and dynamically focused in the image plane.

2 Emerging Applications for Small Animal Models

2.1 High-Frequency Transducer for Mouse

High-frequency ultrasound (>30 MHz) has proven to be a valuable tool for small animal cardiovascular imaging offering high resolution, noninvasiveness, low cost, and portability. Using ultrasound biomicroscopy (UBM), Turnbull et al. [12] reported the observation of mutant phenotyping in the mouse embryo, and more investigations [13–16] were conducted on mice with an improved prototype scanner at a resolution of 40 μm. In addition to B-mode imaging, ultrasound cardiovascular assessment also involves Doppler measurement to obtain quantitative information of blood hemodynamics. A duplex system featured with B-mode imaging and Doppler measurement allows complementary integration of anatomical and physiological information for precise diagnosis. Previous studies show that high-frequency ultrasound is capable of detecting blood flow in small vessels [17, 18], and recent research demonstrated the detection of blood flow in major vessels in mouse [19–21] as well as in mouse microcirculation [22–24].

The high-frequency ultrasound single-element transducer and system for mouse cardiac imaging, capable of B-mode imaging and pulsed-wave Doppler measurement, were reported [25, 26]. B-mode images and Doppler waveforms acquired from the mouse heart are shown in Fig. 2. Figure 2a shows a long-axis parasternal view of a 4-week mouse heart. The chambers of the left ventricle (LV), right atrium (RA), and right ventricle (RV) are able to be resolved. Guided by the B-mode imaging, the pulsed-wave (PW) Doppler waveforms in Fig. 2b display the blood flow in the LV. The sample volume was marked by a long line with two short bars

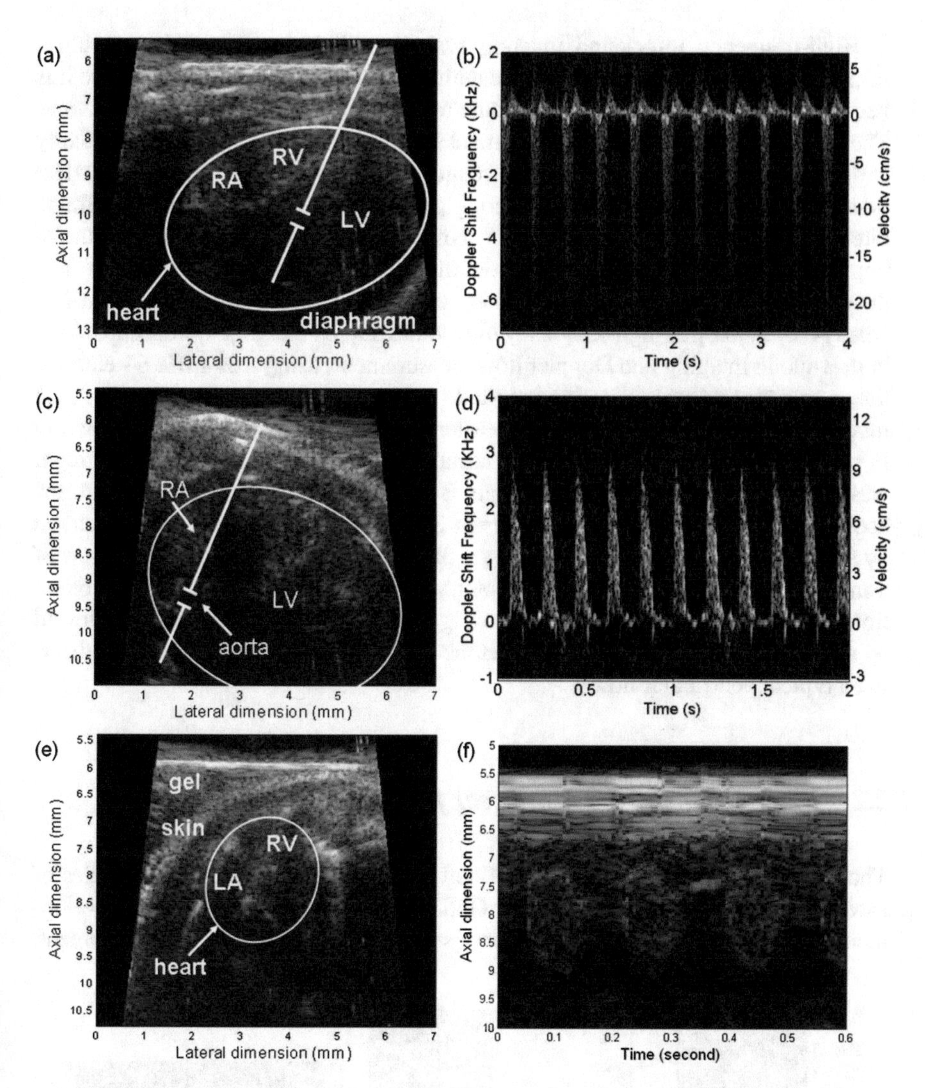

Fig. 2 In vivo images of adult mouse cardiovascular system using the duplex UBM. (**a**) Parasternal long-axis image of a 4-week mouse heart showing RA, RV, and LV. (**b**) Ventricular Doppler waveform from the location marked by two short bars. (**c**) Parasternal long-axis image showing aorta RV, and LV. (**d**) Aortic Doppler waveform. (**e**) Parasternal short-axis image showing LA and RV. (**f**) M-mode images of the mouse LV. (From Sun [25], *RA* right atrium, *RV* right ventricle, *LV* left ventricle, *LA* left atrium)

in Fig. 2a at an estimated Doppler angle of 45°. Figure 2c shows another parasternal long-axis view of cardiac structures, displaying aorta, LV, and RA. Under the assistance of B-mode imaging, the Doppler measurement in the aorta was acquired at the location marked in Fig. 2c. A short-axis view of a mouse heart is demonstrated in Fig. 2e, showing the chambers of left atrium (LA) and RV. M-mode images shown in Fig. 2f were also acquired on the LV.

High-frequency ultrasound imaging systems utilizing a single-element ultrasonic transducer require mechanical scanning to acquire images, and therefore it is very difficult to achieve a very high frame rate (>100 fps) with a wide field of view. The fixed focal point of the single-element transducer also limits the image quality at depths away from the focus. High-frequency ultrasonic linear array (>30 MHz) systems utilizing electronic beam steering provide a better solution for high frame rate imaging. A linear array allows for dynamic focusing in the imaging plane to improve the lateral resolution throughout the depth of view. Significant progress in the development of high-frequency linear array has been achieved in the past few years [4, 5, 27, 28]. A high-frequency ultrasound duplex imaging system capable of both B-mode imaging and Doppler flow measurement, using a 30 MHz 64-element linear array transducer, was also reported [4, 29]. In vivo study on mice, B-mode image acquired from the long-axis parasternal view of a mouse heart is shown in Fig. 3a. Under the ribs were the RA, aorta, and partial LV. The bright diaphragm was located at the bottom of this image. B-mode images and Doppler waveforms acquired in duplex mode are shown in Figs. 3b and 3c. The sample volume is marked by a long line with two short bars. Figure 3b shows a long-axis parasternal view of a mouse heart. The pulsed-wave Doppler waveform in Fig. 3c displays the blood flow acquired at the location marked in Fig. 3b. The large positive inflow followed by a negative outflow pattern suggested the filling and emptying of the LV, indicating a typical ventricular flow.

2.2 *High-Frequency Transducer for Zebrafish*

The zebrafish has emerged as an excellent genetic model organism for studies of cardiovascular development [30–32]. Optical transparency and external development during embryogenesis allow for visual analysis in the early development.

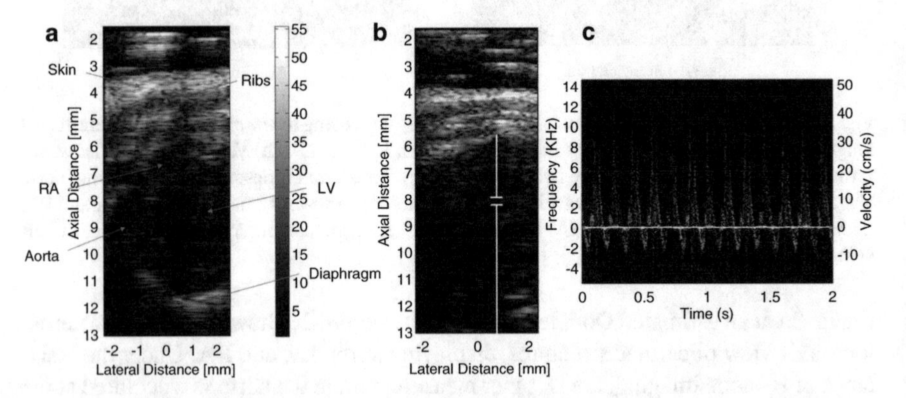

Fig. 3 (**a**) In vivo image of an adult mouse cardiovascular system using the duplex linear array imaging system with a 30 MHz linear array. (**b**) A duplex B-mode imaging of a mouse heart and its corresponding ventricular Doppler waveform from the marked location. (From Zhang [29])

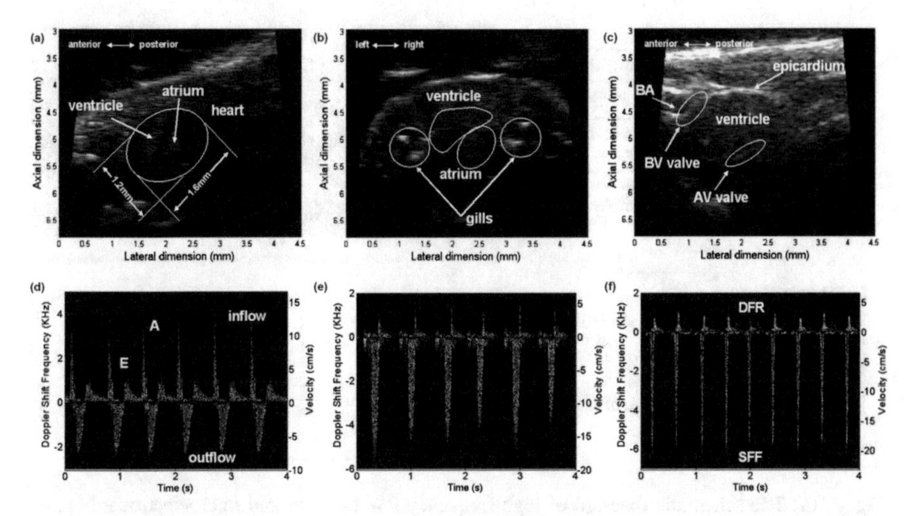

Fig. 4 Adult zebrafish heart at the isovolumic relaxation stage in an upside-down position. (**a**) Sagittal view showing the atrium and ventricle located on the ventral side underneath the skin. The cardiac dimensions were 1.2 × 1.6 mm. (**b**) Transverse views displaying the heart positioned medially between the gills. (**c**) A sagittal view of the adult zebrafish heart at the isovolumic contraction stage showing the ventricle, bulbus arteriosus (BA), bulboventricular (BV) valve, atrioventricular (AV) valve, and epicardium. (**d**) Pulsed-wave Doppler waveform of ventricular inflow (positive) and outflow (negative). The inflow consisted of early diastolic flow (E) and late diastolic flow (A). (**e**) Doppler waveform at medial BA. (**f**) Doppler waveform at posterior BA close to BV valve. In addition to systolic forward flow (SFF), diastolic flow reversal (DFR) appears when the ventricular pressure drops and blood flows backward. (From Sun [37])

However, optical methods are not suitable for the study of adult zebrafish due to opacity beyond its early stage. Histology [33, 34], scanning electron microscopy (SEM), and transmission electron microscopy [34] were applied to evaluate the cardiac morphology in fixed adult zebrafish hearts, without providing in vivo data. Magnetic resonance microscopy [35] was used to examine anatomical structures in adult zebrafish ex vivo and in vivo, but the image acquisition took a long period of time (128 to 480 s). A conventional ultrasonic imaging device was employed to image the heart of adult zebrafish at 7 and 8.5 MHz [36], but the resolution was inadequate, and useful observations of zebrafish cardiac functions were difficult. Therefore, to understand the cardiovascular structures and functions of the adult zebrafish requires a high-resolution, real-time, noninvasive imaging. The high-frequency ultrasound system for adult zebrafish cardiac imaging, capable of 75 MHz B-mode imaging and 45 MHz pulsed-wave Doppler measurement, was reported [37]. In vivo imaging studies showed the identification of the atrium, ventricle, bulbus arteriosus, atrioventricular valve, and bulboventricular valve in real-time images, with cardiac measurement at various stages. Doppler waveforms acquired at the ventricle and the bulbus arteriosus demonstrated the utility of this system to study the zebrafish cardiovascular hemodynamics (Fig. 4).

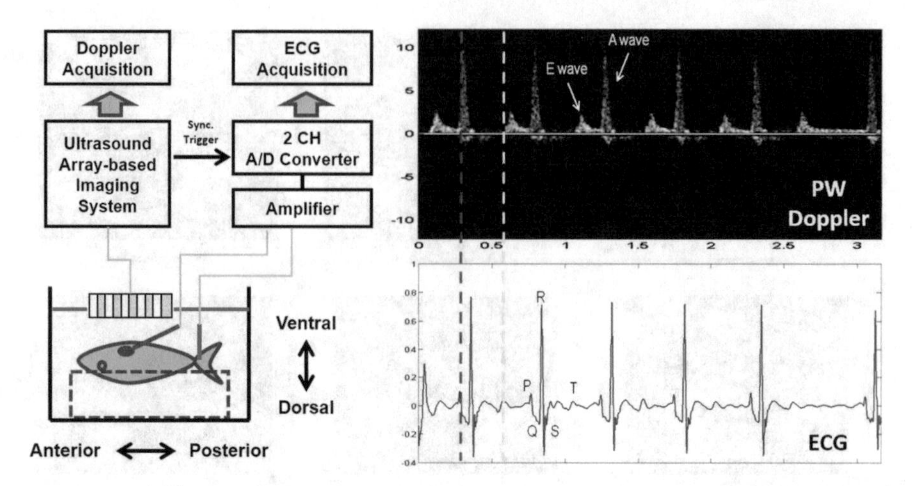

Fig. 5 (**a**) The schematic diagram of high-frequency PW Doppler and microelectrocardiogram (μECG) systems. (**b**) Synchronized ECG features coregistered with the hemodynamic events as captured by PW Doppler: P wave in ECG precedes atrial contraction, resulting in A waves in Doppler; and T wave in ECG preceded ventricular relaxation, resulting in E waves in Doppler. An integration of both modalities revealed ~500 ms delay in electromechanical coupling

A high-frequency ultrasound duplex imaging system capable of both B-mode imaging and Doppler flow measurement, using a 30 MHz 256-element linear array, is also reported [11, 38]. In vivo study on adult zebrafish [39], under the guidance of B-mode imaging, the Doppler gate was positioned inside the ventricle to interrogate both inflow velocities from the atrium and outflow velocities to the aortic valves. Synchronized PW Doppler and ECG recording confirmed the A wave in response to atrial contraction (P wave in ECG) and E wave in response to ventricular relaxation (T wave in ECG) as indicated by the dashed lines in Fig. 5b. PW Doppler signals for ventricular outflow were also detected following the QRS complex.

In humans, the E/A ratio is greater than 1 for the normal ventricular diastolic function. In adult zebrafish, A-wave velocity is greater than compared with the E wave as shown in Fig. 5b, and the E/A ratio is less than 1, suggesting a distinct cardiac physiology in the two-chamber system [40].

References

1. Shung, K. K. (2006). *Diagnostic ultrasound: Imaging and blood flow measurements*. Boca Raton: CRC Press.
2. Lockwood, G. R., Turnball, D. H., Christopher, D. A., & Foster, F. S. (1996). Beyond 30 MHz: Applications of high-frequency ultrasound imaging. *Engineering in Medicine and Biology Magazine, 15*, 60–71.

3. Cannata, J. M., Ritter, T. A., Chen, W. H., Silverman, R. H., & Shung, K. K. (2003). Design of efficient, broadband single-element (20–80 MHz) ultrasonic transducers for medical imaging applications. *IEEE Transactions on Ultrasonics, Ferroelectrics, and Frequency Control, 50*, 1548–1557.
4. Ritter, T. A., Shrout, T. R., Tutwiler, R., & Shung, K. K. (Feb 2002). A 30-MHz piezo-composite ultrasound array for medical imaging applications. *IEEE Transactions on Ultrasonics, Ferroelectrics, and Frequency Control, 49*, 217–230.
5. Cannata, J. M., Williams, J. A., Zhou, Q., Ritter, T. A., & Shung, K. K. (2006). Development of a 35-MHz piezo-composite ultrasound array for medical imaging. *IEEE Transactions on Ultrasonics, Ferroelectrics, and Frequency Control, 53*, 224–236.
6. Ma, T., Yu, M., Li, J., Munding, C. E., Chen, Z., Fei, C., et al. (2015). Multi-frequency intravascular ultrasound (IVUS) imaging. *IEEE Transactions on Ultrasonics, Ferroelectrics, and Frequency Control, 62*, 97–107.
7. Yoon, S., Williams, J., Kang, B. J., Yoon, C., Cabrera-Munoz, N., Jeong, J. S., et al. (Jun 2015). Angled-focused 45 MHz PMN-PT single element transducer for intravascular ultrasound imaging. *Sensors and Actuators A: Physical, 228*, 16–22.
8. Li, X., Wu, W., Chung, Y., Shih, W. Y., Shih, W. H., Zhou, Q., et al. (2011). 80-MHz intravascular ultrasound transducer using PMN-PT free-standing film. *IEEE Transactions on Ultrasonics, Ferroelectrics, and Frequency Control, 58*, 2281–2288.
9. Li, X., Ma, T., Tian, J., Han, P., Zhou, Q., & Shung, K. K. (2014). Micromachined PIN-PMN-PT crystal composite transducer for high-frequency intravascular ultrasound (IVUS) imaging. *IEEE Transactions on Ultrasonics, Ferroelectrics, and Frequency Control, 61*, 1171–1178.
10. Chen, R., Cabrera-Munoz, N. E., Lam, K. H., Hsu, H. S., Zheng, F., Zhou, Q., et al. (2014). PMN-PT single-crystal high-frequency kerfless phased array. *IEEE Transactions on Ultrasonics, Ferroelectrics, and Frequency Control, 61*, 1033–1041.
11. Cannata, J. M., Williams, J. A., Zhang, L., Hu, C. H., & Shung, K. K. (2011). A high-frequency linear ultrasonic array utilizing an interdigitally bonded 2-2 piezo-composite. *IEEE Transactions on Ultrasonics, Ferroelectrics, and Frequency Control, 58*, 2202–2212.
12. Turnbull, D. H., Bloomfield, T. S., Baldwin, H. S., Foster, F. S., & Joyner, A. L. (1995). Ultrasound backscatter microscope analysis of early mouse embryonic brain development. *Proceedings of the National Academy of Sciences of the United States of America, 92*, 2239–2243.
13. Foster, F. S., Pavlin, C. J., Harasiewicz, K. A., Christopher, D. A., & Turnbull, D. H. (2000). Advances in ultrasound biomicroscopy. *Ultrasound in Medicine & Biology, 26*, 1–27.
14. Foster, F. S., Zhang, M. Y., Zhou, Y. Q., Liu, G., Mehi, J., Cherin, E., et al. (2002). A new ultrasound instrument for in vivo microimaging of mice. *Ultrasound in Medicine & Biology, 28*, 1165–1172.
15. Kaufmann, B. A., Lankford, M., Behm, C. Z., French, B. A., Klibanov, A. L., Xu, Y., et al. (2007). High-resolution myocardial perfusion imaging in mice with high-frequency echocardiographic detection of a depot contrast agent. *Journal of the American Society of Echocardiography, 20*, 136–143.
16. Zhou, Y. Q., Foster, F. S., Nieman, B. J., Davidson, L., Chen, X. J., & Henkelman, R. M. (2004). Comprehensive transthoracic cardiac imaging in mice using ultrasound biomicroscopy with anatomical confirmation by magnetic resonance imaging. *Physiological Genomics, 18*, 232–244.
17. Christopher, D. A., Burns, P. N., Starkoski, B. G., & Foster, F. S. (1997). A high-frequency pulsed-wave Doppler ultrasound system for the detection and imaging of blood flow in the microcirculation. *Ultrasound in Medicine & Biology, 23*, 997–1015.
18. Kruse, D. E., Silverman, R. H., Fornaris, R. J., Coleman, D. J., & Ferrara, K. W. (1998). A swept-scanning mode for estimation of blood velocity in the microvasculature. *IEEE Transactions on Ultrasonics, Ferroelectrics, and Frequency Control, 45*, 1437–1440.
19. Phoon, C. K., Aristizabal, O., & Turnbull, D. H. (Oct 2000). 40 MHz Doppler characterization of umbilical and dorsal aortic blood flow in the early mouse embryo. *Ultrasound in Medicine & Biology, 26*, 1275–1283.

20. Reddy, A. K., Jones, A. D., Martono, C., Caro, W. A., Madala, S., & Hartley, C. J. (2005). Pulsed Doppler signal processing for use in mice: Design and evaluation. *IEEE Transactions on Biomedical Engineering, 52*, 1764–1770.
21. Reddy, A. K., Taffet, G. E., Li, Y. H., Lim, S. W., Pham, T. T., Pocius, J. S., et al. (2005). Pulsed Doppler signal processing for use in mice: Applications. *IEEE Transactions on Biomedical Engineering, 52*, 1771–1783.
22. Chérin, E., Williams, R., Needles, A., Liu, G., White, C., Brown, A. S., et al. (2006). Ultrahigh frame rate retrospective ultrasound microimaging and blood flow visualization in mice in vivo. *Ultrasound in Medicine & Biology, 32*, 683–691.
23. Goertz, D. E., Christopher, D. A., Yu, J. L., Kerbel, R. S., Burns, P. N., & Foster, F. S. (2000). High-frequency color flow imaging of the microcirculation. *Ultrasound in Medicine & Biology, 26*, 63–71.
24. Goertz, D. E., Yu, J. L., Kerbel, R. S., Burns, P. N., & Foster, F. S. (2003). High-frequency 3-D color-flow imaging of the microcirculation. *Ultrasound in Medicine & Biology, 29*, 39–51.
25. Sun, L., Xu, X., Richard, W. D., Feng, C., Johnson, J. A., & Shung, K. K. (2008). A high-frame rate duplex ultrasound biomicroscopy for small animal imaging in vivo. *IEEE Transactions on Biomedical Engineering, 55*, 2039–2049.
26. Sun, L., Richard, W. D., Cannata, J. M., Feng, C. C., Johnson, J. A., Yen, J. T., et al. (2007). A high-frame rate high-frequency ultrasonic system for cardiac imaging in mice. *IEEE Transactions on Ultrasonics, Ferroelectrics, and Frequency Control, 54*, 1648–1655.
27. Lukacs, M., Yin, J., Pang, G., Garcia, R. C., Cherin, E., Williams, R., et al. (2006). Performance and characterization of new micromachined high-frequency linear arrays. *IEEE Transactions on Ultrasonics, Ferroelectrics, and Frequency Control, 53*, 1719–1729.
28. Brown, J. A., Foster, F. S., Needles, A., Cherin, E., & Lockwood, G. R. (2007). Fabrication and performance of a 40-MHz linear array based on a 1-3 composite with geometric elevation focusing. *IEEE Transactions on Ultrasonics, Ferroelectrics, and Frequency Control, 54*, 1888–1894.
29. Zhang, L., Xu, X., Hu, C., Sun, L., Yen, J. T., Cannata, J. M., et al. (2010). A high-frequency, high frame rate duplex ultrasound linear array imaging system for small animal imaging. *IEEE Transactions on Ultrasonics, Ferroelectrics, and Frequency Control, 57*, 1548–1557.
30. Chen, J. N., Haffter, P., Odenthal, J., Vogelsang, E., Brand, M., van Eeden, F. J., et al. (1996). Mutations affecting the cardiovascular system and other internal organs in zebrafish. *Development, 123*, 293–302.
31. Serbedzija, G. N., Chen, J. N., & Fishman, M. C. (1998). Regulation in the heart field of zebrafish. *Development, 125*, 1095–1101.
32. Thisse, C., & Zon, L. I. (2002). Organogenesis – heart and blood formation from the zebrafish point of view. *Science, 295*, 457–462.
33. Hu, N., Sedmera, D., Yost, H. J., & Clark, E. B. (2000). Structure and function of the developing zebrafish heart. *The Anatomical Record, 260*, 148–157.
34. Hu, N., Yost, H. J., & Clark, E. B. (2001). Cardiac morphology and blood pressure in the adult zebrafish. *The Anatomical Record, 264*, 1–12.
35. Kabli, S., Alia, A., Spaink, H. P., Verbeek, F. J., & De Groot, H. J. (2006). Magnetic resonance microscopy of the adult zebrafish. *Zebrafish, 3*, 431–439.
36. Ho, Y. L., Shau, Y. W., Tsai, H. J., Lin, L. C., Huang, P. J., & Hsieh, F. J. (2002). Assessment of zebrafish cardiac performance using Doppler echocardiography and power angiography. *Ultrasound in Medicine & Biology, 28*, 1137–1143.
37. Sun, L., Lien, C. L., Xu, X., & Shung, K. K. (2008). In vivo cardiac imaging of adult zebrafish using high frequency ultrasound (45–75 MHz). *Ultrasound in Medicine & Biology, 34*, 31–39.
38. Hu, C., Zhang, L., Cannata, J. M., Yen, J., & Shung, K. K. (2011). Development of a 64 channel ultrasonic high frequency linear array imaging system. *Ultrasonics, 51*, 953–959.
39. Lee, J., Cao, H., Kang, B. J., Jen, N., Yu, F., Lee, C. A., et al. (2014). Hemodynamics and ventricular function in a zebrafish model of injury and repair. *Zebrafish, 11*, 447–454.
40. Grego-Bessa, J., Luna-Zurita, L., del Monte, G., Bolós, V., Melgar, P., Arandilla, A., et al. (2007). Notch signaling is essential for ventricular chamber development. *Developmental Cell, 12*, 415–429.

Minimally Invasive Technologies for Biosensing

Shiming Zhang, KangJu Lee, Marcus Goudie, Han-Jun Kim, Wujin Sun, Junmin Lee, Yihang Chen, Haonan Ling, Zhikang Li, Cole Benyshek, Martin C. Hartel, Mehmet R. Dokmeci, and Ali Khademhosseini

1 Introduction

People of age 65 or older represent a rapidly growing demographic, currently accounting for nearly 10% of the world's population. As world life expectancy increases, the size of this demographic is projected to double by the year of 2050 [1]. Senior citizens are more likely to experience health complications because of degraded organ functionality due to their age. Older people are more vulnerable to

S. Zhang · K. Lee · M. Goudie · H.-J. Kim · W. Sun · J. Lee · C. Benyshek · M. C. Hartel
Department of Bioengineering, University of California-Los Angeles, Los Angeles, CA, USA

Center for Minimally Invasive Therapeutics (C-MIT), University of California-Los Angeles, Los Angeles, CA, USA

California NanoSystems Institute, University of California-Los Angeles, Los Angeles, CA, USA

Y. Chen
Center for Minimally Invasive Therapeutics (C-MIT), University of California-Los Angeles, Los Angeles, CA, USA

California NanoSystems Institute, University of California-Los Angeles, Los Angeles, CA, USA

Department of Materials Science and Engineering, University of California-Los Angeles, Los Angeles, CA, USA

H. Ling
Center for Minimally Invasive Therapeutics (C-MIT), University of California-Los Angeles, Los Angeles, CA, USA

California NanoSystems Institute, University of California-Los Angeles, Los Angeles, CA, USA

Department of Mechanical and Aerospace Engineering, University of California-Los Angeles, Los Angeles, CA, USA

H. Cao et al. (eds.), *Interfacing Bioelectronics and Biomedical Sensing*,
https://doi.org/10.1007/978-3-030-34467-2_9

various diseases including diabetes, arthritis, and osteoporosis, which result in tissue loss and organ failure. Additionally, some acute and fatal diseases such as heart failure and stroke continue to be prevalent threats to one's well-being. To address these issues, devices that enable timely monitoring, diagnosis, and treatment of certain diseases are required.

While the specific methods for detection, diagnosis, and therapy vary for different diseases, current clinical approaches are similar. They seek to develop therapies after the onset of serious complications, rather than addressing the prevention of such diseases. Results often require invasive surgical operations such as artificial joint surgery and tissue/organ transplantation, both of which result in substantial wound trauma after treatment. Chemical techniques including chemotherapy are also frequently used, but they are often accompanied with strong side effects that damage healthy cells. The invasiveness of these approaches contributes to a painful post-therapy recovery period for the patient. Moreover, certain treatments can pose future risks for the patient, such as lethal organ rejection in the case of organ transplantation. The time to fully recover and resume everyday activities often takes very long, and complications may persist for the remainder of the patients life. It is of great urgency to develop new methods for accurate medical diagnoses that result in minimal trauma, thus preventing further complications. The goal of minimally

Z. Li
Department of Bioengineering, University of California-Los Angeles, Los Angeles, CA, USA

Center for Minimally Invasive Therapeutics (C-MIT), University of California-Los Angeles, Los Angeles, CA, USA

California NanoSystems Institute, University of California-Los Angeles, Los Angeles, CA, USA

School of Mechanical Engineering, Xi'an Jiaotong University, Xi'an, China

M. R. Dokmeci
Center for Minimally Invasive Therapeutics (C-MIT), University of California-Los Angeles, Los Angeles, CA, USA

California NanoSystems Institute, University of California-Los Angeles, Los Angeles, CA, USA

Department of Radiology, University of California-Los Angeles, Los Angeles, CA, USA

A. Khademhosseini (✉)
Department of Bioengineering, University of California-Los Angeles, Los Angeles, CA, USA

Center for Minimally Invasive Therapeutics (C-MIT), University of California-Los Angeles, Los Angeles, CA, USA

California NanoSystems Institute, University of California-Los Angeles, Los Angeles, CA, USA

Department of Radiology, University of California-Los Angeles, Los Angeles, CA, USA

Department of Chemical and Biomolecular Engineering, University of California-Los Angeles, Los Angeles, CA, USA
e-mail: khademh@ucla.edu

invasive therapies is to deliver effective treatments with reduced risk of wound infection and other complications, while minimizing the discomfort and recovery times experienced by patients [1, 2].

Another shortcoming of most current disease diagnostic techniques is their inability to provide measurements in a continuous manner. For instance, conventional blood tests require mL levels of blood per test, therefore making continuous sampling and time-dependent readout unlikely. Furthermore, overexposure to techniques such as X-ray or ultrasound, which are crucial and elementary tools in acquiring inner-body information, can cause unavoidable complications. More convenient and less irritative devices are needed to continuously monitor relevant physiological parameters and further lower morbidity. Continuous monitoring can be implemented for a wide array of markers to benefit those affected by multiple ailments, such as the elderly, who suffer higher incidences of having diabetes and hypertension. It is clear that by monitoring typical health instabilities more frequently, corrective action could be taken more quickly. These benefits would apply to both acute and chronic complications, which require diligent monitoring to rectify. However, if based on current medical devices and diagnostic systems, real-time monitoring would remain impossible.

The combination of minimally invasive technology and biosensing provides a promising alternative to realize such benefits [3]. At present, minimally invasive technologies have been used in many fields, including disease detection and biosensing, to diagnose lethal or chronic diseases at earlier stages [2]. Compared to conventional analytical techniques, minimally invasive biosensing is better accepted by patients because it causes much less pain with reduced risk of infection. They offer more direct and effective monitoring for potentially unhealthy status changes, including a broad range of diseases such as heart failure, tumors, and cancer. In recent years, minimally invasive biosensing technologies have proliferated. They have rapidly become an important part of personalized healthcare, which remains a focused research area in biomedical engineering. Minimally invasive devices of various forms have been developed as an important subbranch of point-of-care devices, categorized here as wearable biosensors, edible biosensors, microneedle patches, and smart wound-healing bandages. In this chapter, these different forms of minimally invasive devices are discussed with a focus on how they can contribute to the detection, diagnosis, and treatment of certain diseases. In the conclusion, future challenges, perspectives, and possibilities regarding these technologies are discussed.

2 Point-of-Care Devices

Point of care (POC) or point of patient care is a clinical idea to provide patients with viable and accurate medical diagnostic feedback whenever and wherever they receive care [4]. POC devices seek to reduce the risks of medical errors and to personalize pre- and postsurgical treatments. For practical health monitoring, feedback includes parameters ranging from the temperature and pH of biofluids to the existence or specific concentration of ions and biomarkers, such as glucose and lactate [5]. The design and fabrication of portable, sensitive, and signal-processable POC devices that target these parameters mark the first substantial step to realize and commercialize this idea.

Most POC devices are inherently minimally invasive or require small trauma to the human body since they are often used by nonmedical professionals. Paper-based POC devices have been developed as one of the most accepted branches of practical POC devices [6] (Fig. 1a). They utilize capillary force to drive biofluids within designed routes on paper. After passing functionalized regions, the biomarkers are captured and react to be visualized as a detection line or color change (Fig. 1b). Most paper-based devices have numerous advantages, including low cost, ease of operation, noninvasiveness, and no required power supply [8]. Lateral flow paper-based POC devices represent one of the largest classes of testing paper POC devices. They have rapidly been applied to help determine pregnancy status and infection diagnostics [9]. However, simple output via line or color change inevitably restricts traditional paper POC devices for qualitative applications. The demands for more accurate and informative diagnoses require more quantitative output. Recent research has mainly developed two types of output: complex optical signals and electrical signals. Complex optical sensing can be processed via colorimetric, fluorescent, or luminescent analysis [10]. Yetisen et al. invented a portable paper-based microfluidic tear sensing system that could quantify electrolytes in artificial tear fluid, via smartphone-readable fluorescence outputs (Fig. 1c) [11]. The device

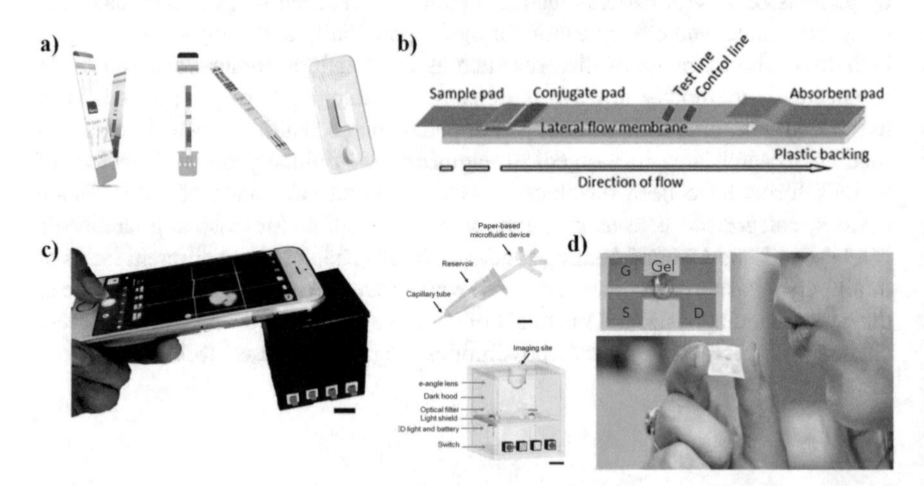

Fig. 1 Examples of point-of-care devices. (**a**) Commercialized testing paper-based POC device for qualitative detection. (Adapted with permission from [6]. Copyright (2013) Royal Society of Chemistry.) (**b**) Conventional POC device based on lateral flow testing. (Adapted with permission from [7]. Copyright (2014) Springer-Verlag Berlin Heidelberg.) (**c**) A smartphone-assisted portable paper-based microfluidic tear sensing system that uses complex optical signals in target molecular quantification. (**d**) A disposable paper-based blood alcohol quantitative breathalyzer using organic transistor

consists of four channels leading to four separate analysis regions for the detection of: Na^+, K^+, Ca^{2+}, and pH. Because the fluorescence intensity of the four sensing regions is highly correlated to their respective concentrations, quantification was achieved with the help of the smartphone's image capturing capabilities. The phone also served as a platform to analyze the image and ultimately provide a more accurate approach to categorize dry eye subclassifications. Electrical sensing is a more common approach to acquire quantitative data, with numerous advantages including high sensitivity and reliability. Bihar et al. have demonstrated a proof-of-concept inexpensive and disposable paper-based blood alcohol quantitative breathalyzer using an organic transistor (Fig. 1d) [12]. Ethanol molecules in breath trigger an enzymatic reaction in the electrolyte solution on the testing paper, during which electron transport affects the conductivity of the transistor, thereby transforming the chemical signals of blood alcohol content into processable electrical signals.

Recent research on POC devices has produced fruitful and impressive achievements, namely, more accurate quantitative diagnostics and more wearable portable devices. However, many challenges remain to be addressed for the field to reach its true potential. These POC devices must be able to sense, respond, and provide feedback in real time. They should also actively transfer wireless signals across farther distances to notify experts in clinics [13].

3 Wearable Biosensors

In recent years, we have witnessed the rise of wearable electronic devices as an emerging branch of POC devices [14] (Fig. 2). Wearable devices are well-known for their flexibility, enabling them to be incorporated into clothing or to be worn on the human body as implants and accessories. As one example, the Apple Watch provides medical functions such as heart rate monitoring to generate a corresponding electrocardiogram. The watch can therefore be regarded as a wearable electronic sensor that is flexible enough to be worn on one's wrist. Wearable devices are considered noninvasive or minimally invasive as they normally don't cause any trauma to human body [15]. However, achieving sufficient softness as well as maintaining electrical functionality has been challenging for the traditional semiconductor industry. Manufacturing procedures for commercial semiconductor devices typically rely on rigid chips which are unable to bend, stretch, or twist in shapes required by human skin or organs. Therefore, most wearable devices are fabricated on soft plastic or elastomers that can tolerate mechanical bending and stretching [16]. However, this poses challenges in materials development and device fabrication as they are not compatible with established industrial processes. Nevertheless, in recent years many revolutionary flexible electronic materials [17], fabrication technologies [18, 19], and discoveries have been reported. Examples include flexible organic photovoltaics [19], transistors [20, 21], and energy harvesters [19].

As flexible device technology rapidly advances, epidermal electronics have emerged as next-generation minimally invasive wearable biosensors [22]. These

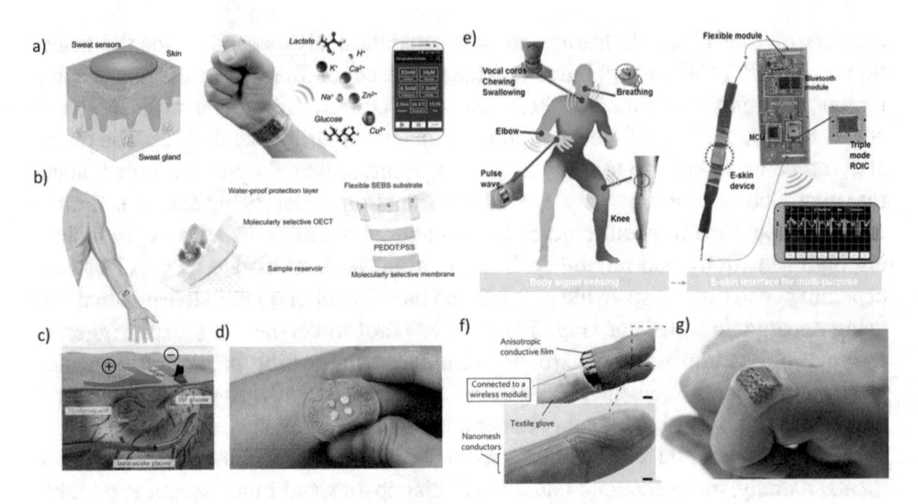

Fig. 2 Minimally invasive sensing of healthcare signals with wearable biosensors: (**a–d**) biomarkers detection. (**a**) glucose and various of ions in sweat [14]. (Adapted with permission from [14]. Copyright (2016) Springer Nature); (**b**) cortisol sensing with wearable organic transistors [29]; (**c**) iontophoresis sensing of glucose in interstitial fluid [30]; and (**d**) colorimetric sensing of ions, glucose, and pH in sweat [31]. (**e–g**) Electrophysiological signals sensing on human body: (**e**) integrated electronic system monitors movement at different places [32]; (**f**) skin-attachable nanomeshes mentoring muscle behavior [33]. (Adapted with permission from [33]. Copyright (2017) Springer Nature.) (**g**) Wearable ultrasound sensor monitors blood pressure inside human body [34]. (Adapted with permission from [34]. Copyright (2018) Springer Nature)

devices are skin-attachable and stretchable, leaving no gap between the sensing units and skin. This allows for more intimate contact with the human body, enabling continuous and precise monitoring of human healthcare signals [20]. Such devices have higher patient compliance due to their noninvasive, ultrathin, and ultralightweight properties. However, because of the uniqueness of our skin's mechanophysiology, fabrication of ultrathin epidermal sensors has been challenging. Compared to flexible electronics, epidermal devices require additional stretchability to minimize the gap between sensor surfaces and the skin. At the same time, they should be ultralightweight and small to be imperceptible and even optically transparent (invisible) to users. To realize conformable biosensing on the user's skin, devices not only need direct pain-free contact with the skin but should also be able to function properly regardless of their location on the body. For example, textile-based sensors meet the first requirement as they only achieve intimate contact with the skin at specific regions [23, 24]. Furthermore, they can only be incorporated into certain types of fabrics, limiting the user's preference [25]. Currently, tattoo-based biosensors [26] and intrinsically stretchable thin film sensors made of conductive polymers [20], elastomers [20], hydrogels [27], and nanomaterials [28] represent the best candidates for epidermal biosensors.

4 Minimally Invasive Sensing with Wearable Biosensors

Generally, biosensing that could reflect the health condition of our body can be divided into three categories: mechanical sensing, biomarker sensing, and electrophysiological sensing. Mechanical sensors are demonstrated to detect and differentiate human movements, including motions of the arms, legs, and fingers, and also monitor pulse, swallowing, and breathing activities. These signals can be detected by strain sensors. In comparison to mechanical sensors, sensing of biomarkers and electrophysiological signals sensing (discussed below) is more important and enables the acquisition of more accurate health information because biomarkers are direct indicators of biological states or conditions of cells, while electrophysiological signals are specific signals generated by electroactive cells such as cardiac cells, muscle cells, and neuron cells.

4.1 Biomarker Sensing

A biomarker is a molecule or substance that indicates some bodily function or phenomenon. These can be important indicators for monitoring diseases or responses to environmental stresses, such as dehydration. Biomarkers must be continuously and accurately monitored for wearable biosensors to reach their full potential as a form of personalized medicine. The major classes of noninvasively sampled biomarker mediums include urine, sweat, blood, interstitial fluid (ISF), tears, breath, saliva, and wound excretions. A wide variety of conditions can be diagnosed by studying biomarkers, so accurately measuring them will allow wearable medical sensors to become the next big breakthrough.

Different biomarkers require different types of wearable sensors, many of which already exist (Fig. 2a–d). For example, sweat sensors have been integrated onto flexible textiles [35], tattoos [36], and plastics [37]. Tear sensors have been put into contact lenses through various means [38], and breath analyzers have been reduced to fingernail-sized patches [39]. Electronic circuits have been combined with chemical sensors that react in the presence of molecular biomarkers within a fluid. The exact electrochemical sensor used depends on the biomarker medium and what specific marker requires measurement. Some biomarkers are easier to measure noninvasively than others. Biomarker fluids like sweat, saliva, and breath are readily obtained from the body's exterior, while others like blood and ISF must be measured indirectly [40]. Wearable devices such as pulse oximeters use noninvasive light rays to detect the composition of an internal fluid or tissue [10]. Medical sensors have been around for decades, but recent advances in nanotechnology have enabled wearable sensors with ultrathin external power sources [41]. Nonetheless, wearable sensors are still very much developing, and as we make improvements in our understanding of biomarkers, their potential only grows.

Biomarkers' true importance stems from their ability to indicate various body phenomena. Through sweat monitoring, electrolyte imbalances, blood ethanol concentrations, kidney failure, and cystic fibrosis can be monitored [42] or diagnosed [43, 44]. Urine monitoring [45] can detect glomerular disease, kidney disease, and healthy menstrual function [46]. Tear sensing [47] can detect ocular diseases as well as glucose levels which are considered as metrics for diabetic patients [48]. Noninvasive blood monitoring can still detect many phenomena, including blood oxidation levels and blood glucose levels [49]. Noninvasive ISF analysis can detect metabolic disorders and assess organ failures [3, 50]. Breath analysis can detect blood ethanol levels and asthma diagnoses, among other things [51]. Saliva monitoring can detect stress levels as well as the presence of some types of cancer. Wound excretions can determine the healing status of the wound and the presence of infections. There are many other uses for sensing these biomarkers as their applications are extremely diverse. Of course, each type of biomarker presents a different limitation. Sensing these components noninvasively can limit the amount and specificity of information available, which further reduces the accuracy of diagnostics in some cases. Additionally, some biomolecules that are analyzed to predict certain ailments are inherently quasi-inaccurate indicators [10]. Thus, as our understanding of biological processes improves, so too will our understanding of indicative biomarkers. Not only will current biomarkers become more diverse in their applications, but new biomarkers will be discovered, expanding the scope of this field even more.

4.2 Electrophysiological Sensing

In addition to biomarker detection, wearable biosensors offer competitive replacements for conventional bulky medical instruments which measure electrophysiological signals including electrocardiograms (ECGs), electromyograms (EMGs), electroculograms (EOGs), and electroencephalogram (EEGs) [52], as shown in Fig. 2e–g. Most likely, wearable biosensors will be integrated with a combination of these functions to obtain a more complete evaluation of one's health in the future [19, 53].

Previous work has demonstrated capacitive sensors. Such devices are normally fabricated with rigid electrodes or conductive fabric meshes covering dielectric materials such as pyre varnish, polydimethylsiloxane (PDMS), or even clothes. Despite that by capping, belting, or taping, sensors with rigid electrodes could be attached to human skin, it remains difficult to achieve conformal contact. Coupling capacitance is vulnerable to inevitable motion-induced deformations and the dynamic mismatch between devices and the skin. Moreover, discomfort accompanying long-term wearing is another issue to solve, since rigid electrodes available are too thick to be naturally attached without irritations [54].

Cardiac and heart rate monitoring devices mainly use optical or electrode-based sensors to obtain physiological signals. These devices are commonly used for clinical cardiac condition diagnoses including arrhythmia and ischemia. However, the use of these devices in mobile and home environments is often limited due to the size and price of facilities and complexity of operation. Dae-Hyeong Kim et al. [55] reported an ECG monitor based on the integration of sensors, amplifiers, and organic light-emitting diodes (OLEDs), which supported continuous and real-time acquisition for readable ECG signals. Proven by cyclic mechanical deformation tests, the ECG monitor could tolerate severe deformation on the skin while retaining electrical stability. Rogers and colleagues [19] reported a low-cost and low-weight flexible wearable cardiac sensor (WiSP) that could be wirelessly read and powered by NFC-enabled smartphone. The device exhibited outstanding accuracy that matched the commercialized product, Polar H7 heart rate monitor. Due to the small size and thin-layered structure, WiSP could also be comfortably attached to human body while decreasing the impedance between the sensors and human, improving the fidelity of measured signals.

Central blood pressure (CBP) is another parameter to represent human health status. Continuous monitoring of CBP has important clinical value in predicting all-cause cardiovascular mortality. However, existing noninvasive methods, such as optical methods (photoplethysmography or volume clamp) and tonometry, are only capable of monitoring the peripheral blood pressure waveform, which is different from CBP induced by the variation of arterial stiffness. Moreover, although current ultrasound technology allows noninvasive deep tissue observation such as ultrasound wall-tracking, rigid ultrasonic probes would introduce gap and prohibit seamless contact with target skin, introducing inevitable impedance instability while operating. Xu and colleagues developed a wearable and stretchable ultrasonic device for continuous and accurate monitoring of the CBP noninvasively [34]. Encapsulated by silicone elastomer, the 240 μm thick device could be naturally attached to skin at different locations on the human body.

5 Edible Biosensors

Beyond epidermal and wearable electronics, edible electronics (digital capsule, ingestible electronics) are emerging as a new minimally invasive medical device for biosensing (Fig. 3). These ingestible devices are developed with biocompatible and dissolvable electronic materials. The concept uses minimized electronic devices such as sensors, cameras, and drug delivery components powered by nontoxic batteries made from biodegradable electronic materials [56, 57]. These devices can be bent and encased in commercially available gelatin capsules, allowing for timed release in the human body and enabling specific targeting, such as within the gastrointestinal tract [56]. As the capsule dissolves in the body, the electronic devices are exposed to the gastrointestinal environment and begin sensing and transmitting data wirelessly. Patients can take these capsules regularly like standard medication, while the devices maintain functionality in the digestive system for 1–2 days. The devices would ultimately be absorbed or passed in the stool.

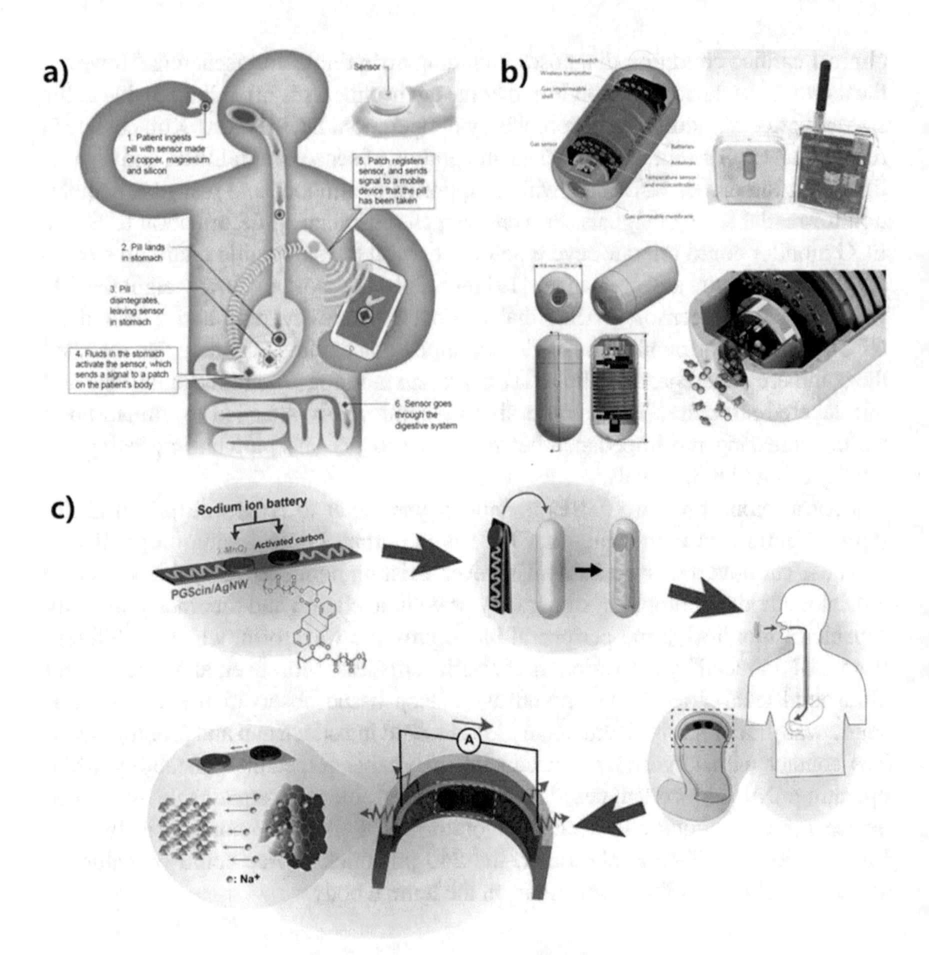

Fig. 3 Examples of edible (capsule) biosensors: (**a**) The journey of an ingestible capsule sensor (Credit: Proteus Digital Health); (**b**) ingestible capsule biosensors capable of sensing various gases in the gut [56]. (Adapted with permission from [56]. Copyright (2018) Springer Nature.) (**c**) Envisioned deployment and operation of edible battery that are compatible with noninvasive deployment strategies and are composed entirely of edible materials [57]. (Adapted with permission from [57]. Copyright (2013) Royal Society of Chemistry)

Edible biosensors have many advantageous properties over other sensing technologies in part because they are minimally invasive and more convenient to use [58]. In addition, "oral medication" of these sensors, with respect to other biosensors, may be accepted by more people because there is almost no discomfort. The health data collected in the gastrointestinal tract is more accurate than those sensed outside the body, such as body temperature. Moreover, this technique provides a means of direct detection that is impossible with conventional techniques, such as detecting biomarkers and gases in the digestive systems, which could provide more information on the body condition [57].

Capsules containing ingestible electronic sensors have been demonstrated to be capable of sensing different gases including oxygen, hydrogen, and carbon dioxide [56]. These minimally invasive ingestible electronic capsules use a combination of thermal conductivity and semiconducting sensors to monitor different gases in the gut, with good selectivity and sensitivity. A non-transparent polyethylene shell was selected as the capsule material. Inside lie all the sensing electronic components including CO_2, H_2, and O_2 sensors that can operate in various anaerobic and aerobic conditions, a temperature sensor, a transmission system, a microcontroller, and button batteries made of silver oxide [56]. The capsule shell is also incorporated with membranes embedded with nanomaterials to allow fast diffusion of dissolved gases. Once the sensors detect the presence of certain gases, they gather information and transmit it in real time to a small receiver carried by the user.

Edible electronic capsule sensors can be also used to monitor the dose of opioids that patients take in to prevent issues like overdose and addiction [56, 59]. The capsule pills are based on standard gelatin capsules, containing an ingestible radiofrequency emitter and a 5 mg oxycodone tablet. Once this capsule is ingested, the gelatin capsule is dissolved, liberating the oxycodone medication. The emitter is then activated by chloride ions and sends a signal to a wearable reader. By collecting the data sent from the frequency emitter, concentrations of drugs such as morphine is obtained.

Apart from edible biosensors which perform detection inside the human body and send signals out to a receiver, actively powered edible devices have also been developed to supply current inside the human body for biosensing metabolic activity, heart rate, and core body temperature [57]. These devices consist of two functional components: flexible composite electrodes and a sodium ion electrochemical cell. The composite electrodes are fabricated by incorporating silver nanowires in a PGScin polymer substrate. This edible current source is packed in gelatin capsules to be ingested orally, and after ingestion, the capsule dissolves, activating the device via hydration. The current flow is initiated from the wet electrochemical cell, thus providing power. There already exist commercially available products using capsule biosensors.

To ensure biocompatibility and degradability of edible biosensors, most of these devices use “green” electronic materials [58]. However, most of these green materials are organic and have poor electrical properties and significant performance degradation over time. For example, conducting polymers are frequently used as active materials in edible biosensors. However, compared with inorganic materials, organic materials have poor electrical performance and poor environmental stability. In addition, edible devices require a wireless electrical unit to transmit the data to mobile devices, but these units must rely on commercial silicon chips which are toxic and corrosive. Therefore, the development of non-corrosive batteries is a must, because otherwise fatal damage to the human body will be inflicted. Even though nontoxic, edible batteries have been developed using biocompatible organic materials such as melanin, a natural compound from the human body, they still have very limited performance. Nevertheless, due to the significant advantages and potential market of edible biosensors, it is envisioned that advanced materials and fabrication technologies will be developed to promote their commercialization for in vivo minimally invasive biosensing.

6 Microneedle-Based Biosensors

The skin is the single largest organ of the human body and contains a lot of information to be analyzed. In the last two decades, microneedles (MNs) have been fully demonstrated to be a good platform for skin applications (Fig. 4). MNs applied to the skin can be made to have sufficient height to penetrate the physical tissue membrane, which acts as a barrier, but short enough to avoid damaging the tissue or nerves. By doing so, MNs do not cause bleeding and are considered minimally invasive. Such lack of pain associated with MNs has been scientifically verified [60]. The degree of pain resulting from the insertion of a 150 μm height silicon MN array was compared to that of a 26G hypodermic needle (negative control) and a silicon plate without MN structures (positive control). After giving informed consent, 12 healthy volunteers (male and female between 18 and 40 year of age) were blindly exposed once to each of the three treatments so that the subjects could rate subsequent pain scores. The resulting sensation caused by the MN array was statistically indistinguishable from a smooth surface, indicating that the MN array treatment was not painful and did not injure the tissue. The pain caused by a hypodermic needle was demonstrated to be substantially higher than that caused by the MN array. Skin resealing after MN application was also demonstrated using an in vivo test in human subjects [61]. A solid correlation between permeability and impedance of the skin was confirmed, where an increase in skin permeability generally corresponded with a decreased impedance of the skin. The resealing properties of the human skin were analyzed using impendence spectroscopy. The results indicated that all the sites treated with MNs recovered barrier properties within 2 hours in the absence of occlusion. Occluded sites resealed slower with a resealing window of 3–40 hours depending on the geometry of the MNs. Furthermore, the safety of MN arrays has been assessed from human volunteers. MNs arrays were able to overcome the physical skin barrier and only cause minimal irritation which lasted less than 2 hours. The study found no biological viability damage of the skin tissue [61]. As verified by many subsequent studies, these minimally invasive characteristics were investigated. The increased skin permeability afforded by MNs was demonstrated to facilitate the delivery of a wide variety of drugs from small to large molecules, proteins, and genes. Beyond harnessing the MNs for drug delivery, MNs can be used for wearable biosensing or long-term health monitoring devices in transdermal field.

ISF is one heavily researched body fluid because its composition follows that of blood quite closely in many instances. Unlike blood, however, ISF is more readily measured noninvasively with the use of MNs. ISF exists in the epidermal tissues less than 1 mm below the skin's surface, so it can either be extracted or analyzed in

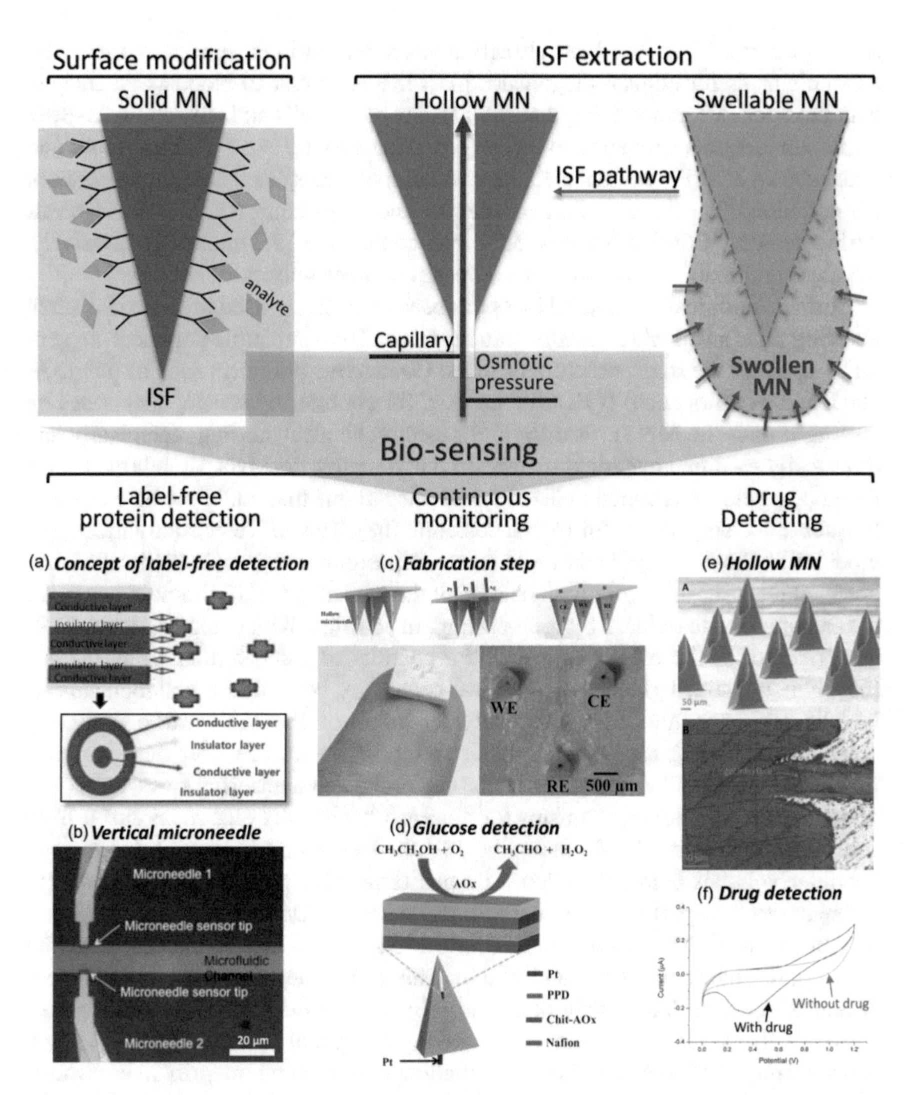

Fig. 4 Schematic of microneedle biosensing. Microneedle can access to ISF via surface-modified solid MN and extract ISF with hollow and swellable MNs. Their representative biosensing applications; (i) label-free protein detection: (**a**) concept of label-free detection by directly inserting (**b**) vertical microneedle to the target site. (Adapted with permission from [62]. Copyright (2013) Elsevier.) (ii) Continuous monitoring patch: (**c**) the fabrication of continuous monitoring electrode consisting of hollow MNs and (**d**) the principle of glucose detection at working electrode. (Adapted with permission from [63]. Copyright (2017) Elsevier.) (iii) Drug detection: (**e**) an image of hollow MN for the electrode and (**f**) drug detection by the difference of cyclic voltammetry. (Adapted with permission from [64]. Copyright (2017) Elsevier)

situ. Therefore, MNs have been attracting much current interest because they offer novel channels for clinical diagnostics [65]. ISF is similar to blood as an analyte, with the main difference being that ISF has few blood cells and only about 25–30% of protein amounts compared to blood [66]. Although the exact concentrations and time profiles of ISF's content can differ, small molecules often equilibrate between ISF and blood. The different time profiles occasionally manifest as lags in ISF levels during sensing. Considering this, MNs can contact the ISF minimally invasively, thereby minimizing the relative lag compared to other sensors.

Surface modification of solid MNs has been extensively used to sense ISF, often allowing them swell when inserted into the tissue. Biocompatible polymers are generally used for the main structure of MNs. Conductive polymers such as poly(3,4-ethylenedioxythiophene) (PEDOT) are used for configuring sensing electrodes by coating it onto the MN surface [67]. For electrochemical sensing, chemically and electrically modified materials are used to functionalize the MNs. Bollela et al. have developed gold MNs functionalized with nanocarbons that could mediate electron transfer to lactate oxidase for lactate detection [68]. The gold electrode surface was modified in three steps: (1) thin film electrodeposition using gold-multiwalled carbon nanotubes, (2) electropolymerization of the mediator, and (3) immobilization of the enzyme lactate oxidase by drop-casting. In addition, Kim et al. developed MNs comprised of cyclic olefin copolymer for continuous glucose monitoring (CGM) [69]. Functionalized glucose oxidase was covalently bonded to a terthiophene carboxylic acid monomer, and electropolymerization on the MN surface coated by gold followed using a potential cycling method. The surface after treatment was protected by Nafion layer to avoid detaching probe and interfering species.

In general, contact mechanisms for biosensing with MN electrodes suffer from insufficient amounts of ISF. Sampling ISF for biosensing is limited due to only nanoliter volumes being extracted. MN platforms have proven to be a promising technique to transport ISF from the skin to electrodes. One mechanism of ISF collection involves the diffusion of ISF into MNs, which can be accomplished by inserting dry hydrogel MNs into the skin. Chang et al. developed a MN patch consisting of methacrylated hyaluronic acid [70]. This swellable MN patch extracted sufficient ISF volumes in minutes without the additional device by using the high water affinity of MeHA. The MN array maintained structural integrity in its swollen state, leaving no residues in the skin. This approach facilitates remarkably timely metabolic analysis.

Leveraging capillary action by using hollow MNs is another method for extracting ISF. Hollow MNs have been fabricated with various techniques ranging from microfabrication to drawing lithography. A press-fitting process has recently been introduced to fabricate hollow MN arrays for ISF collection and sensing. To make an electrochemical transducer, platinum and silver wires, which diameters are 100 μm, were inserted into MN lumen to make channels [63]. ISF was sampled via the hollow MN aperture driven by capillary force. Mishra et al. used 3D printing to fabricate 3 × 3 hollow MN arrays with pyramidal shapes and triangular bases [71]. The MNs were designed to detect a nerve agent, with specific geometries of 1500 μm in height and vertical cylindrical holes of 425 in diameter on the pyramid structure.

An osmotic driving force and a pressure difference between the dermis and skin surface can drive ISF from the dermis through its pores [72]. To generate this osmotic pressure gradient, Samant et al. filled microchannels created by hollow MNs with a solution of elevated ionic strength to drive ISF from the dermis into the microchannels. By lacking a permselective membrane between the microchannels and dermis, solutes in the microchannels could diffuse into the dermis over time, gradually decreasing the osmotic pressure gradient. The pressure difference could be accomplished by suction to create negative pressure on the skin surface at the microchannels. Alternatively, applying positive pressure on the skin surface at sites distant from the microchannels would also create a negative pressure in the skin near the microchannels.

Continuous bio-signal monitoring has been demonstrated by using MN sensors. Continuous glucose monitoring has shown promising clinical evidence by reducing the frequency of hypoglycemic episodes and HbA1c levels in Type 1 diabetes patients. Besides the advantage of minimizing any damage to blood vessels and nerve cells in the dermis layer, noise and sweat contamination can be avoided during sensing. This is because the MNs are not located on the surface of the skin but penetrate through the tissue and remain inside. Jina et al. developed and tested a MN-based CGM system that could measure glucose signals for 3 days with functionalized hollow MNs [73]. They validated the ISF results by also measuring glucose levels in blood for 3 days in ten diabetic patients using conventional fingerstick sampling. The results showed that the MN was more accurate and well tolerated by the subjects compared to conventional fingerstick. In addition to glucose, alcohol levels within the skin have also been detected continuously. Mohan et al. developed a MN sensor for continuous alcohol monitoring in ISF [63]. The MNs used a three-electrode system configured with Pt and Ag wires within the hollow MNs aperture. The Pt wire, as the working electrode, functionalized with alcohol oxidase, permselective chitosan, and a Nafion layer. The results showed high selectivity and stability of MN biosensing in an ex vivo mouse model with an ethanol substrate.

Enzyme-linked immunosorbent assays (ELISA) are the most common technique for quantifying protein concentrations. This technique relies on labeling the target analyte with an antibody and subsequently detecting optical fluorescence. Despite being the clinical gold standard, ELISA is time-consuming and expensive. Impedance sensors have been developed to provide an alternative to ELISA by using rapid, label-free detection in situ with miniaturized systems. The underlying principle of impedance sensors is to measure the impedance changes of the electrode surface after specific binding events occur on the sensor. However, they still suffer from limitations in terms of low sensitivity compared with their fluorescent counterparts. Song et al. developed a micromachined flexible MN impedance sensor for label-free protein detection in situ [74]. The sensor was configured with a microwell array and MN tips for label-free detection. Despite the presence of complex interfering compounds and high concentrations of salt, the MN sensor maintained high sensitivity. As a result, real-time, rapid, and label-free protein detection was achieved in vitro on skin phantom tissue.

In addition to protein concentrations, monitoring dynamic changes in pH provides essential insight into many physiological phenomena. Zhou et al. reported a high-performance electrochemical pH MN for real-time pH monitoring using a brain of rate [75]. A MN sensor was fabricated by assembling layer to layer with disulfide and molybdenum polyaniline for detecting hydrogen ions with high sensitivity. The MN sensor reduced other potential interfering species in the brain and showed clear selectivity with a Nernstian response, indicating 51.2 mV per pH over a range of pH 3–9. Furthermore, MN arrays for pH detection have also been fabricated via injection molding [76]. This sensor was integrated with other electronic parts to scale up a portable device for real-time pH wireless sensing in biological ISF samples. The fabricated MNs were coated with gold and surface functionalized with iridium oxide electrodeposition to sensitize them to dynamic pH changes. Miniaturized electronics were integrated with the MN sensor for wireless data transmission and analog-to-digital conversion, thus enabling the sensor to measure real-time pH within soft and dynamic biological tissue, such as muscle.

MN-based sensors have also been used to detect the therapeutic level of drugs to minimize side effects and overdose. Vazquez et al. developed a hollow MN integrated with a micro-interface of immiscible fluids for the detection of drug concentrations in body fluids [77]. This device could detect ionic species unaffected by redox processes and without ISF extraction. The MN sensor was able to detect propranolol in a physiological phantom solution and in artificial saliva samples with high sensitivity. Using the miniaturized interfaces for analysis, the detection limit was lower compared to previous publications regarding propranolol. The MN sensor also provided stable linear range detection.

Despite the success of MN biosensors in the past years, yet few products are available in the market. From the materials point of view, few of them are Food and Drug Administration (FDA) approved; thus biocompatible and mechanically robust materials should be further developed to mitigate the skin response to the MN and thus reduce skin inflammation. Besides, efficiency degradation and delamination are always associated with MN biosensors with time, which needs to be solved before the practical application. Further integration of drug delivery and biosensing in the same MN that could be wirelessly controlled and self-powered may hold the potential to dominate the MN application in the future, yet such devices require endeavor from convergent engineering works in material development, device fabrication, drug delivery, and electrical engineering.

7 Smart Bandages

Compared to normal human tissue, wounds present a special case for biosensing because wound sites have lost their natural skin barrier. This enables biosensing devices to contact bodily fluids directly, therefore providing more accurate information. Currently, chronic wound management represents an ongoing problem for clinicians. For example, in 2012 it was estimated that chronic wounds care in the

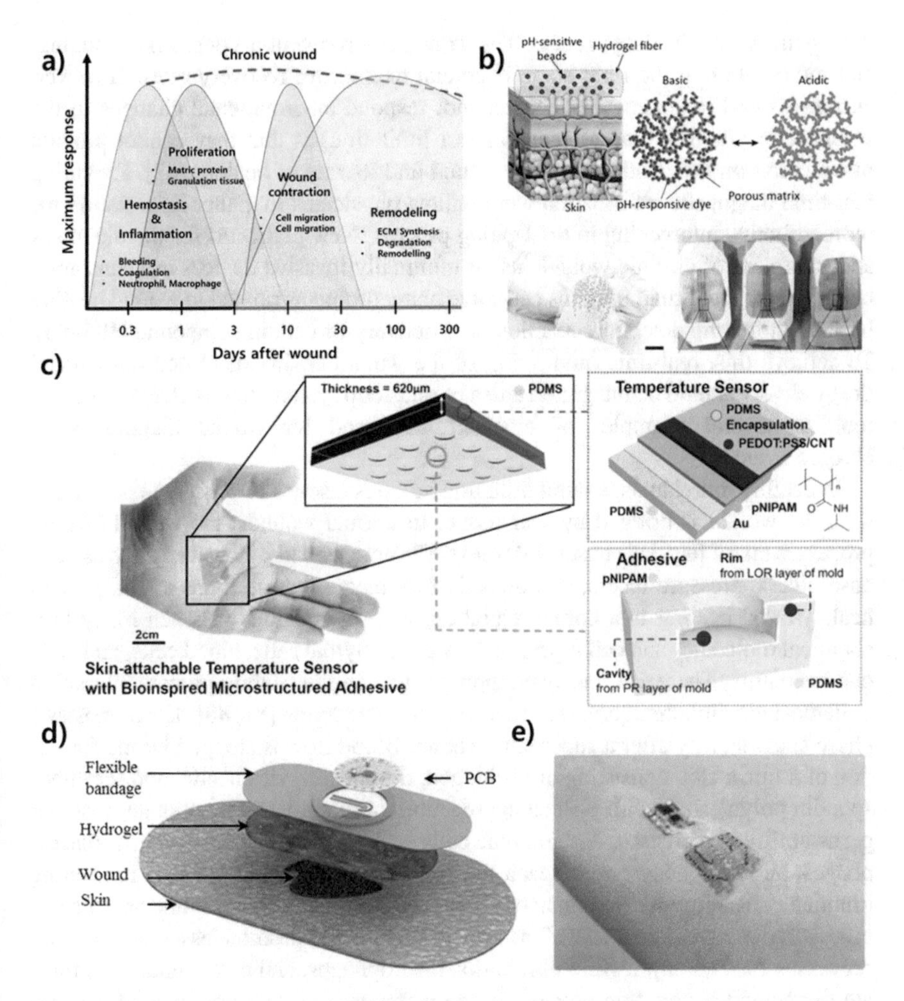

Fig. 5 Smart bandages for wound healing monitor. (**a**) Chronic wound healing stages. (**b**) pH-sensing hydrogel fibers. (Adapted with permission from [93]. Copyright (2016) John Wiley and Sons); highly stretchable potentiometric pH sensor. (Adapted with permission from [94]. Copyright (2017) American Chemical Society.) (**c**) Highly sensitive flexible temperature sensor with a bioinspired octopus-mimicking adhesive. (Adapted with permission from [57, 95]. Copyright (2018) American Chemical Society.) (**d**) Three-dimensional model of wireless flexible smart bandage for real-time data acquisition of oxygen concentration. (Adapted with permission from [96]. Copyright (2015) IEEE.) (**e**) Automated smart bandage. (Adapted with permission from [97]. Copyright (2018) John Wiley and Sons)

United States cost of almost $25 million annually [78]. The misdiagnosis of wound healing status can result in various complications, including infections and scar hyperplasia. In the past years, we have seen the rise of smart bandages as a minimally invasive technology for wound healing, biosensing, and drug delivery (Fig. 5).

Wound diagnostics using minimally invasive bioelectronics on the skin's surface are important for regeneration prognoses and determining treatment direction during the wound healing process. The wound environment is constantly changing, and intervention at the right stage is crucial to improve recovery rates. However, current wound treatment dressings cannot respond environmental changes in the wound cite [79–81]. One of their biggest limitations is that they cannot provide information on the condition of the wound and its rate of healing [82]. Providing real-time diagnostic information would allow physicians to gather information for more effective intervening in the healing process. New platforms for the diagnosis and treatment of chronic wounds using minimally invasive devices are constantly being developed. Smart systems can solve many problems related to wound healing by facilitating the detection, reaction, and delivery of certain compounds [83–86]. To achieve this, real-time monitoring of the wound environment and on-demand drug delivery should be integrated into a closed-loop system. In this chapter, we will look at several examples of bioelectronics used for wound diagnostics in chronic wounds.

In healthy individuals, wound infliction initiates a series of processes performed until the wound is bridged by scar tissue. In normal wounds, the wound healing processes take a few days to several weeks. Chronic wounds, however, such as varicose ulcers, pressure ulcers, and diabetic foot ulcers require more than a year to heal. Wound healing is a complex process with interactions between many biochemical molecules and cell types, such as parenchymal cells, blood cells, and various mediators. There are four overlapping stages in the healing process, classified as hemostatic, inflammatory, proliferative, and remodeling [87, 88]. The hemostasis phase starts shortly after a superficial wound. Blood flow is stopped by the formation of a fibrin clot, consisting of fibrinogen, fibronectin, vitronectin, and thrombospondin polymerized with platelet aggregation. In the inflammatory stage, vascular permeability is increased, and immune cells such as enzymes, neutrophils, macrophages, and lymphocytes can reach the damaged area [87, 89]. These infiltrating immune cells help remove damaged tissue and dead cells. Neutrophils and macrophages secrete mediators as well as growth factors that promote essential recovery processes for 2–3 days, while fibroblasts, macrophages, and endothelial cells initiate the wound contraction process. In the worst case of a chronic wound that has stopped in the inflammatory phase, there may be a healing delay up to several months after the initial injury [82, 90]. During the proliferative stage, fibroblasts secrete a transient extracellular matrix (ECM) that provides substrates for cell movement, facilitates changes in cell behavior, and provides structure to reestablish the integrity and function of the tissue. Particularly, cell migration in such a wound site allows the epidermal cells to cause wound contraction. In addition to the ECM, angiogenesis, contraction, and keratinocyte migration are essential components for cell migration and proliferation. The stimulation of these ECMs and angiogenic growth factors leads to the observation of granulation tissue composed of new connective tissue and small blood vessels. Endothelial cells enter the fast growth phase, while granular tissue experiences angiogenesis to supply this active metabolic region. This helps collagen and wound edge contraction. During the final phase of remodeling, the vascular network is retracted, while the ECM is extensively replaced and modified. Fibroblasts continue to stimulate collagen synthesis and have a unique triple helix shape, allowing intracellular and intermolecular interactions to form very stable and strong collagen meshes [90–92].

Many of the existing wound care technologies rely on passive mechanisms for drug release to prevent infection and do not respond to environmental variations in the wound. These dressings can release anti-inflammatory drugs, antibiotics, anti-bacterial compounds, angiogenic factors, or uptake excessive exudate. More advanced dressings passively release biological factors and compounds to facilitate tissue healing. However, a key limitation of current wound care products is the lack of the mobility to monitor the status of the wound bed and its healing rate, which could be used by physicians to make real-time decisions on treatment.

Infection is the most common complication associated with decreased wound healing, of which staphylococci and streptococci are the two most common pathogens found in superficial wounds [98, 99]. While late-stage infection can be observed from symptoms such as localized increases in temperature, swelling, puss, odor, pain, or sepsis, the goal is to detect or prevent infection so interventions may occur, increasing the safety of the patient [100].

Currently, the extent of infection is determined using a swab from the wound site, which can be analyzed in a microbiology laboratory for bacterial growth. Growth is determined with ratings from "scanty" to "heavy"; however motile and fast-growing organisms are more easily detected than anaerobes, which may be underrepresented [101]. Also, this method is only capable of detecting pathogens at the surface of the wound and fails to represent what is happening deeper in the tissue. False positives are also common, showing bacterial growth not associated with infection. Therefore, the ability to detect infection early will guide physicians on when treatment is or is not required, reducing costs and the overuse of antibiotics associated with the emergence of antimicrobial resistant bacteria. In general, being able to quantify and capture the wound status would provide critical insights on the status of the wound, which is even more difficult for chronic wounds. For example, diagnosing infection or abnormalities in diabetic wounds is more difficult as differences from the healthy state are less severe. Here, various methods and approaches for the engineering of these smart bandages are reviewed, with the aim at providing more information or better treatment of wounds for clinical practice.

Preventing infection has been a primary goal for both an increased rate in wound healing and as a means of preventing mortality and costs associated with hospital stays. The development of antimicrobial wound dressings has been a broad effort in biomedical engineering. A large variety of approaches have been taken to integrate antimicrobial characteristics into wound dressings, such by using metallic ions [102, 103], nitric oxide [104–107], and antibiotics, all of which have been reviewed extensively [79, 108–110]. While these materials are highly applicable for developing wound dressings to prevent infection, this section will focus on technologies associated with real-time monitoring and the treatment of wounds. As the wound healing process progresses, the need for intervention could be different at different healing stages and time points.

As referenced above, the ability to monitor pH throughout the wound healing process can give insight into the progression of healing. A cost-effective method for determining pH is by integrating dyes that absorb different wavelengths of light depending on the pH into the dressing. This method is attractive as the sensors are generally easy to use and robust and may not require additional electronics. However, methods to prevent the leaching the pH-responsive dyes are typically needed. Secondly, the dye selection must show significant responses over the pH ranges observed during the wound healing process. One example of these dyes is a series of hydroxyl-substituted azobenzene derivatives [111, 112]. Colorimetric pH sensors have also been used where pH is determined based on image processing; however, if there is a significant color change, viewing it with the naked eye can be used for the estimation of the pH value or identification of wound status, but method leaves errors associated with human interpretation. In one method to decrease leaching, the pH-responsive dyes were incorporated into polyester beads before incorporation into microfluidic-generated alginate-based hydrogel fibers. These fibers were then used in conjunction with transparent medical tape for long-term monitoring of cutaneous wounds [93]. Response time of the fibers for various levels of pH was shown to vary with fiber diameter. Changes in the pH of the wound site could be monitored using smartphones; however, image analysis can be susceptible to errors introduced by differences in lighting and image quality. Hydrogel dressing with pH-responsive dyes has been used to measure pH using an integrated photodiode with wireless communication [113]. One strategy for monitoring pH is the detection of uric acid within the wound. Uric acid can be observed at concentrations up to 4100 $mmol \cdot L^{-1}$ in the wound fluid [114]. A wireless pH monitor that can be imbedded within wound dressing to continuously read pH was shown using a pH-selective hydrogel between two micromachined wire coils [115].

Damage to the vasculature associated with wounds can also lead to acute hypoxia and is particularly dangerous in a chronic wound where longer periods of hypoxia may result in unnecessary tissue loss [116]. Therefore, the measurement of oxygen at the wound site can be of particular interest, not as a means of determining infection but confirmation that the wound is recovering. In healthy subcutaneous tissues, partial pressures of O_2 are 30–50 mmHg, while chronic wounds may exhibit decreased O_2 levels near 5–20 mmHg. A flexible and wireless smart bandage with an oxygen sensor was shown to monitor the wound bed in real time [96]. The oxygen sensor was constructed using silver and electroplated zinc electrodes on a flexible parylene C substrate, generating a proportional voltage associated with oxygen concentration. One advantage of this system was an electronic system that allowed for an adjustable gain from the sensor, which allowed for large ranges of O_2 concentrations to be quantified. Parylene, as the base dressing material, is also impermeable to oxygen, limiting noise from atmospheric O_2. A liquid bandage has also been developed with integrated oxygen sensing porphyrin dendrimers in New Skin® liquid bandage [117]. Quantification and mapping of O_2 levels was observed through the level of phosphorescence of the applied bandage. These bandages were applied to ex vivo and in vivo porcine models to monitor the progression of burn wounds. While it is important for oxygen sensors to distinguish between the wound and atmospheric O_2 conditions, the use of impermeable membranes to limit sensor noise may negatively affect the wound healing process.

The moisture content in the wound environment is also critical for proper healing: too much moisture can result in maceration, while too little can lead to the wound drying out [118]. While changing wound dressings regularly may seem logical, frequent removal of the dressing can lead to repeated damage to the wound site, further delaying wound healing. To control this moisture content, engineers have looked to use a variety of materials (alginates, gelatin derivatives) as well as various physical structures (films, hydrocolloids, foams, and hydrogels) [79, 100]. While hydrophilic materials can retain high levels of moisture within the dressing, they may also draw moisture from the wound array, whereas hydrophobic materials can create a more substantial barrier to the wound (keeping moisture in) but have less flexibility in controlling moisture through the dressing. McColl et al. have developed an array of eight moisture sensors that allow for special resolution within the wound dressing [119]. Moisture levels were measured by electrical impedance between silver/silver chloride electrodes insulated by a thin layer of silicone. This system was used to evaluate the performance of six various wound dressings of foams, films, and hydrofibers in vitro. The WoundSense™ is a commercially available moisture sensor by Ohmedics, which was approved in 2011 [120]. These sensors utilize low-current impedance and allow for the sensor to be integrated under the wound dressing. Readouts from the sensor go to a handheld meter and have been used in observational studies to help clinicians indicate when dressing changes are needed. Mehmood et al. demonstrated a multifunctional smart bandage, where moisture content was determined using Honeywell HIH4030 piezo-electric and Multicomp's HCZ-D5 piezo-resistive moisture sensors [121]. These dressings included wireless communication capable of continuous monitoring for ~ 2 days and could be constructed for approximately $40.

Active drug delivery systems enable the controlled delivery of therapeutics to the wound and are typically governed by the response to changes in the environment (those described above). Mostafalu et al. utilized drug-loaded particles in Poly N-isopropylacrilamide (PNIPAM) an alginate hydrogel layer and controlled the drug delivery rate using an external microheater and microcontroller [122]. The microheater was fabricated by 20 nm chromium and 200 nm gold deposition onto a parylene substrate. The entire construct was attached to a transparent medical tape to form a wearable platform that was less than 3 mm thick. One advantage of this platform is the low-cost microheater is disposable, while the microcontroller components could be reused. One consideration moving forward will be the development of non-fouling sensors, as the protein-rich environment of the wound can lead to loss in sensor precision over time.

8 Conclusion and Outlook

Minimally invasive technologies for human healthcare monitoring have received great research attention. The influx of health data from continuous monitoring allows clinicians to make earlier diagnoses and better treatment guidelines [123].

Technological innovations in materials, fabrication processes, and data transmission are enabling the development of more miniaturized, painless, user-friendly, and continuous health monitoring systems. These technologies are all contributing to a worldwide shift towards preventative health monitoring and new treatment approaches. This chapter discussed several state-of-the-art minimally invasive detection devices, including point-of-care sensors, wearable, edible, and microneedle biosensors, and wound healing bandages. The working principles, fabrication methods, materials, applications, and current research status of these biosensors were introduced in detail. These portable, wearable biosensors can function in many physiological mediums, including urine, sweat, blood, ISF, tears, breath, saliva, and wound exudate. The noninvasive sampling of such key fluids has enabled the continual monitoring of many major biomarkers and physiological vital signs, like blood glucose fluctuations, blood ethanol concentrations, stress levels, and electrolyte imbalances; monitoring these parameters enables biosensors to diagnose a wide variety of health conditions. Although significant progress has been made in this field, many challenges remain to be addressed before their full potential and commercial applications can be achieved. Some of these challenges include obtaining more effective biocompatibility, flexibility, sensitivity, selectivity, long-term stability, power consumption, self-healing, data transmission, and system-level integration. Nonetheless, the use of these noninvasive detection devices is expected to grow rapidly in the upcoming decades.

For biosensors attached to or implanted in human bodies, biocompatibility is the most vital requirement. Long-term toxicity analyses of their active layers and flexible substrates should be implemented before they are used for practical applications. To address this, more work needs to be done on material development, surface modifications, and interface engineering as well as the devices' fabrication methods and packaging technologies [124–126]. Furthermore, the development of new biocompatible materials and fabrication processes will extend the scope and viability of minimally invasive biosensing technologies.

Another major obstacle for wearable biosensors is being sufficiently flexible without compromising their sensitivity. The large-scale deformations tend to decrease a device's sensitivity; thus, greater flexibility decreases sensitivity. Efforts should focus on the development of new materials like intrinsic soft organic conjugated polymers [21, 127, 128] or on new macroscopic designs of traditional inorganic semiconductors. Moreover, entirely novel processes or innovative working mechanisms might be necessary for researchers to solve this challenge [123, 124, 129, 130]. Optimally, devices and systems should eventually match the mechanical properties of the target tissue, and further work must be done to make that possible.

Ensuring high sensor selectivity towards the target biomarker represents another on-going challenge. The large amount of interfering species existing in complex measurement mediums makes this especially challenging. Developing new analyte-specific materials such as enzymes, conducting polymers, and other biomolecules can be used to address problems associated with low selectivity. In addition, utilizing multi-sensor arrays and multivariate analysis represent promising approaches to optimizing analytical solutions [131, 132].

For healthcare applications in particular, biosensors should be very stable over long time periods. In some cases, sensors are needed to conduct continuous detection in harsh environments for months or years without maintenance and recalibration, posing a large issue in many instances. Currently, failing to mitigate this in all the different biofluid environments, such as tears, ISF, sweat, and wound exudate, is one of the main factors preventing biosensors from being used clinically. Though it remains speculative, solutions to the stability problem may be found by developing novel materials and transduction mechanisms, adjusting surface chemistry, and designing calibration-free devices [131, 133].

Decreasing the power consumption is also needed before the devices can be put into practice for long-term monitoring. Some researchers have aimed to solve this by developing low-power micro-pumps, passive textile-based microfluidics, and low-power transmitting circuits that decrease power usage in peripheral components [129]. Various modalities have been demonstrated based on novel biofuel cells, energy harvesting, motion storage systems, solar energy, and ultrasound.

Self-healing is an incredibly appealing property for wearable biosensors to have [134]. Unfortunately, the properties of currently available self-healing polymers, such as electrical conductivity, chemical sensitivity, mechanical flexibility, and photochemical properties, still cannot match those of comparable non-healing materials. More work is needed to develop self-healing biosensors that use self-healing polymers, to integrate self-healing sensors into circuit boards, and to develop other similar advanced processes [135, 136].

Additionally, wireless transmission and system-level integration clearly impact multiple performance parameters. In order to analyze and display the results from wearable biosensors, signals must be transmitted and processed. Although Bluetooth low energy (BLE) and near-field communication (NFC) have been widely used for biosensors to obtain real-time data transmission, they are not efficient for high-density data processing. Advanced transmission mechanisms, more efficient algorithms, and user-friendly display platforms are needed to achieve their full potential. Additionally, more efforts should be focused on the integration of biosensors, power sources, wireless communication modules, and signal process circuits in compact devices for practical applications [10, 137, 138].

Despite these challenges, wearable biosensing devices continue to make great improvements. For example, devices such as a 1-mm, nontoxic intracranial pressure sensor; a resorbable, microscale arterial blood pressure sensor; a water-stable, flexible microelectrode array nerve stimulator; an ultrathin, wirelessly powered photodynamic therapy patch; a 128-channel, wireless neural activity modulator; and a wireless, dynamic MRI signal transducer have all been successfully developed in recent years [139]. Even more efficient devices are being reported each year, exemplifying the speed at which these devices are approaching being fit for use on and within human bodies.

Minimally invasive biosensors have become an integral part of personal health monitoring and treatment. These minimally invasive biosensing technologies not only promise to help people better monitor their health but also provide continuous medical data for personalized diagnoses and treatments. Multidisciplinary cooperation is required to integrate the key developments in biosensing, biomaterials, manufacturing, wireless communication, energy harvesting, and sample processing to overcome the remaining technological challenges.

Acknowledgments K. L, M. G, and H. K contributed equally to this work. The authors declare no conflict of interests in this work. This work has been supported by National Institutes of Health (1R01HL140951-01A1, 1R01GM126571-01, 1R01GM126831-01, 1R01EB023052-01A1).

References

1. Ashammakhi, N., Ahadian, S., Darabi, M. A., El Tahchi, M., Lee, J., Suthiwanich, K., Sheikhi, A., Dokmeci, M. R., Oklu, R., & Khademhosseini, A. (2019). Minimally invasive and regenerative therapeutics. *Advanced Materials, 31*(1), 1804041.
2. Daddona, P. E., Fieldson, G. T., Nat, A. S., & Lin, W.-Q. (2000). *Minimally invasive detecting device*. Google Patents.
3. Bandodkar, A. J., & Wang, J. (2014). Non-invasive wearable electrochemical sensors: a review. *Trends in Biotechnology, 32*(7), 363–371.
4. Wang, J. (2006). Electrochemical biosensors: towards point-of-care cancer diagnostics. *Biosensors and Bioelectronics, 21*(10), 1887–1892.
5. Corstjens, A. M., Ligtenberg, J. J., van der Horst, I. C., Spanjersberg, R., Lind, J. S., Tulleken, J. E., Meertens, J. H., & Zijlstra, J. G. (2006). Accuracy and feasibility of point-of-care and continuous blood glucose analysis in critically ill ICU patients. *Critical Care, 10*(5), R135.
6. Yetisen, A. K., Akram, M. S., & Lowe, C. R. (2013). Paper-based microfluidic point-of-care diagnostic devices. *Lab on a Chip, 13*(12), 2210–2251.
7. Teerinen, T., Lappalainen, T., & Erho, T. (2014). A paper-based lateral flow assay for morphine. *Analytical and Bioanalytical Chemistry, 406*(24), 5955–5965.
8. Vashist, S. K., Luppa, P. B., Yeo, L. Y., Ozcan, A., & Luong, J. H. T. (2015). Emerging technologies for next-generation point-of-care testing. *Trends in Biotechnology, 33*(11), 692–705.
9. Chin, C. D., Linder, V., & Sia, S. K. (2012). Commercialization of microfluidic point-of-care diagnostic devices. *Lab on a Chip, 12*(12), 2118–2134.
10. Yang, Y., & Gao, W. (2019). Wearable and flexible electronics for continuous molecular monitoring. *Chemical Society Reviews, 48*(6), 1465–1491.
11. Yetisen, A. K., Jiang, N., Tamayol, A., Ruiz-Esparza, G. U., Zhang, Y. S., Medina-Pando, S., Gupta, A., Wolffsohn, J. S., Butt, H., Khademhosseini, A., & Yun, S.-H. (2017). Paper-based microfluidic system for tear electrolyte analysis. *Lab on a Chip, 17*(6), 1137–1148.
12. Bihar, E., Deng, Y., Miyake, T., Saadaoui, M., Malliaras, G. G., & Rolandi, M. (2016). A Disposable paper breathalyzer with an alcohol sensing organic electrochemical transistor. *Scientific Reports, 6*, 27582.
13. Sun, W., Lee, J., Zhang, S., Benyshek, C., Dokmeci, M. R., & Khademhosseini, A. (2018). Engineering precision medicine. *Advanced Science, 0*(0), 1801039.
14. Gao, W., Emaminejad, S., Nyein, H. Y. Y., Challa, S., Chen, K., Peck, A., Fahad, H. M., Ota, H., Shiraki, H., & Kiriya, D. (2016). Fully integrated wearable sensor arrays for multiplexed in situ perspiration analysis. *Nature, 529*(7587), 509.
15. Windmiller, J. R., & Wang, J. (2013). Wearable electrochemical sensors and biosensors: a review. *Electroanalysis, 25*(1), 29–46.
16. Miyamoto, A., Lee, S., Cooray, N. F., Lee, S., Mori, M., Matsuhisa, N., Jin, H., Yoda, L., Yokota, T., & Itoh, A. (2017). Inflammation-free, gas-permeable, lightweight, stretchable on-skin electronics with nanomeshes. *Nature Nanotechnology, 12*(9), 907.

17. Oh, J. Y., Rondeau-Gagné, S., Chiu, Y.-C., Chortos, A., Lissel, F., Wang, G.-J. N., Schroeder, B. C., Kurosawa, T., Lopez, J., & Katsumata, T. (2016). Intrinsically stretchable and healable semiconducting polymer for organic transistors. *Nature, 539*(7629), 411.
18. Sessolo, M., Khodagholy, D., Rivnay, J., Maddalena, F., Gleyzes, M., Steidl, E., Buisson, B., & Malliaras, G. G. (2013). Easy-to-fabricate conducting polymer microelectrode arrays. *Advanced Materials, 25*(15), 2135–2139.
19. Park, S., Heo, S. W., Lee, W., Inoue, D., Jiang, Z., Yu, K., Jinno, H., Hashizume, D., Sekino, M., & Yokota, T. (2018). Self-powered ultra-flexible electronics via nano-grating-patterned organic photovoltaics. *Nature, 561*(7724), 516.
20. Wang, S., Xu, J., Wang, W., Wang, G.-J. N., Rastak, R., Molina-Lopez, F., Chung, J. W., Niu, S., Feig, V. R., & Lopez, J. (2018). Skin electronics from scalable fabrication of an intrinsically stretchable transistor array. *Nature, 555*(7694), 83.
21. Zhang, S., Hubis, E., Tomasello, G., Soliveri, G., Kumar, P., & Cicoira, F. (2017). Patterning of stretchable organic electrochemical transistors. *Chemistry of Materials, 29*(7), 3126–3132.
22. Krishnan, S. R., Ray, T. R., Ayer, A. B., Ma, Y., Gutruf, P., Lee, K., Lee, J. Y., Wei, C., Feng, X., & Ng, B. (2018). Epidermal electronics for noninvasive, wireless, quantitative assessment of ventricular shunt function in patients with hydrocephalus. *Science Translational Medicine, 10*(465), eaat8437.
23. Enzo Pasquale, S., Lorussi, F., Mazzoldi, A., & Rossi, D. D. (2003). Strain-sensing fabrics for wearable kinaesthetic-like systems. *IEEE Sensors Journal, 3*(4), 460–467.
24. Yang, Y.-L., Chuang, M.-C., Lou, S.-L., & Wang, J. (2010). Thick-film textile-based amperometric sensors and biosensors. *Analyst, 135*(6), 1230–1234.
25. Chuang, M.-C., Windmiller, J. R., Santhosh, P., Ramírez, G. V., Galik, M., Chou, T.-Y., & Wang, J. (2010). Textile-based Electrochemical Sensing: Effect of fabric substrate and detection of nitroaromatic explosives. *Electroanalysis, 22*(21), 2511–2518.
26. Bandodkar, A. J., Molinnus, D., Mirza, O., Guinovart, T., Windmiller, J. R., Valdés-Ramírez, G., Andrade, F. J., Schöning, M. J., & Wang, J. (2014). Epidermal tattoo potentiometric sodium sensors with wireless signal transduction for continuous non-invasive sweat monitoring. *Biosensors and Bioelectronics, 54*, 603–609.
27. Yuk, H., Lu, B., & Zhao, X. (2019). Hydrogel bioelectronics. *Chemical Society Reviews, 48*(6), 1642–1667.
28. Wang, J. (2005). Nanomaterial-based electrochemical biosensors. *Analyst, 130*(4), 421–426.
29. Parlak, O., Keene, S. T., Marais, A., Curto, V. F., & Salleo, A. (2018). Molecularly selective nanoporous membrane-based wearable organic electrochemical device for noninvasive cortisol sensing. *Science Advances, 4*(7), eaar2904.
30. Chen, Y., Lu, S., Zhang, S., Li, Y., Qu, Z., Chen, Y., Lu, B., Wang, X., & Feng, X. (2017). Skin-like biosensor system via electrochemical channels for noninvasive blood glucose monitoring. *Science Advances, 3*(12), e1701629.
31. Koh, A., Kang, D., Xue, Y., Lee, S., Pielak, R. M., Kim, J., Hwang, T., Min, S., Banks, A., Bastien, P., Manco, M. C., Wang, L., Ammann, K. R., Jang, K.-I., Won, P., Han, S., Ghaffari, R., Paik, U., Slepian, M. J., Balooch, G., Huang, Y., & Rogers, J. A. (2016). A soft, wearable microfluidic device for the capture, storage, and colorimetric sensing of sweat. *Science Translational Medicine, 8*(366).
32. Kim, S.-W., Lee, Y., Park, J., Kim, S., Chae, H., Ko, H., & Kim, J. (2018). A triple-mode flexible e-skin sensor interface for multi-purpose wearable applications. *Sensors, 18*(1), 78.
33. Miyamoto, A., Lee, S., Cooray, N. F., Lee, S., Mori, M., Matsuhisa, N., Jin, H., Yoda, L., Yokota, T., Itoh, A., Sekino, M., Kawasaki, H., Ebihara, T., Amagai, M., & Someya, T. (2017). Inflammation-free, gas-permeable, lightweight, stretchable on-skin electronics with nanomeshes. *Nature Nanotechnology, 12*, 907.
34. Wang, C., Li, X., Hu, H., Zhang, L., Huang, Z., Lin, M., Zhang, Z., Yin, Z., Huang, B., Gong, H., Bhaskaran, S., Gu, Y., Makihata, M., Guo, Y., Lei, Y., Chen, Y., Wang, C., Li, Y., Zhang, T., Chen, Z., Pisano, A. P., Zhang, L., Zhou, Q., & Xu, S. (2018). Monitoring of the central blood pressure waveform via a conformal ultrasonic device. *Nature Biomedical Engineering, 2*(9), 687–695.

35. Rose, D. P., Ratterman, M. E., Griffin, D. K., Hou, L., Kelley-Loughnane, N., Naik, R. R., Hagen, J. A., Papautsky, I., & Heikenfeld, J. C. (2015). Adhesive RFID sensor patch for monitoring of sweat electrolytes. *IEEE Transactions on Biomedical Engineering, 62*(6), 1457–1465.
36. Bandodkar, A. J., Hung, V. W. S., Jia, W., Valdés-Ramírez, G., Windmiller, J. R., Martinez, A. G., Ramírez, J., Chan, G., Kerman, K., & Wang, J. (2013). Tattoo-based potentiometric ion-selective sensors for epidermal pH monitoring. *Analyst, 138*(1), 123–128.
37. Gao, W., Nyein, H. Y. Y., Shahpar, Z., Fahad, H. M., Chen, K., Emaminejad, S., Gao, Y., Tai, L.-C., Ota, H., Wu, E., Bullock, J., Zeng, Y., Lien, D.-H., & Javey, A. (2016). Wearable microsensor array for multiplexed heavy metal monitoring of body fluids. *ACS Sensors, 1*(7), 866–874.
38. Yao, H., Shum, A. J., Cowan, M., Lähdesmäki, I., & Parviz, B. A. (2011). A contact lens with embedded sensor for monitoring tear glucose level. *Biosensors and Bioelectronics, 26*(7), 3290–3296.
39. Borini, S., White, R., Wei, D., Astley, M., Haque, S., Spigone, E., Harris, N., Kivioja, J., & Ryhänen, T. (2013). Ultrafast graphene oxide humidity sensors. *ACS Nano, 7*(12), 11166–11173.
40. Bandodkar, A. J., Jia, W., Yardımcı, C., Wang, X., Ramirez, J., & Wang, J. (2015). Tattoo-based noninvasive glucose monitoring: A proof-of-concept study. *Analytical Chemistry, 87*(1), 394–398.
41. Tehrani, Z., Korochkina, T., Govindarajan, S., Thomas, D. J., O'Mahony, J., Kettle, J., Claypole, T. C., & Gethin, D. T. (2015). Ultra-thin flexible screen printed rechargeable polymer battery for wearable electronic applications. *Organic Electronics, 26*, 386–394.
42. Sonner, Z., Wilder, E., Heikenfeld, J., Kasting, G., Beyette, F., Swaile, D., Sherman, F., Joyce, J., Hagen, J., Kelley-Loughnane, N., & Naik, R. (2015). The microfluidics of the eccrine sweat gland, including biomarker partitioning, transport, and biosensing implications. *Biomicrofluidics, 9*(3), 031301.
43. Sato, K. (1977). The physiology, pharmacology, and biochemistry of the eccrine sweat gland. In R. H. Adrian, E. Helmreich, H. Holzer, R. Jung, K. Kramer, O. Krayer, R. J. Linden, F. Lynen, P. A. Miescher, J. Piiper, H. Rasmussen, A. E. Renold, U. Trendelenburg, K. Ullrich, W. Vogt, & A. Weber (Eds.), *Reviews of physiology, biochemistry and pharmacology* (Vol. 79, pp. 51–131). Berlin, Heidelberg: Springer Berlin Heidelberg.
44. Al-Tamer, Y. Y., Hadi, E. A., & Al-Badrani, I. e. I. (1997). Sweat urea, uric acid and creatinine concentrations in uraemic patients. *Urological Research, 25*(5), 337–340.
45. Varghese, S. A., Powell, T. B., Budisavljevic, M. N., Oates, J. C., Raymond, J. R., Almeida, J. S., & Arthur, J. M. (2007). Urine biomarkers predict the cause of glomerular disease. *Journal of the American Society of Nephrology, 18*(3), 913.
46. Tesch, G. H. (2010). Serum and urine biomarkers of kidney disease: A pathophysiological perspective. *Nephrology, 15*(6), 609–616.
47. Holly, F. J., & Lemp, M. A. (1977). Tear physiology and dry eyes. *Survey of Ophthalmology, 22*(2), 69–87.
48. von Thun und Hohenstein-Blaul, N., Funke, S., & Grus, F. H. (2013). Tears as a source of biomarkers for ocular and systemic diseases. *Experimental Eye Research, 117*, 126–137.
49. Garg, S. K., Schwartz, S., & Edelman, S. V. (2004). Improved glucose excursions using an implantable real-time continuous glucose sensor in adults with type 1 diabetes. *Diabetes Care, 27*(3), 734–738.
50. Degim, I. T., Ilbasmis, S., Dundaroz, R., & Oguz, Y. (2003). Reverse iontophoresis: a non-invasive technique for measuring blood urea level. *Pediatric Nephrology, 18*(10), 1032–1037.
51. Risby, T. H., & Solga, S. F. (2006). Current status of clinical breath analysis. *Applied Physics B, 85*(2), 421–426.
52. Rogers, J. A. (2017). Nanomesh on-skin electronics. *Nature Nanotechnology, 12*, 839.
53. Zhang, S., & Cicoira, F. (2018). Flexible self-powered biosensors. *Nature, 561*(7724), 466.
54. Jeong, J.-W., Kim, M. K., Cheng, H., Yeo, W.-H., Huang, X., Liu, Y., Zhang, Y., Huang, Y., & Rogers, J. A. (2014). Capacitive epidermal electronics for electrically safe, long-term electrophysiological measurements. *Advanced Healthcare Materials, 3*(5), 642–648.

55. Koo, J. H., Jeong, S., Shim, H. J., Son, D., Kim, J., Kim, D. C., Choi, S., Hong, J.-I., & Kim, D.-H. (2017). Wearable electrocardiogram monitor using carbon nanotube electronics and color-tunable organic light-emitting diodes. *ACS Nano, 11*(10), 10032–10041.
56. Kalantar-Zadeh, K., Berean, K. J., Ha, N., Chrimes, A. F., Xu, K., Grando, D., Ou, J. Z., Pillai, N., Campbell, J. L., Brkljača, R., Taylor, K. M., Burgell, R. E., Yao, C. K., Ward, S. A., McSweeney, C. S., Muir, J. G., & Gibson, P. R. (2018). A human pilot trial of ingestible electronic capsules capable of sensing different gases in the gut. *Nature Electronics, 1*(1), 79–87.
57. Kim, Y. J., Chun, S.-E., Whitacre, J., & Bettinger, C. J. (2013). Self-deployable current sources fabricated from edible materials. *Journal of Materials Chemistry B, 1*(31), 3781–3788.
58. Bettinger, C. J. (2015). Materials advances for next-generation ingestible electronic medical devices. *Trends in Biotechnology, 33*(10), 575–585.
59. Chai, P. R., Carreiro, S., Innes, B. J., Chapman, B., Schreiber, K. L., Edwards, R. R., Carrico, A. W., & Boyer, E. W. (2017). Oxycodone ingestion patterns in acute fracture pain with digital pills. *Anesthesia & Analgesia, 125*(6), 2105–2112.
60. Kaushik, S., Hord, A. H., Denson, D. D., McAllister, D. V., Smitra, S., Allen, M. G., & Prausnitz, M. R. (2001). Lack of pain associated with microfabricated microneedles. *Anesthesia and Analgesia, 92*(2), 502–504.
61. Gupta, J., Gill, H. S., Andrews, S. N., & Prausnitz, M. R. (2011). Kinetics of skin resealing after insertion of microneedles in human subjects. *Journal of Controlled Release, 154*(2), 148–155.
62. Esfandyarpour, R., Esfandyarpour, H., Javanmard, M., Harris, J. S., & Davis, R. W. (2013). Microneedle biosensor: A method for direct label-free real time protein detection. *Sensors and Actuators B-Chemical, 177*, 848–855.
63. Mohan, A. M. V., Windmiller, J. R., Mishra, R. K., & Wang, J. (2017). Continuous minimally-invasive alcohol monitoring using microneedle sensor arrays. *Biosensors & Bioelectronics, 91*, 574–579.
64. Vazquez, P., Herzog, G., O'Mahony, C., O'Brien, J., Scully, J., Blake, A., O'Mathuna, C., & Galvin, P. (2014). Microscopic gel-liquid interfaces supported by hollow microneedle array for voltammetric drug detection. *Sensors and Actuators B-Chemical, 201*, 572–578.
65. Miller, P. R., Narayan, R. J., & Polsky, R. (2016). Microneedle-based sensors for medical diagnosis. *Journal of Materials Chemistry B, 4*(8), 1379–1383.
66. Foghandersen, N., Altura, B. M., Altura, B. T., & Siggaardandersen, O. (1995). Composition of interstitial fluid. *Clinical Chemistry, 41*(10), 1522–1525.
67. Keum, D. H., Jung, H. S., Wang, T., Shin, M. H., Kim, Y. E., Kim, K. H., Ahn, G. O., & Hahn, S. K. (2015). Microneedle biosensor for real-time electrical detection of nitric oxide for in situ cancer diagnosis during endomicroscopy. *Advanced Healthcare Materials, 4*(8), 1153–1158.
68. Bollella, P., Sharma, S., Cass, A. E. G., & Antiochia, R. (2019). Microneedle-based biosensor for minimally-invasive lactate detection. *Biosensors & Bioelectronics, 123*, 152–159.
69. Kim, L. W., KB, Cho, C. H., Park, D. S., Cho, S. J., & Shim, Y. B. (2019). Continuous glucose monitoring using a microneedle array sensor coupled with a wireless signal transmitter. *Sensors and Actuators B: Chemical, 281*(15), 14–21.
70. Chang, H., Zheng, M. J., Yu, X. J., Than, A., Seeni, R. Z., Kang, R. J., Tian, J. Q., Khanh, D. P., Liu, L. B., Chen, P., & Xu, C. J. (2017). A swellable microneedle patch to rapidly extract skin interstitial fluid for timely metabolic analysis. *Advanced Materials, 29*(37), 1702243.
71. Mishra, R. K., Mohan, A. M. V., Soto, F., Chrostowski, R., & Wang, J. (2017). A microneedle biosensor for minimally-invasive transdermal detection of nerve agents. *Analyst, 142*(6), 918–924.
72. Samant, P. P., & Prausnitz, M. R. (2018). Mechanisms of sampling interstitial fluid from skin using a microneedle patch. *Proceedings of the National Academy of Sciences of the United States of America, 115*(18), 4583–4588.

73. Jina, M. J. T. A., Tamada, J. A., McGill, S., Desai, S., Chua, B., Chang, A., & Christiansen, M. (2014). Design, development, and evaluation of a novel microneedle array-based continuous glucose monitor. *Journal of Diabetes Science and Technology, 8*, 483–487.
74. Naixin Song, P. X., Shen, W., Javanmard, M., Allen, M. G. (2018). Microwell-array on a flexible needle: A transcutaneous insertable impedance sensor for label-free cytokine detection. In 2018 IEEE Micro Electro Mechanical Systems (MEMS) (pp. 392–395). IEEE.
75. Zhou, J. X., Ding, F., Tang, L. N., Li, T., Li, Y. H., Zhang, Y. J., Gong, H. Y., Li, Y. T., & Zhang, G. J. (2018). Monitoring of pH changes in a live rat brain with MoS2/PAN functionalized microneedles. *Analyst, 143*(18), 4469–4475.
76. Mirza, K. B., Zuliani, C., Hou, B., Ng, F. S., Peters, N. S., & Toumazou, C. (2017). Injection moulded microneedle sensor for real-time wireless pH monitoring. In *2017 39th Annual International Conference of the IEEE Engineering in Medicine and Biology Society (EMBC)* (pp. 189–192). IEEE.
77. Vazquez, P., O'Mahony, C., O'Brien, J., Scully, J., Blake, A., O'Mathuna, C., Galvin, P., Herzog, G. (2014). Microneedle sensor for voltammetric drug detection in physiological fluids. In *SENSORS, 2014 IEEE* (pp. 1768–1771). IEEE.
78. O'Connor, J. (2012). Higher wound care costs are driving treatment research. *McKnight's*.
79. Dhivya, S., Padma, V. V., & Santhini, E. (2015). Wound dressings–a review. *Biomedicine, 5*(4), 22.
80. Lochno, P., Kraus-Pfeiffer, G., & Jacobs, P. (2013). Review and basics of frequently used wound dressings. *MMW Fortschritte der Medizin, 155*(10), 59.
81. Sarabahi, S. (2012). Recent advances in topical wound care. *Indian Journal of Plastic Surgery: Official Publication of the Association of Plastic Surgeons of India, 45*(2), 379.
82. Frykberg, R. G., & Banks, J. (2015). Challenges in the treatment of chronic wounds. *Advances in Wound Care, 4*(9), 560–582.
83. Derakhshandeh, H., Kashaf, S. S., Aghabaglou, F., Ghanavati, I. O., & Tamayol, A. (2018). Smart bandages: The future of wound care. *Trends in Biotechnology, 36*(12), 1259–1274.
84. Gianino, E., Miller, C., & Gilmore, J. (2018). Smart wound dressings for diabetic chronic wounds. *Bioengineering, 5*(3), 51.
85. Andreu, V., Mendoza, G., Arruebo, M., & Irusta, S. (2015). Smart dressings based on nanostructured fibers containing natural origin antimicrobial, anti-inflammatory, and regenerative compounds. *Materials, 8*(8), 5154–5193.
86. Boateng, J. S., Matthews, K. H., Stevens, H. N., & Eccleston, G. M. (2008). Wound healing dressings and drug delivery systems: a review. *Journal of Pharmaceutical Sciences, 97*(8), 2892–2923.
87. Velnar, T., Bailey, T., & Smrkolj, V. (2009). The wound healing process: an overview of the cellular and molecular mechanisms. *Journal of International Medical Research, 37*(5), 1528–1542.
88. Brown, A. (2015). Phases of the wound healing process. *Nursing Times, 111*(46), 12–13.
89. Eming, S. A., Martin, P., & Tomic-Canic, M. (2014). Wound repair and regeneration: mechanisms, signaling, and translation. *Science Translational Medicine, 6*(265), 265sr266.
90. Zhu, X., Zhang, Y., & Hu, C. (2001). Study on the molecular mechanisms involved in the increased collagen synthesis by platelet-derived wound healing factors during wound healing in alloxaninduced diabetic rat. *Zhongguo xiu fu chong jian wai ke za zhi= Zhongguo xiufu chongjian waike zazhi= Chinese Journal of Reparative and Reconstructive Surgery, 15*(4), 223–226.
91. Epstein, F. H., Singer, A. J., & Clark, R. A. (1999). Cutaneous wound healing. *New England Journal of Medicine, 341*(10), 738–746.
92. Falanga, V. (2005). Wound healing and its impairment in the diabetic foot. *The Lancet, 366*(9498), 1736–1743.
93. Tamayol, A., Akbari, M., Zilberman, Y., Comotto, M., Lesha, E., Serex, L., Bagherifard, S., Chen, Y., Fu, G., & Ameri, S. K. (2016). Flexible pH-sensing hydrogel fibers for epidermal applications. *Advanced Healthcare Materials, 5*(6), 711–719.

94. Rahimi, R., Ochoa, M., Tamayol, A., Khalili, S., Khademhosseini, A., & Ziaie, B. (2017). Highly stretchable potentiometric pH sensor fabricated via laser carbonization and machining of Carbon–Polyaniline composite. *ACS Applied Materials & Interfaces, 9*(10), 9015–9023.
95. Oh, J. H., Hong, S. Y., Park, H., Jin, S. W., Jeong, Y. R., Oh, S. Y., Yun, J., Lee, H., Kim, J. W., & Ha, J. S. (2018). Fabrication of high-sensitivity skin-attachable temperature sensors with bioinspired microstructured adhesive. *ACS Applied Materials & Interfaces, 10*(8), 7263–7270.
96. Mostafalu, P., Lenk, W., Dokmeci, M. R., Ziaie, B., Khademhosseini, A., & Sonkusale, S. R. (2015). Wireless flexible smart bandage for continuous monitoring of wound oxygenation. *IEEE Transactions on Biomedical Circuits and Systems, 9*(5), 670–677.
97. Mostafalu, P., Tamayol, A., Rahimi, R., Ochoa, M., Khalilpour, A., Kiaee, G., Yazdi, I. K., Bagherifard, S., Dokmeci, M. R., Ziaie, B., Sonkusale, S. R., & Khademhosseini, A. (2018). Smart bandage for monitoring and treatment of chronic wounds. *Small, 14*(33), 1703509.
98. Dowd, S. E., Wolcott, R. D., Sun, Y., McKeehan, T., Smith, E., & Rhoads, D. (2008). Polymicrobial nature of chronic diabetic foot ulcer biofilm infections determined using bacterial tag encoded FLX amplicon pyrosequencing (bTEFAP). *PLoS One, 3*(10), e3326.
99. Davies, C. E., Wilson, M. J., Hill, K. E., Stephens, P., Hill, C. M., Harding, K. G., & Thomas, D. W. (2001). Use of molecular techniques to study microbial diversity in the skin: chronic wounds reevaluated. *Wound Repair and Regeneration, 9*(5), 332–340.
100. Dargaville, T. R., Farrugia, B. L., Broadbent, J. A., Pace, S., Upton, Z., & Voelcker, N. H. (2013). Sensors and imaging for wound healing: a review. *Biosensors and Bioelectronics, 41*, 30–42.
101. Healy, B., & Freedman, A. (2006). ABC of wound healing: Infections. *BMJ: British Medical Journal, 332*(7545), 838.
102. Burd, A., Kwok, C. H., Hung, S. C., Chan, H. S., Gu, H., Lam, W. K., & Huang, L. (2007). A comparative study of the cytotoxicity of silver-based dressings in monolayer cell, tissue explant, and animal models. *Wound Repair and Regeneration, 15*(1), 94–104.
103. Nasajpour, A., Ansari, S., Rinoldi, C., Rad, A. S., Aghaloo, T., Shin, S. R., Mishra, Y. K., Adelung, R., Swieszkowski, W., & Annabi, N. (2018). A multifunctional polymeric periodontal membrane with osteogenic and antibacterial characteristics. *Advanced Functional Materials, 28*(3), 1703437.
104. Pant, J., Goudie, M., Brisbois, E., & Handa, H. (2016). Nitric oxide-releasing polyurethanes. In S. L. Cooper & J. Guan (Eds.), *Advances in polyurethane biomaterials* (pp. 471–550). Duxford: Woodhead Publishing.
105. Brisbois, E. J., Bayliss, J., Wu, J., Major, T. C., Xi, C., Wang, S. C., Bartlett, R. H., Handa, H., & Meyerhoff, M. E. (2014). Optimized polymeric film-based nitric oxide delivery inhibits bacterial growth in a mouse burn wound model. *Acta Biomaterialia, 10*(10), 4136–4142.
106. Masters, K. S. B., Leibovich, S. J., Belem, P., West, J. L., & Poole-Warren, L. A. (2002). Effects of nitric oxide releasing poly (vinyl alcohol) hydrogel dressings on dermal wound healing in diabetic mice. *Wound Repair and Regeneration, 10*(5), 286–294.
107. Carpenter, A. W., & Schoenfisch, M. H. (2012). Nitric oxide release: Part II. Therapeutic applications. *Chemical Society Reviews, 41*(10), 3742–3752.
108. Saghazadeh, S., Rinoldi, C., Schot, M., Kashaf, S. S., Sharifi, F., Jalilian, E., Nuutila, K., Giatsidis, G., Mostafalu, P., & Derakhshandeh, H. (2018). Drug delivery systems and materials for wound healing applications. *Advanced Drug Delivery Reviews, 127*, 138–166.
109. Simões, D., Miguel, S. P., Ribeiro, M. P., Coutinho, P., Mendonça, A. G., & Correia, I. J. (2018). Recent advances on antimicrobial wound dressing: A review. *European Journal of Pharmaceutics and Biopharmaceutics, 127*, 130–141.
110. Han, G., & Ceilley, R. (2017). Chronic wound healing: a review of current management and treatments. *Advances in Therapy, 34*(3), 599–610.
111. Trupp, S., Alberti, M., Carofiglio, T., Lubian, E., Lehmann, H., Heuermann, R., Yacoub-George, E., Bock, K., & Mohr, G. (2010). Development of pH-sensitive indicator dyes for the preparation of micro-patterned optical sensor layers. *Sensors and Actuators B: Chemical, 150*(1), 206–210.

112. Mohr, G. J., Müller, H., Bussemer, B., Stark, A., Carofiglio, T., Trupp, S., Heuermann, R., Henkel, T., Escudero, D., & González, L. (2008). Design of acidochromic dyes for facile preparation of pH sensor layers. *Analytical and Bioanalytical Chemistry, 392*(7–8), 1411–1418.
113. Kassal, P., Kim, J., Kumar, R., de Araujo, W. R., Steinberg, I. M., Steinberg, M. D., & Wang, J. (2015). Smart bandage with wireless connectivity for uric acid biosensing as an indicator of wound status. *Electrochemistry Communications, 56*, 6–10.
114. James, T. J., Hughes, M. A., Cherry, G. W., & Taylor, R. P. (2003). Evidence of oxidative stress in chronic venous ulcers. *Wound Repair and Regeneration, 11*(3), 172–176.
115. Sridhar, V., & Takahata, K. (2009). A hydrogel-based passive wireless sensor using a flex-circuit inductive transducer. *Sensors and Actuators A: Physical, 155*(1), 58–65.
116. Sen, C. K. (2009). Wound healing essentials: let there be oxygen. *Wound Repair and Regeneration, 17*(1), 1–18.
117. Li, Z., Roussakis, E., Koolen, P. G., Ibrahim, A. M., Kim, K., Rose, L. F., Wu, J., Nichols, A. J., Baek, Y., & Birngruber, R. (2014). Non-invasive transdermal two-dimensional mapping of cutaneous oxygenation with a rapid-drying liquid bandage. *Biomedical Optics Express, 5*(11), 3748–3764.
118. Winter, G. D. (1962). Formation of the scab and the rate of epithelization of superficial wounds in the skin of the young domestic pig. *Nature, 193*(4812), 293.
119. McColl, D., Cartlidge, B., & Connolly, P. (2007). Real-time monitoring of moisture levels in wound dressings in vitro: An experimental study. *International Journal of Surgery, 5*(5), 316–322.
120. Milne, S. D., Seoudi, I., Al Hamad, H., Talal, T. K., Anoop, A. A., Allahverdi, N., Zakaria, Z., Menzies, R., & Connolly, P. (2016). A wearable wound moisture sensor as an indicator for wound dressing change: an observational study of wound moisture and status. *International Wound Journal, 13*(6), 1309–1314.
121. Mehmood, N., Hariz, A., Templeton, S., & Voelcker, N. H. (2015). A flexible and low power telemetric sensing and monitoring system for chronic wound diagnostics. *Biomedical Engineering Online, 14*(1), 17.
122. Mostafalu, P., Tamayol, A., Rahimi, R., Ochoa, M., Khalilpour, A., Kiaee, G., Yazdi, I. K., Bagherifard, S., Dokmeci, M. R., & Ziaie, B. (2018). Smart bandage for monitoring and treatment of chronic wounds. *Small, 14*(33), 1703509.
123. Bandodkar, A. J., Jeerapan, I., & Wang, J. (2016). Wearable chemical sensors: Present challenges and future prospects. *Acs Sensors, 1*(5), 464–482.
124. Han, S. T., Peng, H., Sun, Q., Venkatesh, S., Chung, K. S., Lau, S. C., Zhou, Y., & Roy, V. (2017). An overview of the development of flexible sensors. *Advanced Materials, 29*(33), 1700375.
125. Choi, S., Lee, H., Ghaffari, R., Hyeon, T., & Kim, D. H. (2016). Recent advances in flexible and stretchable bio-electronic devices integrated with nanomaterials. *Advanced Materials, 28*(22), 4203–4218.
126. Dias, D., & Paulo Silva Cunha, J. (2018). Wearable health devices—vital sign monitoring, systems and technologies. *Sensors, 18*(8), 2414.
127. Wang, Y., Zhu, C., Pfattner, R., Yan, H., Jin, L., Chen, S., Molina-Lopez, F., Lissel, F., Liu, J., & Rabiah, N. I. (2017). A highly stretchable, transparent, and conductive polymer. *Science Advances, 3*(3), e1602076.
128. Boubée de Gramont, F., Zhang, S., Tomasello, G., Kumar, P., Sarkissian, A., & Cicoira, F. (2017). Highly stretchable electrospun conducting polymer nanofibers. *Applied Physics Letters, 111*(9), 093701.
129. Yang, T., Xie, D., Li, Z., & Zhu, H. (2017). Recent advances in wearable tactile sensors: Materials, sensing mechanisms, and device performance. *Materials Science and Engineering: R: Reports, 115*, 1–37.
130. Amjadi, M., Kyung, K. U., Park, I., & Sitti, M. (2016). Stretchable, skin-mountable, and wearable strain sensors and their potential applications: A review. *Advanced Functional Materials, 26*(11), 1678–1698.

131. Kassal, P., Steinberg, M. D., & Steinberg, I. M. (2018). Wireless chemical sensors and biosensors: A review. *Sensors and Actuators B: Chemical, 266*, 228–245.
132. Wu, W., & Haick, H. (2018). Materials and wearable devices for autonomous monitoring of physiological markers. *Advanced Materials, 30*(41), 1705024.
133. Pappa, A.-M., Parlak, O., Scheiblin, G., Mailley, P., Salleo, A., & Owens, R. M. (2018). Organic electronics for point-of-care metabolite monitoring. *Trends in Biotechnology, 36*(1), 45–59.
134. Zhang, S., & Cicoira, F. (2017). Water-enabled healing of conducting polymer films. *Advanced Materials, 29*(40), 1703098.
135. Huynh, T. P., Sonar, P., & Haick, H. (2017). Advanced materials for use in soft self-healing devices. *Advanced Materials, 29*(19), 1604973.
136. Huynh, T. P., & Haick, H. (2018). Autonomous flexible sensors for health monitoring. *Advanced Materials, 30*(50), 1802337.
137. Xu, M., Obodo, D., & Yadavalli, V. K. (2019). The design, fabrication, and applications of flexible biosensing devices–A review. *Biosensors and Bioelectronics, 124*, 96–114.
138. Lee, S. P., Klinker, L. E., Ptaszek, L., Work, J., Liu, C., Quivara, F., Webb, C., Dagdeviren, C., Wright, J. A., & Ruskin, J. N. (2015). Catheter-based systems with integrated stretchable sensors and conductors in cardiac electrophysiology. *Proceedings of the IEEE, 103*(4), 682–689.
139. Zhou, A., Santacruz, S. R., Johnson, B. C., Alexandrov, G., Moin, A., Burghardt, F. L., et al. (2019). A wireless and artefact-free 128-channel neuromodulation device for closed-loop stimulation and recording in non-human primates. *Nature Biomedical Engineering, 3*(1), 15.

Index

A
Acoustic energy, 102
Action potentials (AP)
  cardiomyocyte, 121
  pacemaker, 120
Active drug delivery systems, 213
Actively multiplexed, flexible electrode arrays
  capacitively coupled arrays, 10–14
  concept, 9, 10
  rationale, 9, 10
Acute coronary syndrome, 137
Acute respiratory distress syndrome (ARDS), 171
Agent Orange, 118
Air-filled lung, 169
American National Standards Institute (ANSI), 132
Amorphous silicon (a-Si), 42
Amplifier, 11
Amplifier blanking circuits, 4
Antimicrobial wound dressings, 211
Artifact, 2
Artifact-free recording, 5
Artificial intelligence, 135–137
Artificial neural networks (ANNs), 136, 137, 181
Association for the Advancement of Medical Instrumentation (AAMI), 132
Atherosclerosis
  cholesterols, 143
  detection techniques, 144
  oxidative stress, 144
  oxLDL, 143, 144
  serum biomarkers, 144
  unstable plaque, 143, 144
Atomic layer deposition (ALD), 53
Atrioventricular (AV) node, 122
Au-nanorod (NR), 64

B
Bandwidth, 14, 15
Batteries, 56, 57
Battery-based stimulation, 106, 107
Batteryless direct stimulation, 105, 106
Beer-Lambert law, 16
Bench-top test, 6
Bioactive glasses, 47
Biodegradable bioelectronics, 73
Biodegradable electronics, 31
Biodegradable materials, 93
Biodegradable passive components, 73
Biodegradable photoluminescent polymer (BPLP), 50
Biodegradable supercapacitors, 41
Bioelectronics, 39
Biofuel harvesting, 102
Biomarker fluids, 199
Biomarkers, 199, 200
Biomedical engineering
  bioresorbable systems, 56
  biosensing, 59
  energy supply, 54–58
Biomedical implants, 88, 90, 92
Bioresorbable electronics
  active and passive components, 30
  bulk materials, 31
  CMOS, 31, 32
  controlled/on-demand transience, 32
  economical/scalable manufacturing processes, 32
  e-waste, 29

H. Cao et al. (eds.), *Interfacing Bioelectronics and Biomedical Sensing*,
https://doi.org/10.1007/978-3-030-34467-2

Bioresorbable electronics (*cont.*)
flexible electronic production, 31
flexible organic electronics, 31
implants, 29
OFETs, 31
OLEDs, 31
OTFTs, 31
small-scale material deposition techniques, 31
TFTs, 31
thin film organic photovoltaic cells, 31
water-soluble electronics, 31
Bioresorbable electronic stent (BES), 92
Bioresorbable materials
biocompatibility, 34
conducting, 34
conductor
inorganic, 34–41
device components, constituent materials and dissolution behaviors, 34–37
dielectric, 34
electrical conductivity, 33
functioning components, 32, 33
insulator (*see* Insulator)
mechanical and chemical properties, 34
semiconducting, 33–34
semiconductor
organic, 43, 44
valence and conduction bands, 33
Bioresorbable silicon electrodes, 88
Bioresorbable silicon electronic sensors, 88
Bioresorbable triboelectric generators, 58
Biosensing, 32
chemical sensing, 60, 61
ECoG and EEG monitoring, 59
EP monitors, 59
motion, 60, 61
pressure, 60, 61
temperature, 60, 61
therapeutics
heat-stimulated drug release, 62–63
multifunctional, 64
tissue regeneration, 63, 64
transient electronic, 59
Bipolar stimulation, 4
Bit error rate (BER), 17, 20
Bluetooth-low-energy (BLE), 133, 215
Boltzmann's constant, 18
Bone-skin interface, 17
Boron-doped Si-NM encapsulating layers, 52
Borosilicate glass, 47
Brain function, 1
Brain-instrument interfaces, 29
Brain-machine interface (BMI), 14
Breath analysis, 200
Buffer, 9
Bulk materials, 31

C
Capacitively coupled arrays
amplifier, 11
characteristics, 10
concept, 11
encapsulation material, 10
encapsulation methods, 10
fabrication process, 13
flexible devices, 10
NMOS transistors, 12
organic/inorganic-based multilayer, 10
output characteristics, 13
photolithography, 12, 13
rat auditory cortex, 13, 14
semiconducting channels, 11
Si NM transistor, 12, 13
SOI wafer, 12
thin flexible films, 10
ultrathin thermal SiO_2, 10, 11
Capsule shell, 203
Carboxymethyl cellulose, 48–49
Cardiac and heart rate monitoring devices, 200
Cardiac pacemakers, 101
Cardiomyocyte AP, 121
Cardiopulmonary monitoring
ARDS, 171
CF, 172
COPD, 171
mechanical ventilation, 170
motivation, 169
pulmonary perfusion, 170
Cardiovascular disease (CVD), 117, 118
Cardioverter defibrillator, 29
Causative genes, 117
Cellulose, 48
Central blood pressure (CBP), 201
Central unit (CU), 20
Ceria nanoparticles (ceria NPs), 92
Cerium dioxide nanoparticles, 64
Chemical-mechanical planarization (CMP), 82
Chemical vapor deposition (CVD), 41
Chemotherapy, 194
Chlorophyll, 43
Chopper amplifier techniques, 4
Chronic electrophysiology, 10–14
Chronic obstructive pulmonary diseases (COPD), 171
Circuit techniques, 5
Clinical translation, 177–179

Cloud-based monitor, 134
CMOS inverter device, 49
Coil inductance, 103
Colorimetric pH sensors, 212
Commercial off-the-shelf (COTS) components, 82
Complementary metal-oxide-semiconductor (CMOS), 31, 32
Complete electrode model (CEM), 165
Compressed sensing circuitry, 15
Concentric bipolar electrodes (CBE)
- advantages, 150
- Ag/AgCl electrode, 148
- atherosclerosis features, 150
- EIS implementation, atherosclerosis detection, 148
- electrical currents, 150
- electrochemical characterization, 151
- electrochemical insights, 151
- equivalent circuit model, 150
- in vitro, 150
- in vivo animal study, 151, 153

Conducting materials, 34
Conduction pathway, 41
Conductive polymers, 73, 206
Conductive wax (C-wax), 93–95
Conductor, 75, 77
- inorganic materials
  - in bioresorbable electronics, 34
  - bulk materials, 38
  - corrosion, 34
  - electrode potential, 38
  - extrinsic factors, 38
  - Gibbs free energy, 34, 38
  - intrinsic factors, 38
  - iron (Fe), 39
  - magnesium (Mg), 38, 39
  - molybdenum (Mo), 40
  - Nernst equation, 38
  - transient electronics, 34
  - tungsten, 40, 41
  - zinc (Zn), 40
- organic materials, 41

Constant phase element (CPE), 146
Contact electrode
- dry, 125–126
- wet, 123–125

Continuous bio-signal monitoring, 207
Continuous glucose monitoring (CGM), 206
Conventional ultrasonic imaging systems, 189
Conventional ultrasound imaging systems, 185
Coupling capacitance, 200
Coupling coefficient, 103
Cross-linked urethane-doped polyester (CUPE), 51
Cystic fibrosis (CF), 172

D

Data encryption, 93
Data rate requirement, 14, 15
DBS, 4, 8
Decoding, 1
Deep neural networks (DNNs), 136
De-multiplexers (DEMUXs), 22
Density functional theory (DFT), 75
Design challenges
- applications, 1
- electrode array fabrication, 1
- focalized stimulation, 7–8
- recording scheme, 2
- stimulation artifact cancellation (*see* Stimulation artifact cancellation)

Device components and systems, 92
Device-grade silicon, 73
Diabetic wounds, 211
Dialysis machines, 29
Dielectric materials, 34, 73
- inorganic, 44
- organic, 46
- in parallel plate capacitors, 44

Digital processing unit (DSP), 15
Digital signal processing, 4, 5
Diphenylalanine (FF), 78
Doping, 42
Doppler
- B-mode images, 188
- flow measurement, 188, 190
- measurement, 186
- waveforms, 186–189

DRAM memory architecture, 9
Drug delivery devices, 90–91
Drug delivery system, 92
Drug-loaded silk, 90
Dry electrode, 125–126

E

Edible biosensors
- advantageous properties, 202
- biocompatibility and degradability, 203
- capsule shell, 203
- capsules containing ingestible electronic sensors, 203
- conducting polymers, 203
- detection inside, human body, 203
- electronic capsule sensors, 203

Edible biosensors (*cont.*)
gelatin capsules, 201
minimally invasive medical device, 201
nontoxic, edible batteries, 203
e-field distribution, 8
e-field intensity, 7
EIDORS open-source tools, 168
Electric stimulation technologies, 8
Electrical conductivity, 33
Electrical impedance tomography (EIT)
cardiopulmonary monitoring (*see* Cardiopulmonary monitoring)
EIDORS open-source tools, 168
functional lung imaging, 173
inverse problem, 165, 180, 181
liver fat infiltration (*see* Nonalcoholic fatty liver disease (NAFLD))
noninvasive medical imaging technique, 163
nonlinear inverse problem, 165–167
principle, 164, 165
respiratory monitoring, 163
Electrical sensing, 197
Electrical stimulation, 32
Electrocardiogram (ECG)
acquisition, zebrafish, 127
cardiac system, 122
cardiomyocyte AP, 121
contact electrode, 123–126
electrophysiological pathway, 122
in vital sign monitoring, 119
mammalian hearts, 118
MEA membranes, 128, 130
monitoring in humans, 132, 133
NCEs, 126, 127
pacemaker AP, 120
wearable technology, 132
zebrafish, 118, 119, 130, 131
Electrochemical impedance spectroscopy (EIS)
CBE, 148–151
current density distribution, 145
distinct material properties, 145
electric field, 145
electrochemical impedance, 145
equivalent circuit model, 146, 147
four-point EIS, 147, 148
in vivo animal study, CBE, 151, 153
intrinsic electrochemical properties, 145
lipid-free tissue, 145
3-D EIS interrogation, NZW rabbit model, 155–157
2-point EIS, 147, 148
2-point symmetric configuration, 153–155
Electroconvulsive therapy (ECT), 8
Electrocorticography (ECoG), 41, 88
Electrode array fabrication, 1
Electromagnetic (EM) fields, 15, 16
Electron beam (e-beam) deposition, 45
Electron mobilities, 44
Electronic fabrication methods, 41
Electronic healthcare records, 29
Electronic materials, 41
Electronic mobilities, 42
Electronic waste (e-waste), 29
Electrophysiological (EP) monitors, 59
Electrophysiological (EP) sensor, 62, 92
Electrophysiological signals
capacitive sensors, 200
cardiac and heart rate monitoring devices, 200, 201
CBP, 201
coupling capacitance, 200
Encapsulation layer
bioresorbable device, 52
organic, 53
swelling/nonuniform dissolution, 52
transient electronic devices, 52
End-expiratory lung volume (EELV), 170
Energy budget, 15, 16
Energy-storage unit (ESU), 107
Energy supply
batteries, 56, 57
MEHs, 57, 58
MSCs, 58
self-sustainable, 54
wireless energy supplies, 54
Enhanced bioimaging, 29
Environmental sensing, 31
Enzyme-linked immunosorbent assays (ELISA), 207
Equivalent circuit model, 146
Equivalent series resistance (ESR), 58
Extracellular matrix (ECM), 210

F

Far-field WPT, 110–112
Fat volume fraction (FVF), 179
Fine motor control, 9
Finite element model (FEM), 174
Flexible 2-point electrodes, 154
Flexible 3-D interrogation, 157
Flexible and stretchable electronics, 155
Flexible device technology, 197
Flexible electrodes, 130
Flexible electronics, 31, 54, 151, 155
Flexible organic electronics, 31
Flexible property, 92

Flow fraction reserve (FFR), 158
Focalized stimulation, 7–8
Food and Drug Administration (FDA), 47, 208
Fractional flow reserve (FFR), 144
Fresnel equations, 17
Fully integrated neuromuscular electrical stimulation system (FITNESS), 105
Functional biodegradable materials, 85–87
Functional transformation and active control, 85–87
Functionalized glucose oxidase, 206

G
Gastrointestinal (GI)
 monitor, 1
 motility disorders, 1
Gate/interlayer dielectrics, 93
Gaussian processes, 5
Gelatin, 78
Gelatin capsules, 201
Genetics, 136
Genome-wide association studies (GWAS), 118
Genomics, 136
Gibbs free energy, 34, 38
Gigabit wireless telemetry, 18–20
Gold electrode surface, 206
Gulf War Veterans (GWVs), 118

H
Hafnium oxide (HfO_2), 79
Hampel identifier filters, 5
Heart rate (HR), 132
Heart regeneration, 118
Heat-stimulated drug release, 62–63
Heterojunction bipolar transistors (HBTs), 49
Hexamethylene diisocyanate (HDI), 51
High-data-rate wireless link, 15
High-definition transcranial direct current stimulation (HD-tDCS), 7, 8
High-density electrode array
 actively multiplexed, 9–14
 flexible, 9–14
 neural interfaces, 8, 9
 system architecture, 22, 23
High-density gigabit wireless neural recording system, 20, 21
High-density polyethylene (HDPE), 63
High-density stimulation system, 7
High-frequency PECVD (PECVD-HF), 45
High-frequency transducer
 mouse, 186–188
 zebrafish, 188–190
High-frequency ultrasound duplex imaging system, 190
High-frequency ultrasound imaging systems, 188
High-order analog filter, 4
High-resolution computed tomography (HRCT), 172
Holter monitors, 132
Hydrophilic polymers, 48

I
ImageJ, 58
Implantable actuators, 90–91
Implantable devices
 bioresorbable triboelectric generators, 58
 bioresorption, 31
 brain-instrument interfaces, 29
 cardioverter defibrillator, 29
 degradation mechanisms, 32
 and drug delivery, 48
 flexibility and elasticity, 50
 flexible nonelectronic biodegradable, 29
 ventricular assist devices, 29
Implantable flexible electrode arrays, 155
Implantable sensors, 88–90, 92
Independent component analysis (ICA) techniques, 5
Indigo, 43
Inductive power transfer (IPT) system, 102, 104, 105, 107, 108
Infusion pumps, 29
Innovative power harvesting methods, 102
Inorganic dissolvable bioelectronics
 application, 73
 architecture, 73
 biodegradable passive components, 73
 biomedical implants, 88, 90, 92
 conductors, 75, 77
 functional transformation and active control, 85–87
 insulating polymers, 73
 insulators, 78, 79
 manufacturing processes, 79–82
 material selection, 76–77
 physical invariant performance, 73
 power supply components, 83–85
 semiconductors, 74, 75
Inorganic materials
 conductor, 34
 dielectric (insulator), 44
 substrates (insulator), 46, 47
Insulating polymers, 73

Insulators, 78, 79
- dielectric
  - inorganic, 44–46
  - organic, 46
- encapsulation layer
  - organic, 53
- substrate
  - inorganic, 46, 47
  - organic, 47–51

Internet-of-Things (IoT), 136
Interstitial fluid (ISF), 199, 204–208, 214, 215
Intracranial pressure (ICP) sensor, 53
Intracranial pressure monitoring, 32
Intravascular detection, 144
Intravascular ultrasound (IVUS), 143, 144, 154
Invasive coronary angiography, 157
Iron (Fe), 39
IR-UWB transceiver, 18

L

LC tank circuit, 104
Lead-related complications, 101
Linearly constrained minimum variance (LCMV), 7
Lipid-rich plaque, 144
Liver MR images, 179
Low-frequency PECVD (PECVD-LF), 45
Low-pressure chemical vapor deposition (LPCVD), 45
Lung protective ventilation (LPV), 170

M

Machine learning algorithm, 8
Machine learning (ML), 135–137
Magnesium (Mg), 38, 39
Magnetic field, 102, 103
Magnetic resonance electrical impedance tomography (MREIT), 181
Magnetic resonance microscopy, 189
Manufacturing processes, 79–82, 93, 95
Max intensity method, 8
Maximum intensity, 7
Mechanical energy, 102
Mechanical energy harvesters (MEHs), 57, 58
Mechanical sensors, 199
Mechanical ventilation, 170
Medical device development, 101
Medical devices, 32, 49, 51, 65
Medtronic's Activa RC+S device, 3
Metal foils, 46, 47
Metal insulator metal (MIM) capacitor, 52
Methylcellulose, 48
Mg spiral inductor, 52
Microelectrode array (MEA) membranes, 119, 128, 130
Microelectronics, 44
Microheater, 213
Microneedles (MNs)
- and biosensing, 208
- capillary action, 206
- continuous bio-signal monitoring, 207
- drug delivery, 204, 208
- ELISA, 207
- FDA, 208
- gold electrode surface, 206
- ISF, 204–207
- micromachined flexible, 207
- miniaturized electronics, 208
- pain, 204
- PEDOT, 206
- pH detection, 208
- physical skin barrier, 204
- propranolol, 208
- skin applications, 204
- surface modification, 206
- therapeutic level, drugs, 208
- three-electrode system, 207

Microsupercapacitors (MSCs), 53, 58, 85
Midfield WPT, 112, 113
Minimally invasive technologies
- biocompatibility, 214
- and biosensing, 195
- chemical techniques, 194
- continuous monitoring, 195
- conventional analytical techniques, 195
- disease diagnostic techniques, 195
- edible biosensors, 201–204
- healthcare applications, 215
- healthcare monitoring, 213
- healthcare signals, wearable biosensors, 198
- MNs (*see* Microneedles (MNs))
- personal health monitoring, 195, 215
- POC, 195–197
- smart bandages (*see* Smart bandages)
- treatment, 215
- wearable (*see* Wearable biosensors)

Molybdenum (Mo), 40
Mono-Si, 43
Mono-Si-NM encapsulating layers, 52
Mono-SiNMs, 43
MOSFETs, 82
Multichannel recording system, 16
Multi-coil stimulation, 108–110
Multidisciplinary cooperation, 215
Multifunctional sensing, 64
Multifunctional vascular stents, 32

Multiplexed flexible silicon transistors, 10–14
Multiplexers (MUXs), 9, 11, 22

N
Nanocellulose paper, 48
Nanogenerator, 58
Natural organic materials, 48, 49
Natural/organic semiconductors, 43
n-channel metal oxide semiconductor (nMOS) transistor, 12, 31
n-channel metal-oxide-semiconductor field-effect transistors (n-MOSFETs), 46
Near-field communication (NFC), 93–95, 215
Near-field radio frequency (RF) power transfer modules, 83
Near-field WPT
  battery-based stimulation, 106, 107
  batteryless direct stimulation, 105, 106
  coil inductance, 103
  components, 104
  coupling coefficient, 103
  LC tank circuit, 104
  magnetic field, 102, 103
  multi-coil stimulation, 108–110
  power transfer efficiency (PTE), 102, 104
  quality factor, 103
  radiative zone, 102
  reactive zone, 102
  remote-controlled stimulation, 107–109
  SAR, 104
  tissue energy absorption, 104
  transmitting and receiving coils, 102
Near-infrared (NIR), 64, 88
Nernst equation, 38
Neural diseases
  closed-loop therapy, 2
Neural interfaces, 8, 9
Neural stimulation
  amplifier design, 4
  artifact cancellation (*see* Stimulation artifact cancellation)
  and recording, 1
Neuromodulation devices, 101
Noise figure (NF), 17
Nonalcoholic fatty liver disease (NAFLD)
  average tissue conductivities, 174
  characterization, 180
  clinical translation, 177–179
  conductivity values, 177, 180
  low-cost screening tool, 180
  motivation, 173
  NZW Rabbit model, 176, 177, 179, 180
  simulation, 174–176
Noncontact electrodes (NCEs), 124, 126, 127
Noninvasive applications, 8
Noninvasive blood monitoring, 200
Noninvasive ISF analysis, 200
Non-transparent polyethylene shell, 203
Nonvolatile memory devices, 44
NZW Rabbit Model, 176, 177

O
On-demand drug release, 32
One-dimensional (1-D) model, 77
On-off keying (OOK) modulator, 18
On-site multiplexing, 9
Open-circuit voltage, 64
Optical coherence tomographic (OCT), 158
Optical energy, 102
Optimization method, 8
Organic electrochemical transistor (OECT), 59
Organic/inorganic-based multilayer, 10
Organic light-emitting devices (OLEDs), 31
Organic materials
  conductor, 41
  dielectric (insulator), 46
  encapsulation layer (insulator), 53
  semiconductors, 43, 44
  substrates (insulator), 47–51
Organic thin-film transistors (OTFTs), 31
Orthosilicic (silicic) acid ($Si(OH)_4$), 42
Oxidative stress, 144
Oxidized low-density lipoprotein (oxLDL), 143, 144
Oxygen gas, 45

P
Pacemaker AP, 120
Pacemaker cells, 120, 122
Paper-based POC devices, 196
Parallel LC tank, 104
Particle swarm optimization (PSO), 181
Parylene C (PAC), 153
PECVD-deposited materials, 53
PEDOT-doped polyanion poly(styrene sulfonate) (PEDOT:PSS), 41
Perylenediimide (PDI), 43, 44
Phosphate buffered saline (PBS), 13
Photo-acid generator (PAG), 87
Photodetectors, 31, 42
Photolithography, 12, 13
Photothermal activation, 92
Photothermal therapy, 32

Physical invariant performance, 73
Piezoelectric energy harvesters, 57
Plaques, 143
Plasma-enhanced chemical vapor deposition (PECVD), 45
Point-of-care devices (POC)
  accurate quantitative diagnostics, 197
  clinical idea, 195
  complex optical sensing, 196
  electrical sensing, 197
  paper-based POC devices, 196
  reduce, risks of medical errors, 195
  wearable portable devices, 197
Poly(1,8-octanediol-co-citrate (POC), 92
Poly(3,4-ethylenedioxythiophene) (PEDOT), 206
Poly(caprolactone) (PCL), 49
Poly(ethylene glycol) (PEG), 51
Poly(glycerol sebacate) (PGS), 46, 51
Poly(lactic-co-glycolic acid) (PLGA), 46, 58, 78
Poly(thiophenes) like poly(3,4-ethylenedioxythiophene) (PEDOT), 41
Poly(vinyl alcohol) (PVA), 46, 51, 78
Poly(vinylpyrrolidone) (PVP), 51
Polyanhydride (PA), 51, 53, 85
Polycaprolactone (PCL), 78
Polycrystalline silicon (p-Si), 42
Polydimethylsiloxane (PDMS), 130, 155
Polyfluorene (PF), 78
Polyglycolic acid (PGA), 78
Polyhydroxyalkanoates, 50
Polyimide (PI), 12
Polylactic acid (PLA), 53, 78
Polynomial curve fitting, 5
POMC, 50
Portable patient unit (PPU), 133
Positron emission tomography (PET), 169
Power constraint, 15, 16
Power supply components, 83–85
Power transfer efficiency (PTE), 102, 104
Principal component analysis (PCA), 5
Printed circuit board (PCB), 82
Pulmonary perfusion, 170

Q

Quality factor, 103

R

Radiative zone, 102
Radio frequency (RF), 90, 102
Rat auditory cortex, 13, 14
Reactive oxygen species (ROS), 64
Reactive zone, 102
Real-time EIT images, 179
Recording units (RU), 20, 21
Reference electrode (RE), 128
Refractive index, 17
Remote-controlled stimulation, 107–109
Remote device, 102
Resistive random-access memory (RRAM), 45, 64
Resistor, 90
Respiratory monitor, 163
Rice paper, 49
ROI, 176, 178

S

Saliva monitor, 200
Scalable manufacturing methods, 41
Scanning electron microscopy (SEM), 189
Self-degrading data security systems, 31
Self-healing, 215
Semiconducting, 73
Semiconducting materials, 34
Semiconductors, 74, 75
  organic, 43, 44
Sericin, 48
Series LC tank, 104
Serum biomarkers, 144
Shellac, 78
Si NM transistor, 12, 13
Si_3N_4 exhibits, 44
SiGe dissolutions, 43
Signal post-processing, 4, 5
Signal processing techniques, 5
Signal-to-noise ratio (SNR), 17, 123–125, 127, 128, 174
Silicate-based SOGs, 53
Silicon-based high-performance electronics, 78
Silicon-on-insulator (SOI) wafer, 12, 82
Silk, 48, 78
Single-element ultrasound transducers, 185
Single-photon emission computerized tomography (SPECT), 170
Single-walled carbon nanotubes (SWNTs), 93
Sinoatrial (SA) node, 122
Skin-air interface, 17
Small-scale material deposition techniques, 31
Smart bandages
  active drug delivery systems, 213
  antimicrobial characteristics, 211
  chronic wound management, 208
  colorimetric pH sensors, 212
  damage, vasculature, 212
  ECMs, 210

infection, 211
limitations, 210
microheater, 213
misdiagnosis, wound healing status, 209
moisture levels, 213
oxygen sensor, 212
pH, 211
preventing infection, 211
real-time diagnostic information, 210
wounds (*see* Wounds)
WoundSense™, 213
Smartphone-based ECG devices, 133
Smartphones, 133
Sodium carboxymethylcellulose (Na-CMC) substrates, 81
Solar cells, 31
Solvent-based processes, 95
Specific absorption rate (SAR), 104
Spinal cord stimulation (SCS), 8
State-of-the-art gigabit wireless telemetry, 18–20
Stimulation and recording units (SRUs), 22
Stimulation artifact cancellation
ADC and DAC, 6
advantage, 3
bench-top test, 6
circuit design, 3
circuit techniques, 5
contamination, 2
digital processing, 6
digital signal processing, 4, 5
in electrodes, 5
methods, 3
signal processing techniques, 5
signals, 2
switched capacitor, 5
system architecture, 22, 23
system design, 5
ultrahigh-density bidirectional neural interfaces, 2
ultrahigh-density neural interface, 2
Stretchable pH sensor, 62
Substrates
inorganic, 46, 47
organic
FDA, 47
limitations, 47
mechanisms, 47
natural, 48, 49
polymers, 47, 48
synthetic, 49–51
Support vector machines, 136
Switched capacitor, 5
Synaptic plasticity, 9
Synthetic and flexible polymers, 46
Synthetic organic materials
aliphatic polyesters, 49
BPLP, 50
copolymer, 49
degradation rates, 49
elastomers, 50
flexibility and elasticity, 50
HDI, 51
PCL, 49
PEG, 51
PGS, 51
PLA, 49
PLGA, 49
POC, 50
polyanhydrides, 51
polyhydroxyalkanoates, 50
POMC, 50
PVA, 51
PVP, 51
Synthetic polymers, 78

T

Tattoo-based biosensors, 198
Tear sensors, 199, 200
Temperature monitoring, 32
Textile-based sensors, 198
Thermal SiO_2 encapsulation approach, 11
Thermal therapy, 90
Thin-cap fibroatheromas (TCFA), 143, 158
Thin-film MgO, 44
Thin-film transistors (TFTs), 31, 42, 44, 75
Thin flexible films, 10
Tissue energy absorption, 104
Tissue regeneration, 63, 64
Trace element, 38
Transient electronics, 30–32, 34, 38, 43, 47, 48, 50, 52, 54, 59, 62, 64, 65
Transient electronics implants
application, 54
Transient N-MOSFET device, 52
Transmission distance, wireless link, 16–18, 20
Transmission electron microscopy, 189
Transmitter (Tx), 18
Triboelectric generator, 64
Tungsten, 40, 41
Two-point symmetric configuration, 153–155

U

Ultrahigh-density bidirectional neural interfaces, 2
Ultrahigh-density neural interface, 2

Ultrasound, 185
Ultrasound biomicroscopy (UBM), 186
Ultrasound imaging systems, 185
Ultrathin thermal SiO_2, 10, 11
Uniform deposition methods, 53
Unity-gain antenna, 17
Uplink operation, 22
Urethane-doped biodegradable photoluminescent polymers (UBPLPs), 51
Urine monitor, 200

V

Valence and conduction bands, 33
Ventricular assist devices, 29
Voltage controlled oscillator (VCO), 18, 19

W

Water-soluble cellulose ether, 48
Water-soluble electronics, 31
Water vapor, 45
Wavelet transforms, 5
Wearable biosensors
  Apple watch, 197
  biomarker, 199, 200
  challenges, 215
  electrophysiological sensing, 200, 201
  fabrication, 197
  flexible, 197, 214
  flexible device technology, 197
  mechanical sensors, 199
  noninvasive/minimally invasive, 197
  physiological mediums, 214
  POC devices, 197
  self-healing, 215
  system-level integration, 215
  tattoo-based biosensors, 198
  textile-based sensors, 198
  wireless transmission, 215
Wearable shirt (WS), 133
Weighted least square method, 7
Wet electrodes, 123–125
Wireless communication, 29
Wireless data link
  bandwidth/data rate requirement, 14, 15
  design, 13
  high-density gigabit wireless neural recording system, 20, 21
  in high-density neural interfaces, 13
  power constraint, 15, 16
  state of the art, 18–20
  transmission distance, 16–18
Wireless medical devices, 101, 102, 104, 106, 111, 114
Wireless microfluidic components, 93
Wireless neural recording system, 20, 21
  artifact cancellation (*see* Stimulation artifact cancellation)
  system architecture, 22, 23
Wireless pH monitor, 212
Wireless power transfer (WPT)
  far-field, 110–112
  IMDs, 105
  lead-related complications, 101
  medical device development, 101
  midfield, 112, 113
  near-field (*see* Near-field WPT)
  remote device, 102
  RF, 102
Working electrodes (WE), 128
Wound care products, 211
Wounds
  chronic, 210
  diagnostics, 209
  dressing, 211, 213
  environment, 210, 213
  healing, 209–212
  hemostasis phase, 210
WoundSense™, 213

X

X-ray computed tomography (CT), 169

Z

Zebrafish, 118, 119, 188
Zinc (Zn), 40

Printed in the United States
By Bookmasters